Special Triangles

Name	*Characteristic*	*Examples*
Right Triangle	Triangle has a right angle.	
Isosceles Triangle	Triangle has two equal sides.	$AB = BC$
Equilateral Triangle	Triangle has three equal sides.	$AB = BC = CA$
Similar Triangles	Corresponding angles are equal; corresponding sides are proportional.	$A = D, B = E, C = F$ $\frac{AB}{DE} = \frac{AC}{DF} = \frac{BC}{EF}$

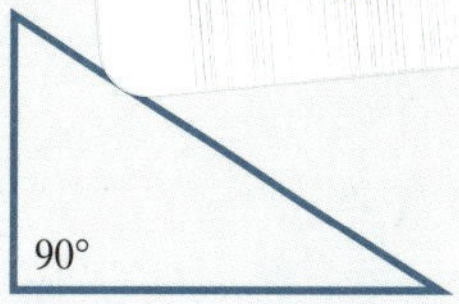

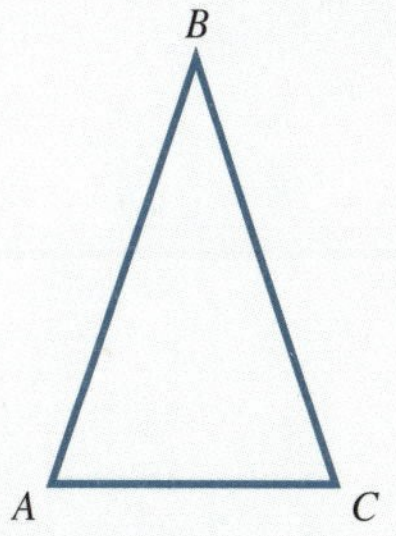

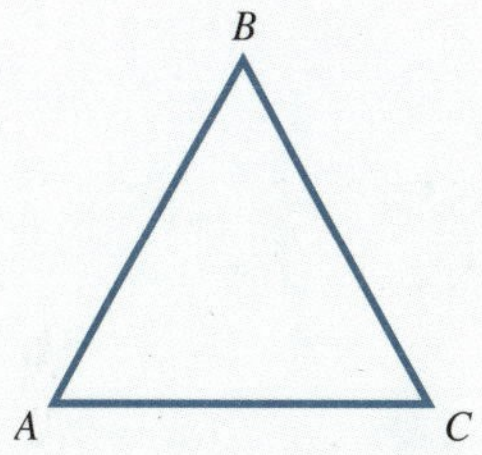

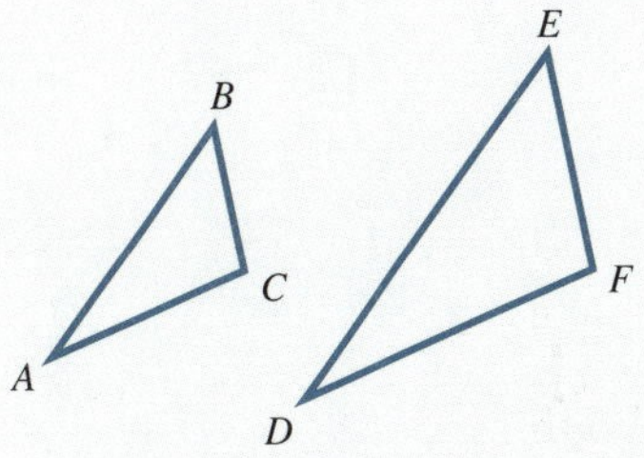

BEGINNING ALGEBRA

EIGHTH EDITION

BEGINNING ALGEBRA

EIGHTH EDITION

Margaret L. Lial

American River College

John Hornsby

University of New Orleans

ADDISON-WESLEY

An Imprint of Addison Wesley Longman, Inc.

Reading, Massachusetts • Menlo Park, California • New York • Harlow, England
Don Mills, Ontario • Sydney • Mexico City • Madrid • Amsterdam

Publisher: Jason Jordan

Acquisitions Editor: Jennifer Crum

Project Manager: Kari Heen

Developmental Editor: Terry McGinnis

Managing Editor: Ron Hampton

Production Supervisor: Kathleen A. Manley

Production Services: Elm Street Publishing Services, Inc.

Compositor: Typo-Graphics

Art Editor: Jennifer Bagdigian

Art Development: Meredith Nightingale

Artists: Precision Graphics, Jim Bryant, and Darwen and Vally Hennings

Marketing Manager: Craig Bleyer

Prepress Buyer: Caroline Fell

Manufacturing Coordinator: Evelyn Beaton

Text and Cover Designer: Susan Carsten

Cover Illustration: © Peter Siu/SIS

Library of Congress Cataloging-in-Publication Data

Lial, Margaret L.
Beginning algebra.—8th ed./Margaret L. Lial, John Hornsby.
p. cm.
Includes index.
ISBN 0-321-03644-1
1. Algebra. I. Hornsby, E. John. II. Title.
QA152.2.L5 1999
512.9—DC21 99-25871
CIP

Reprinted with corrections, November 1999

Printed in the U.S.A.

23456789-VH-02 01 00 99

Contents

CHAPTER 3 Linear Equations in Two Variables 175

CHAPTER 4 Polynomials and Exponents 223

CHAPTER 5 Factoring and Applications 288

Preface

The eighth edition of *Beginning Algebra* is designed for college students who have never studied algebra or who want to review the basic concepts of algebra before taking additional courses in mathematics, science, business, nursing, or computer science. The primary objective of this text is to familiarize students with mathematical symbols and operations in order to solve first- and second-degree equations and applications that lead to these equations.

This revision of *Beginning Algebra* reflects our ongoing commitment to creating the best possible text and supplements package using the most up-to-date strategies for helping students succeed. One of these strategies, evident in our new Table of Contents and consistent with current teaching practices, involves the early introduction of functions and graphing lines in a rectangular coordinate system. We believe that this pedagogy has a great deal of merit as it provides students with the important "input-output" concept that will be an integral part of later mathematics courses. This organization also allows an early treatment of interesting interpretations of data in the form of line and bar graphs—two pictorial representations that students already see on a daily basis in magazines and newspapers. Chapter 3 introduces three sections on linear equations in two variables, ordered pairs, graphing, and slope, with a gentle introduction to the function concept in the form of input-output relationships. This allows students to read graphs in the chapters that immediately follow and to slowly develop an understanding of the basic idea of a function. Chapter 7 introduces the more involved concepts of forms of equations of lines and inequalities in two variables. The function concept is addressed again here, this time with a discussion of domain, range, and function notation.

If for any reason you choose not to cover these topics as our new edition suggests, it will not be difficult to defer Chapter 3 and combine it with Chapter 7 as in previous editions. You will need to skip the last objective in Section 4.1 (Graphing Simple Polynomials). Also, one or two applied problems in an example or exercise in Chapters 4–6 may refer to the function concept, but even those problems can be used without actually working through Chapter 3.

Other up-to-date pedagogical strategies to foster student success include a strong emphasis on vocabulary and problem solving, an increased number of real world applications in both examples and exercises, and a focus on relevant industry themes throughout the text.

Another strategy for student success, an exciting new CD-ROM called "Pass the Test," debuts with this edition of *Beginning Algebra.* Directly correlated to the text's content, "Pass the Test" helps students master concepts by providing interactive pretests, chapter tests, section reviews, and InterAct Math tutorial exercises. To support an increased emphasis on graphical manipulation, the CD-ROM also includes a graphing tool that can be used for open-ended, student-directed exploration of number lines and coordinate graphs, as well as for exercises relevant to the graphing content throughout the book.

The *Student's Study Guide and Journal,* redesigned and enhanced with an optional journal feature for those who would like to incorporate more writing in their mathematics curriculum, provides an additional strategy for student success.

Although *Beginning Algebra,* Eighth Edition, integrates many new elements, it also retains the time-tested features of previous editions: learning objectives for each section, careful exposition, fully developed examples, Cautions and Notes, and design features that highlight important definitions, rules, and procedures. Since the hallmark of any mathematics text is the quality of its exercise sets, we have carefully developed exercise sets that provide ample opportunity for drill and, at the same time, test conceptual understanding. In preparing this edition we have also addressed the standards of the National Council of Teachers of Mathematics and the American Mathematical Association of Two-Year Colleges, incorporating many new exercises focusing on concepts, writing, graph interpretation, technology use, collaborative work, and analysis of data from a wide variety of sources in the world around us.

CONTENT CHANGES

We have fine-tuned and polished presentations of topics throughout the text based on user and reviewer feedback. Some of the content changes you may notice include the following:

- Operations with real numbers are consolidated from four sections to two sections in Chapter 1.
- We consistently emphasize problem solving using a six-step problem-solving strategy, first introduced in Section 2.3 and continually reinforced in examples and the exercise sets throughout the text. Section 2.6 contains a comprehensive discussion of problem solving.
- New Chapter 3 introduces graphing and slope, along with an intuitive introduction to functions using input-output. Equations with two variables are presented earlier so that the applications used in later chapters (for example, with polynomials and rational expressions) can be more realistic and relevant. Both graphing and functions are continued in later chapters.
- New material is included on graphing parabolas in Section 4.1.
- Division of polynomials is consolidated in Section 4.6.
- Solution set and interval notation are now introduced in *Beginning Algebra,* consistent with the approach in *Intermediate Algebra.*

NEW FEATURES

We believe students and instructors will welcome the following new features:

Industry Themes To help motivate the material, each chapter features a particular industry that is presented in the chapter opener and revisited in examples and exercises in the chapter. Identified by special icons, these examples and exercises incorporate sourced data, often in the form of graphs and tables. Featured industries include business, health care, entertainment, sports, transportation, and others. (See pages 1, 88, 175, and 428.)

New Examples and Exercises We have added 25% more real application problems with data sources. These examples and exercises often relate to the industry themes. They are designed to show students how algebra is used to describe and interpret data in everyday life. (See pages 9, 112, 132, 434, and 492.)

The Olympic Committee has come to rely more and more on television rights and major corporate sponsors to finance the games. The pie charts show the funding plans for the first Olympics in Athens and the 1996 Olympics in Atlanta. Use proportions and the figures to answer the questions in Exercises 35 and 36.

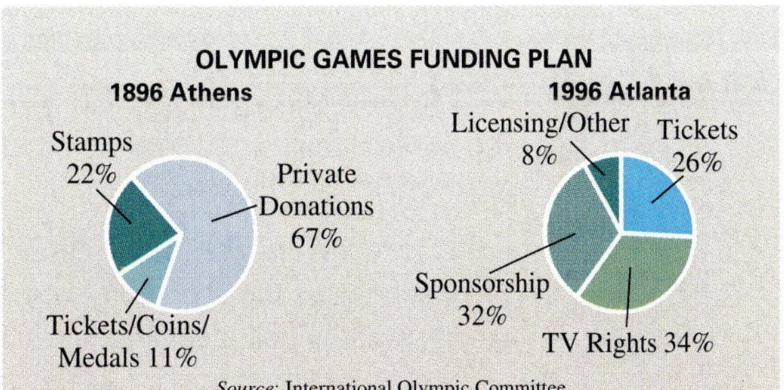

Source: International Olympic Committee.

35. In the 1996 Olympics, total revenue of $350 million was raised. There were 10 major sponsors.
 (a) Write a proportion to find the amount of revenue provided by tickets. Solve it.
 (b) What amount was provided by sponsors? Assuming the sponsors contributed equally, how much was provided per sponsor?
 (c) What amount was raised by TV rights?

36. Suppose the amount of revenue raised in the 1896 Olympics was equivalent to the $350 million in 1996.
 (a) Write a proportion for the amount of revenue provided by stamps and solve it.
 (b) What amount (in dollars) would have been provided by private donations?
 (c) In the 1988 Olympics, there were 9 major sponsors, and the total revenue was $95 million. What is the ratio of major sponsors in 1988 to those in 1996? What is the ratio of revenue in 1988 to revenue in 1996?

Technology Insights Exercises Technology is part of our lives, and we assume that all students of this text have access to scientific calculators. *While graphing calculators are not required for this text,* it is likely that students will go on to courses that use them. For this reason, we have included Technology Insights exercises in selected exercise sets. These exercises illustrate the power of graphing calculators and provide an opportunity for students to interpret typical results seen on graphing calculator screens. (See pages 212, 274, 324, and 506.)

Mathematical Journal Exercises While we continue to include conceptual and writing exercises that require short written answers, new journal exercises have been added that ask students to fully explain terminology, procedures, and methods, document their understanding using examples, or make connections between concepts. Instructors who wish to incorporate a journal component in their classes will find these exercises especially useful. For the greatest possible flexibility, both writing exercises and journal exercises are indicated with icons in the Annotated Instructor's Edition, but not in the Student Edition. (See pages 35, 103, 247, and 415.)

Group Activities Appearing at the end of each chapter, these activities allow students to apply the industry theme of the chapter to its mathematical content in a collaborative setting. (See pages 162, 458, and 506.)

Test Your Word Power To help students understand and master mathematical vocabulary, this new feature has been incorporated at the end of each chapter. Key terms from the chapter are presented with four possible definitions in multiple-choice format. Answers and examples illustrating each term are provided at the bottom of the appropriate page. (See pages 163, 277, and 343.)

HALLMARK FEATURES

We have retained the popular features of previous editions of the text. Some of these features are as follows:

Learning Objectives Each section begins with clearly stated numbered objectives, and material in the section is keyed to these objectives. In this way students know exactly what is being covered in each section.

OBJECTIVES

1 Learn the definition of *factor.*

2 Write fractions in lowest terms.

3 Multiply and divide fractions.

4 Add and subtract fractions.

5 Solve applied problems that involve fractions.

6 Interpret data in a circle graph.

CAUTION NOTE

Cautions and Notes We often give students warnings of common errors and emphasize important ideas in Cautions and Notes that appear throughout the exposition.

Connections Retained from the previous edition, Connections boxes have been streamlined and now often appear at the beginning or the end of the exposition in selected sections. They continue to provide connections to the real world or to other mathematical concepts, historical background, and thought-provoking questions for writing or class discussion. (See pages 105, 201, 226, and 322.)

Problem Solving Increased emphasis has been given to our six-step problem-solving method to aid students in solving application problems. This method is continually reinforced in examples and exercises throughout the text. (See pages 105, 107, 330, 409, and 497.)

Ample and Varied Exercise Sets Students in beginning algebra require a large number and variety of practice exercises to master the material. This text contains approximately 5800 exercises, including about 1600 review exercises, plus numerous conceptual and writing exercises, journal exercises, and challenging exercises that go beyond the examples. More illustrations, diagrams, graphs, and tables now accompany exercises. Multiple-choice, matching, true/false, and completion exercises help to provide variety. Exercises suitable for calculator use are marked with a calculator icon in both the Student Edition and the Annotated Instructor's Edition. (See pages 34, 183, 445, and 473.)

Relating Concepts Previously titled Mathematical Connections, these sets of exercises often appear near the end of selected sections. They tie together topics and highlight the relationships among various concepts and skills. For example, they may show how algebra and geometry are related, or how a graph of a linear equation in two variables is related to the solution of the corresponding linear equation in one variable. Instructors have told us that these sets of exercises make great collaborative activities for small groups of students. (See pages 69, 210, 232, 381, and 490.)

Ample Opportunity for Review Each chapter concludes with a Chapter Summary that features Key Terms and Symbols, Test Your Word Power, and a Quick Review of each section's content. Chapter Review Exercises keyed to individual sections are included as well as mixed review exercises and a Chapter Test. Following every chapter after Chapter 1, there is a set of Cumulative Review Exercises that covers material going back to the first chapter. Students always have an opportunity to review material that appears earlier in the text, and this provides an excellent way to prepare for the final examination in the course. (See pages 214–222 and 459–466)

SUPPLEMENTS

Our extensive supplements package includes the Annotated Instructor's Edition, testing materials, study guides, solutions manuals, CD-ROM software, videotapes, and a Web site. For more information on these and other helpful supplements, contact your Addison Wesley Longman sales representative.

FOR THE INSTRUCTOR

Annotated Instructor's Edition (ISBN 0-321-04128-3)
For immediate access, the Annotated Instructor's Edition provides answers to all text exercises and Group Activities in color in the margin or next to the corresponding exercise, as well as Chalkboard Examples and Teaching Tips. To assist instructors in assigning homework, additional icons not shown in the Student Edition indicate journal exercises, writing exercises, and challenging exercises ▲.

CHALKBOARD EXAMPLE

Write 90 as the product of prime factors.

Answer: $2 \cdot 3^2 \cdot 5$

TEACHING TIP The term *fraction bar* may be unfamiliar to some students.

Exercises designed for calculator use , are indicated in both the Student Edition and the Annotated Instructor's Edition.

Instructor's Solutions Manual (ISBN 0-321-06193-4)
The *Instructor's Solutions Manual* provides solutions to all exercises, including answer art, and lists of all writing, journal, challenging, Relating Concepts, and calculator exercises.

Answer Book (ISBN 0-321-06194-2)
The *Answer Book* contains answers to all exercises and lists of all writing, journal, challenging, Relating Concepts, and calculator exercises. Instructors may ask the bookstore to order multiple copies of the *Answer Book* for students to purchase.

Printed Test Bank (ISBN 0-321-06192-6)
The *Printed Test Bank* contains short answer and multiple-choice versions of a placement test and final exam; six forms of chapter tests for each chapter, including four open-response (short answer) and two multiple-choice forms; 10 to 20 additional exercises per objective for instructors to use for extra practice, quizzes, or tests; answer keys to all of the above listed tests and exercises; and lists of all writing, journal, challenging, Relating Concepts, and calculator exercises.

TestGen-EQ with QuizMaster EQ (ISBN 0-321-06132-2)
This fully networkable software presents a friendly graphical interface which enables professors to build, edit, view, print and administer tests. Tests can be printed or easily exported to HTML so they can be posted to the Web for student practice.

For the Student

Student's Study Guide and Journal (ISBN 0-321-06196-9)
The *Student's Study Guide and Journal* contains a "Chart Your Progress" feature for students to track their scores on homework assignments, quizzes, and tests, additional practice for each learning objective, section summary outlines that give students additional writing opportunities and help with test preparation, and self-tests with answers at the end of each chapter. A manual icon at the beginning of each section in the Student Edition identifies section coverage.

Student's Solutions Manual (ISBN 0-321-06195-0)
The *Student's Solutions Manual* provides solutions to all odd-numbered exercises (journal and writing exercises included). A manual icon at the beginning of each section in the Student Edition identifies section coverage.

InterAct Math Tutorial Software (ISBN 0-321-06140-3 (Student Version))
This tutorial software correlates with every odd-numbered exercise in the text. The program is highly interactive with sample problems and interactive guided solutions accompanying every exercise. The program recognizes common student errors and provides customized feedback with sophisticated answer recognition capabilities. The management system (InterAct Math Plus) allows instructors to create, administer, and track tests, and to monitor student performance during practice sessions.

"Real to Reel" Videotapes (0-321-05662-0)
This videotape series provides separate lessons for each section in the book. A videotape icon at the beginning of each section identifies section coverage. All objectives, topics, and problem-solving techniques are covered and content is specific to *Beginning Algebra,* Eighth Edition.

"Pass the Test" Interactive CD-ROM (ISBN 0-321-06204-3)
This CD helps students to master the course content by providing interactive pre-tests, chapter tests, section reviews, and InterAct tutorial exercises. After studying a chapter in class, students take a pre-test to determine what areas in that chapter need additional work. They are then directed to section reviews and tutorial exercises for continued practice. Students continue to take chapter tests and practice their skills until they have mastered the chapter. A unique graphing tool is provided for exploring the relationship between graphs and their algebraic representation.

World Wide Web Supplement (www.LialAlgebra.com)
Students can visit the Web site to explore additional real world applications related to the chapter themes, look up words in a complete glossary, and work through graphing calculator tutorials. The tutorials consist of step-by-step procedures as well as practice exercises for mastering basic graphing calculator skills.

MathXL (http://www.mathxl.com)
Available on-line with a pre-assigned ID and password by ordering a new copy of *Beginning Algebra,* Eighth Edition, with ISBN 0-201-68155-2, MathXL helps students prepare for tests by allowing them to take practice tests that are similar to the chapter tests in their text. Students also get a personalized study plan that identifies strengths and pinpoints topics where more review is needed. For more information on subscriptions, contact your Addison Wesley Longman sales representative.

Math Tutor Center
Available free to any student who purchases a new Lial/Hornsby, *Beginning Algebra,* Eighth Edition text ordered with ISBN 0-201-66341-4, the Addison Wesley Longman Math Tutor Center is staffed by qualified mathematics instructors who provide students with tutoring on text examples, exercises, and problems. Tutoring assistance is provided by telephone, fax, and e-mail and is available five days a week, seven hours a day. If a student purchases a used book, a registration number for the tutoring service may be obtained for a fee by calling our toll-free customer service number, 1-800-447-2226 and requesting ISBN 0-201-44461-5. Registration for the service is active for one or more of the following time periods depending on the course duration: Fall (8/31–1/31), Spring (1/1–6/30), or Summer (5/1–8/31). The Math Tutor Center service is also available for other Addison Wesley Longman textbooks in developmental math, precalculus math, liberal arts math, applied math, applied calculus, calculus, and introductory statistics. For more information, please contact your Addison Wesley Longman sales representative.

Spanish Glossary (ISBN 0-321-01647-5)
This book includes math terms that would be encountered in Basic Math through College Algebra.

Acknowledgments

For a textbook to last through eight editions, it is necessary for the authors to rely on comments, criticisms, and suggestions of users, nonusers, instructors, and students. We are grateful for the many responses that we have received over the years. We wish to thank the following individuals who reviewed this edition of the text:

Josette Ahlering, *Central Missouri State*
Vickie Aldrich, *Dona Ana Branch Community College*
Lisa Anderson, *Ventura College*
Robert B. Baer, *Miami University–Hamilton*
Julie R. Bonds, *Sonoma State University*
Beverly R. Broomell, *Suffolk Community College*
Cheryl V. Cantwell, *Seminole Community College*
Stanley Carter, *Central Missouri State*
Jeff Clark, *Santa Rosa Junior College*
Ted Corley, *Glendale Community College*
Lisa Delong Cuneo, *Penn State–Dubois Campus*
Marlene Demerjian, *College of the Canyons*
Richard N. Dodge, *Jackson Community College*
Linda Franko, *Cuyahoga Community College*
Linda L. Galloway, *Macon State College*
Theresa A. Geiger, *St. Petersburg Junior College*
Martha Haehl, *Maple Woods Community College*
Melissa Harper, *Embry Riddle Aeronautical University*
W. Hildebrand, *Montgomery College*
Matthew Hudock, *St. Philips College*
Dale W. Hughes, *Johnson County Community College*
Linda Hurst, *Central Texas College*
Nancy Johnson, *Broward Community College–North*
Robert Kaiden, *Lorain County Community College*
Michael Karelius, *American River College*
Margaret Kimbell, *Texas State Technical College*
Linda Kodama, *Kapiolani Community College*
Jeff A. Koleno, *Lorain County Community College*
William R. Livingston, *Missouri Southern State College*
Doug Martin, *Mt. San Antonio College*
Larry Mills, *Johnson County Community College*
Mary Ann Misko, *Gadsden State Community College*
Joanne V. Peeples, *El Paso Community College*
Janice Rech, *University of Nebraska–Omaha*
Joyce Saxon, *Morehead University*
Richard Semmler, *Northern Virginia Community College*
LeeAnn Spahr, *Durham Technical Community College*

No author can complete a project of this magnitude without the help of many other individuals. Our sincere thanks go to Jenny Crum of Addison Wesley Longman who coordinated the package of texts of which this book is a part. Other dedicated staff at Addison Wesley Longman who worked long and hard to make this revision a success include Jason Jordan, Kari Heen, Susan Carsten, Meredith Nightingale, and Kathy Manley.

While Terry McGinnis has assisted us for many years "behind the scenes" in producing our texts, she has contributed far more to these revisions than ever. There is no question that these books are improved because of her attention to detail and consistency, and we are most grateful for her work above and beyond the call of duty. Kitty Pellissier continues to do an outstanding job in checking the answers to exercises. Many thanks to Jenny Bagdigian who coordinated the art programs for the books.

Cathy Wacaser of Elm Street Publishing Services provided her usual excellent production work. She is indeed one of the best in the business. As usual, Paul Van Erden created an accurate, useful index. Becky Troutman prepared the Index of Applications. We are also grateful to Tommy Thompson who made suggestions for the feature "For the Student: 10 Ways to Succeed with Algebra," to Vickie Aldrich and Lucy Gurrola who wrote the Group Activity features, and Janis Cimperman of St. Cloud University and Steve C. Ouellette of the Walpole Massachusetts State Public Schools.

To these individuals and all the others who have worked on these books for 30 years, remember that we could not have done it without you. We hope that you share with us our pride in these books.

Margaret L. Lial
John Hornsby

An Introduction to Calculators

There is little doubt that the appearance of handheld calculators nearly three decades ago and the later development of scientific and graphing calculators have changed the methods of learning and studying mathematics forever. Where the study of computations with tables of logarithms and slide rules made up an important part of mathematics courses prior to 1970, today the widespread availability of calculators make their study a topic only of historical significance.

Most consumer models of calculators are inexpensive. At first, however, they were costly. One of the first consumer models available was the Texas Instruments SR-10, which sold for about $150 in 1973. It could perform the four operations of arithmetic and take square roots, but could do very little more.

Today calculators come in a large array of different types, sizes, and prices. *For the course for which this textbook is intended, the most appropriate type is the scientific calculator,* which costs $10–$20.

In this introduction, we explain some of the features of scientific and graphing calculators. However, remember that calculators vary among manufacturers and models, and that while the methods explained here apply to many of them, they may not apply to your specific calculator. For this reason, it is important to remember that *this introduction is only a guide, and is not intended to take the place of your owner's manual.* Always refer to the manual in the event you need an explanation of how to perform a particular operation.

Scientific Calculators

Scientific calculators are capable of much more than the typical four-function calculator that you might use for balancing your checkbook. Most scientific calculators use *algebraic logic.* (Models sold by Texas Instruments, Sharp, Casio, and Radio Shack, for example, use algebraic logic.) A notable exception is Hewlett Packard, a company whose calculators use *Reverse Polish Notation* (RPN). In this introduction, we explain the use of calculators with algebraic logic.

Arithmetic Operations

To perform an operation of arithmetic, simply enter the first number, press the operation key ([+], [−], [×], or [÷]), enter the second number, and then press the [=] key. For example, to add 4 and 3, use the following keystrokes.

[4] [+] [3] [=]   7

Change Sign Key

The key marked [±] allows you to change the sign of a display. This is particularly useful when you wish to enter a negative number. For example, to enter −3, use the following keystrokes.

[3] [±]   −3

Memory Key

Scientific calculators can hold a number in memory for later use. The label of the memory key varies among models; two of these are [M] and [STO]. [M+] and [M−] allow you to add to or subtract from the value currently in memory. The memory recall key, labeled MR, RM, or RCL, allows you to retrieve the value stored in memory.

Suppose that you wish to store the number 5 in memory. Enter 5, then press the key for memory. You can then perform other calculations. When you need to retrieve the 5, press the key for memory recall.

If a calculator has a constant memory feature, the value in memory will be retained even after the power is turned off. Some advanced calculators have more than one memory. It is best to read the owner's manual for your model to see exactly how memory is activated.

Clearing/Clear Entry Keys

These keys allow you to clear the display or clear the last entry entered into the display. They are usually marked C and CE. In some models, pressing the C key once will clear the last entry, while pressing it twice will clear the entire operation in progress.

Second Function Key

This key is used in conjunction with another key to activate a function that is printed *above* an operation key (and not on the key itself). It is usually marked 2nd. For example, suppose you wish to find the square of a number, and the squaring function (explained in more detail later) is printed above another key. You would need to press 2nd before the desired squaring function can be activated.

Square Root Key

Pressing the square root key, $\sqrt{x}$, will give the square root (or an approximation of the square root) of the number in the display. For example, to find the square root of 36, use the following keystrokes.

3 6 $\sqrt{x}$ → 6

The square root of 2 is an example of an irrational number (Chapter 9). The calculator will give an approximation of its value, since the decimal for $\sqrt{2}$ never terminates and never repeats. The number of digits shown will vary among models. To find an approximation of $\sqrt{2}$, use the following keystrokes.

2 $\sqrt{x}$ → 1.4142136 An approximation

Squaring Key

This key, x^2, allows you to square the entry in the display. For example, to square 35.7, use the following keystrokes.

3 5 . 7 x^2 → 1274.49

The squaring key and the square root key are often found on the same key, with one of them being a second function (that is, activated by the second function key, described above).

Reciprocal Key

The key marked $1/x$ is the reciprocal key. (When two numbers have a product of 1, they are called *reciprocals*. See Chapter 1.) Suppose that you wish to find the reciprocal of 5. Use the following keystrokes.

5 $1/x$ → 0.2

INVERSE KEY

Some calculators have an inverse key, marked [INV]. Inverse operations are operations that "undo" each other. For example, the operations of squaring and taking the square root are inverse operations. The use of the [INV] key varies among different models of calculators, so read your owner's manual carefully.

EXPONENTIAL KEY

The key marked [x^y] or [y^x] allows you to raise a number to a power. For example, if you wish to raise 4 to the fifth power (that is, find 4^5, as explained in Chapter 4), use the following keystrokes.

[4] [x^y] [5] [=] [1024]

ROOT KEY

Some calculators have this key specifically marked [$\sqrt[x]{y}$] or [$\sqrt[y]{x}$]; with others, the operation of taking roots is accomplished by using the inverse key in conjunction with the exponential key. Suppose, for example, your calculator is of the latter type and you wish to find the fifth root of 1024. Use the following keystrokes.

[1] [0] [2] [4] [INV] [x^y] [5] [=] [4]

Notice how this "undoes" the operation explained in the exponential key discussion above.

PI KEY

The number π is an important number in mathematics. It occurs, for example, in the area and circumference formulas for a circle. By pressing the [π] key, you can display the first few digits of π. (Because π is irrational, the display shows only an approximation.) One popular model gives the following display when the [π] key is pressed: [3.1415927].

METHODS OF DISPLAY

When decimal approximations are shown on scientific calculators, they are either *truncated* or *rounded.* To see how a particular model is programmed, evaluate 1/18 as an example. If the display shows .0555555 (last digit 5), it truncates the display. If it shows .0555556 (last digit 6), it rounds off the display.

When very large or very small numbers are obtained as answers, scientific calculators often express these numbers in scientific notation (Chapter 4). For example, if you multiply 6,265,804 by 8,980,591, the display might look like this:

[5.6270623 13].

The "13" at the far right means that the number on the left is multiplied by 10^{13}. This means that the decimal point must be moved 13 places to the right if the answer is to be expressed in its usual form. Even then, the value obtained will only be an approximation: 56,270,623,000,000.

GRAPHING CALCULATORS

Graphing calculators are becoming increasingly popular in mathematics classrooms. While you are not expected to have a graphing calculator to study from this book, we do include a feature in many exercise sets called *Technology Insights* that asks you to interpret typical graphing calculator screens. These exercises can help to prepare you for future courses where graphing calculators may be recommended or even required.

Basic Features

Graphing calculators provide many features beyond those found on scientific calculators. In addition to the typical keys found on scientific calculators, they have keys that can be used to create graphs, make tables, analyze data, and change settings. One of the major differences between graphing and scientific calculators is that a graphing calculator has a larger viewing screen with graphing capabilities. The screens below illustrate the graphs of $y = x$ and $y = x^2$.

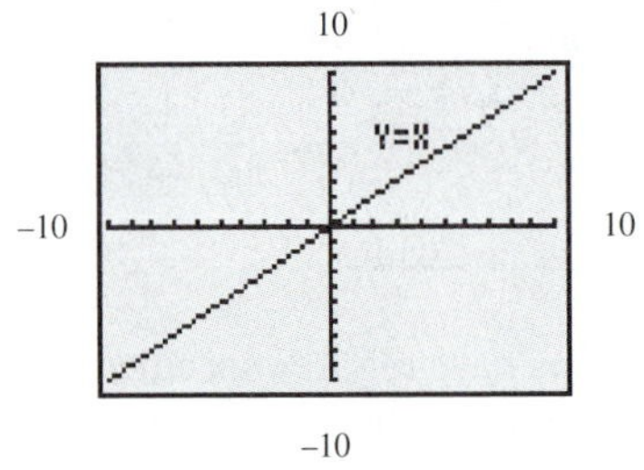

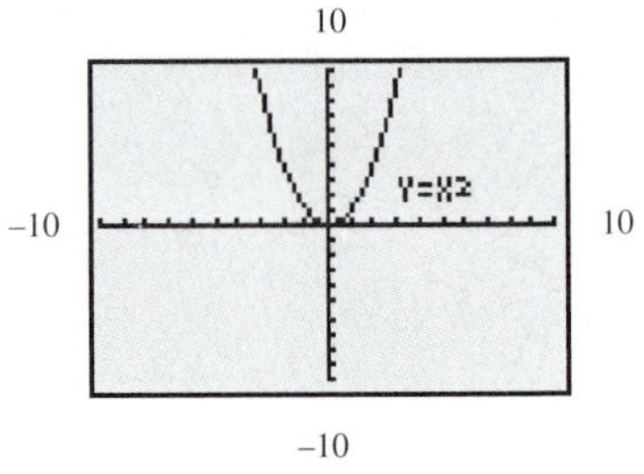

If you look closely at the screens, you will see that the graphs appear to be "jagged" rather than smooth, as they should be. The reason for this is that graphing calculators have much lower resolution than a computer screen. Because of this, graphs generated by graphing calculators must be interpreted carefully.

Editing Input

The screen of a graphing calculator can display several lines of text at a time. This feature allows you to view both previous and current expressions. If an incorrect expression is entered, an error message is displayed. The erroneous expression can be viewed and corrected by using various editing keys, much like a word-processing program. You do not need to enter the entire expression again. Many graphing calculators can also recall past expressions for editing or updating. The screen on the left below shows how two expressions are evaluated. The final line is entered incorrectly, and the resulting error message is shown in the screen on the right.

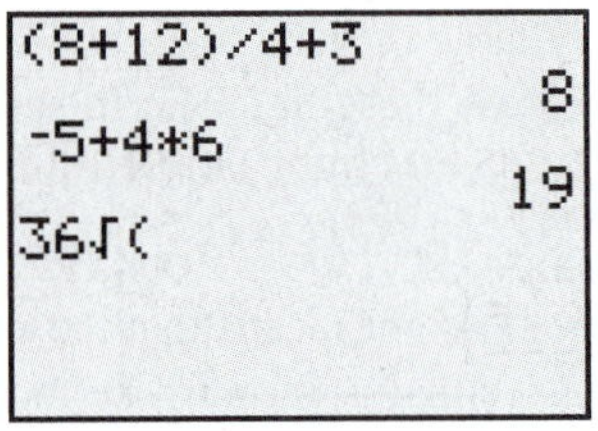

Order of Operations

Arithmetic operations on graphing calculators are usually entered as they are written in mathematical equations. For example, to evaluate $\sqrt{36}$ on a typical scientific calculator, you would first enter 36 and then press the square root key. As seen above, this is not the correct syntax for a graphing calculator. To find this root, you would first press the square root key, and then enter 36. See the screen on the left at the top of the next page. The order of operations on a graphing calculator is also important, and current models assist the user by inserting parentheses when typical errors might occur. The open parenthesis that follows the square root symbol is automatically entered by the calculator, so that an expression such as $\sqrt{2 \times 8}$ will not be calculated incorrectly as $\sqrt{2} \times 8$. Compare the two entries and their results in the screen on the right.

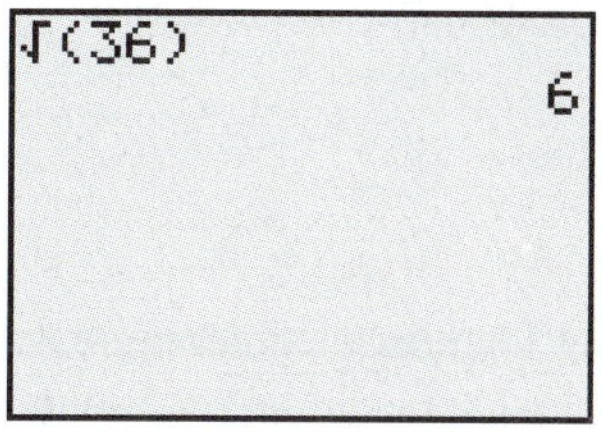

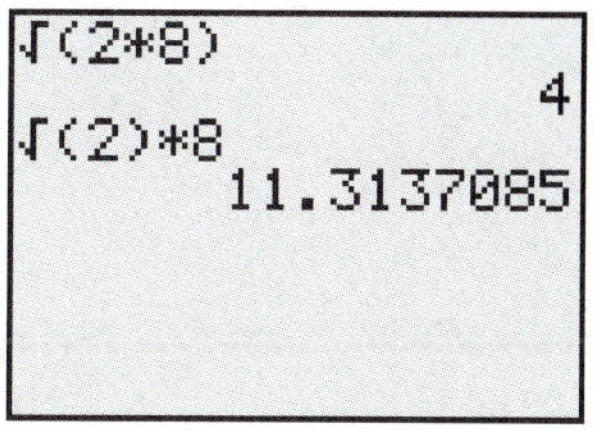

Viewing Windows

The viewing window for a graphing calculator is similar to the viewfinder in a camera. A camera usually cannot take a photograph of an entire view of a scene. The camera must be centered on some object and can only capture a portion of the available scenery. A camera with a zoom lens can photograph different views of the same scene by zooming in and out. Graphing calculators have similar capabilities. The *xy*-coordinate plane is infinite. The calculator screen can only show a finite, rectangular region in the plane, and it must be specified before the graph can be drawn. This is done by setting both minimum and maximum values for the *x*- and *y*-axes. The scale (distance between tick marks) is usually specified as well. Determining an appropriate viewing window for a graph is often a challenge, and many times it will take a few attempts before a satisfactory window is found.

The screen on the left shows a "standard" viewing window, and the graph of $y = 2x + 1$ is shown on the right. Using a different window would give a different view of the line.

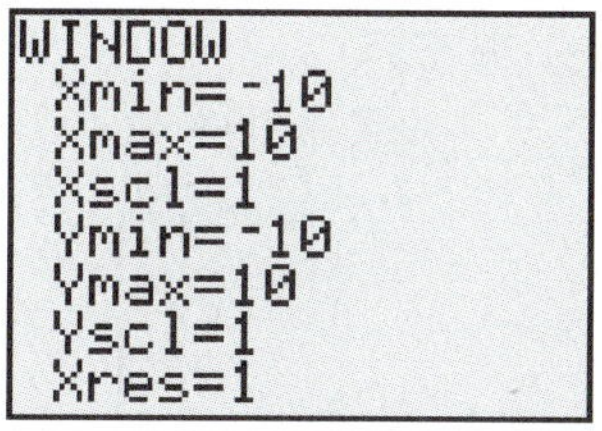

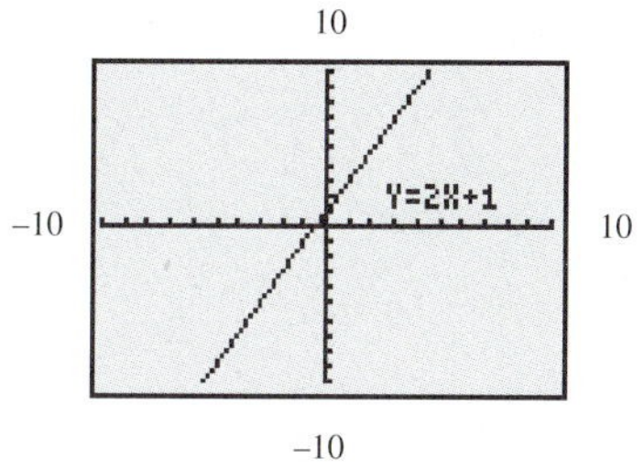

Locating Points on a Graph: Tracing and Tables

Graphing calculators allow you to trace along the graph of an equation, and, while doing this, display the coordinates of points on the graph. See the screen on the left at the top of the next page, which indicates that the point (2, 5) lies on the graph of $y = 2x + 1$. Tables for equations can also be displayed. The screen on the right shows a partial table for this same equation. Note the middle of the screen, which indicates that when $x = 2$, $y = 5$.

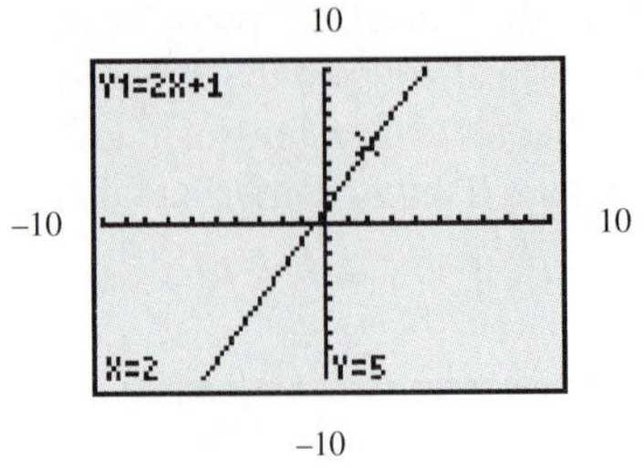

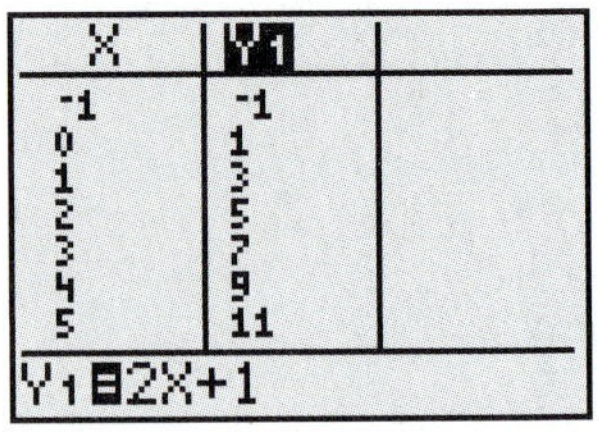

Additional Features

There are many features of graphing calculators that go far beyond the scope of this book. These calculators can be programmed, much like computers. Many of them can solve equations at the stroke of a key, analyze statistical data, and perform symbolic algebraic manipulations. Mathematicians from the past would have been amazed by today's calculators. Many important equations in mathematics cannot be solved by hand. However, their solutions can often be approximated using a calculator. Calculators also provide the opportunity to ask "What if . . . ?" more easily. Values in algebraic expressions can be altered and conjectures tested quickly.

Final Comments

Despite the power of today's calculators, they cannot replace human thought. **In the entire problem-solving process, your brain is the most important component.** Calculators are only tools, and like any tool, they must be used appropriately in order to enhance our ability to understand mathematics. Mathematical insight may often be the quickest and easiest way to solve a problem; a calculator may neither be needed nor appropriate. By applying mathematical concepts, you can make the decision whether or not to use a calculator.

Beginning Algebra

EIGHTH EDITION

The Real Number System 1

Growth in productivity—the amount of goods and services produced for each hour of work—is the most important factor in improving living standards of our country's population. In the latter part of the 1990s, the U.S. construction industry finally began to revive after a long period of decline. Having survived many years of downsizing and layoffs, manufacturing-related industries are again contributing to economic growth. This bar graph shows the increase in construction spending in the latter half of 1996 and all of 1997. During which month and year was construction spending highest? Several exercises in Section 1.2 use this graph. Throughout this chapter, we will see other specific examples and exercises that relate to construction and manufacturing.

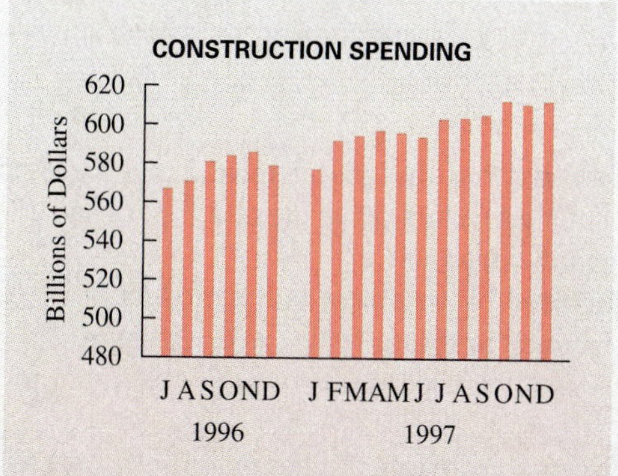

Source: Commerce Department.

Construction/Manufacturing

INTERNET Visit our Web site at www.LialAlgebra.com

1.1 Fractions

OBJECTIVES

1. Learn the definition of *factor.*
2. Write fractions in lowest terms.
3. Multiply and divide fractions.
4. Add and subtract fractions.
5. Solve applied problems that involve fractions.
6. Interpret data in a circle graph.

FOR EXTRA HELP

SSG Sec. 1.1
SSM Sec. 1.1
Pass the Test Software
InterAct Math Tutorial Software
Video 1

As preparation for the study of algebra, this section begins with a brief review of arithmetic. In everyday life the numbers seen most often are the **natural numbers,**

$$1, 2, 3, 4, \ldots,$$

the **whole numbers,**

$$0, 1, 2, 3, 4, \ldots,$$

and **fractions,** such as

$$\frac{1}{2}, \quad \frac{2}{3}, \quad \text{and} \quad \frac{15}{7}.$$

The parts of a fraction are named as follows.

$$\text{Fraction bar} \rightarrow \quad \frac{4}{7} \quad \left(\frac{a}{b} = a \div b\right) \quad \begin{matrix} \leftarrow \text{Numerator} \\ \leftarrow \text{Denominator} \end{matrix}$$

As we will see later, the fraction bar represents division $\left(\frac{a}{b} = a \div b\right)$ and also serves as a grouping symbol.

CONNECTIONS

A common use of fractions is to measure dimensions of tools, amounts of building materials, and so on. We need to add, subtract, multiply, and divide fractions in order to solve many types of measurement problems.

FOR DISCUSSION OR WRITING

Discuss some situations in your experience where you have needed to perform the operations of addition, subtraction, multiplication, or division on fractions. (*Hint:* To get you started, think of art projects, carpentry projects, adjusting recipes, and working on cars.)

OBJECTIVE 1 Learn the definition of *factor.* In the statement $2 \times 9 = 18$, the numbers 2 and 9 are called **factors** of 18. Other factors of 18 include 1, 3, 6, and 18. The result of the multiplication, 18, is called the **product.**

The number 18 is **factored** by writing it as the product of two or more numbers. For example, 18 can be factored in several ways, as $6 \cdot 3$, or $18 \cdot 1$, or $9 \cdot 2$, or $3 \cdot 3 \cdot 2$. In algebra, a raised dot $\cdot$ is often used instead of the $\times$ symbol to indicate multiplication.

A natural number (except 1) is **prime** if it has only itself and 1 as factors. "Factors" are understood here to mean natural number factors. (By agreement, the number 1 is not a prime number.) The first dozen primes are

$$2, 3, 5, 7, 11, 13, 17, 19, 23, 29, 31, 37.$$

A natural number (except 1) that is not prime is a **composite** number.

It is often useful to find all the **prime factors** of a number—those factors that are prime numbers. For example, the only prime factors of 18 are 2 and 3.

EXAMPLE 1 Factoring Numbers

Write the number as the product of prime factors.

(a) 35

Write 35 as the product of the prime factors 5 and 7, or as

$$35 = 5 \cdot 7.$$

(b) 24

One way to begin is to divide by the smallest prime, 2, to get

$$24 = 2 \cdot 12.$$

Now divide 12 by 2 to find factors of 12.

$$24 = 2 \cdot 2 \cdot 6$$

Since 6 can be written as $2 \cdot 3$,

$$24 = 2 \cdot 2 \cdot 2 \cdot 3,$$

where all factors are prime.

NOTE It is not necessary to start with the smallest prime factor, as shown in Example 1(b). In fact, no matter which prime factor we start with, we will *always* obtain the same prime factorization.

OBJECTIVE 2 Write fractions in lowest terms. We use prime numbers to write fractions in *lowest terms.* A fraction is in **lowest terms** when the numerator and denominator have no factors in common (other than 1). By the **basic principle of fractions,** if the numerator and denominator of a fraction are multiplied or divided by the *same* nonzero number, the value of the fraction is unchanged. To write a fraction in lowest terms, use these steps.

Writing a Fraction in Lowest Terms

Step 1 Write the numerator and the denominator as the product of prime factors.

Step 2 Divide the numerator and the denominator by the **greatest common factor,** the product of all factors common to both.

EXAMPLE 2 Writing Fractions in Lowest Terms

Write the fraction in lowest terms.

(a) $\dfrac{10}{15} = \dfrac{2 \cdot 5}{3 \cdot 5} = \dfrac{2 \cdot 1}{3 \cdot 1} = \dfrac{2}{3}$

Since 5 is the greatest common factor of 10 and 15, dividing both numerator and denominator by 5 gives the fraction in lowest terms.

(b) $\dfrac{15}{45} = \dfrac{3 \cdot 5}{3 \cdot 3 \cdot 5} = \dfrac{1 \cdot 1}{3 \cdot 1 \cdot 1} = \dfrac{1}{3}$

The factored form shows that 3 and 5 are the common factors of both 15 and 45. Dividing both 15 and 45 by $3 \cdot 5 = 15$ gives $\frac{15}{45}$ in lowest terms as $\frac{1}{3}$.

We can simplify this process by finding the greatest common factor in the numerator and denominator by inspection. For instance, in Example 2(b), we can use 15 rather than $3 \cdot 5$.

$$\frac{15}{45} = \frac{15}{3 \cdot 15} = \frac{1}{3 \cdot 1} = \frac{1}{3}$$

CAUTION

Errors may occur when writing fractions in lowest terms if the factor 1 is not included. To see this, refer to Example 2(b). In the equation

$$\frac{3 \cdot 5}{3 \cdot 3 \cdot 5} = \frac{?}{3},$$

if 1 is not written in the numerator when dividing common factors, you may make an error. The ? should be replaced by 1.

OBJECTIVE 3 Multiply and divide fractions. The basic operations on whole numbers, addition, subtraction, multiplication, and division, also apply to fractions. We multiply two fractions by first multiplying their numerators and then multiplying their denominators. This rule is written in symbols as follows.

Multiplying Fractions

If $\frac{a}{b}$ and $\frac{c}{d}$ are fractions, then $\frac{a}{b} \cdot \frac{c}{d} = \frac{a \cdot c}{b \cdot d}$.

EXAMPLE 3 Multiplying Fractions

Find the product of $\frac{3}{8}$ and $\frac{4}{9}$, and write it in lowest terms.

First, multiply $\frac{3}{8}$ and $\frac{4}{9}$.

$$\frac{3}{8} \cdot \frac{4}{9} = \frac{3 \cdot 4}{8 \cdot 9}$$ Multiply numerators; multiply denominators.

It is easiest to write a fraction in lowest terms while the product is in factored form. Factor 8 and 9 and then divide out common factors in the numerator and denominator.

$$\frac{3 \cdot 4}{8 \cdot 9} = \frac{3 \cdot 4}{2 \cdot 4 \cdot 3 \cdot 3} = \frac{1 \cdot 3 \cdot 4}{2 \cdot 4 \cdot 3 \cdot 3}$$ Factor. Introduce a factor of 1.

$$= \frac{1}{2 \cdot 3}$$ 3 and 4 are common factors.

$$= \frac{1}{6}$$ Lowest terms

Two fractions are **reciprocals** of each other if their product is 1. For example, $\frac{3}{4}$ and $\frac{4}{3}$ are reciprocals since

$$\frac{3}{4} \cdot \frac{4}{3} = \frac{12}{12} = 1.$$

Also, $\frac{7}{11}$ and $\frac{11}{7}$ are reciprocals of each other. We use the reciprocal to divide fractions. To *divide* two fractions, multiply the first fraction by the reciprocal of the second fraction.

Dividing Fractions

For the fractions $\frac{a}{b}$ and $\frac{c}{d}$, $\frac{a}{b} \div \frac{c}{d} = \frac{a}{b} \cdot \frac{d}{c}$.

(To divide by a fraction, multiply by its reciprocal.)

The reason this method works will be explained in Chapter 6. The answer to a division problem is called a **quotient.** For example, the quotient of 20 and 10 is 2, since $20 \div 10 = 2$.

EXAMPLE 4 Dividing Fractions

Find the following quotients, and write them in lowest terms.

(a) $\frac{3}{4} \div \frac{8}{5} = \frac{3}{4} \cdot \frac{5}{8} = \frac{3 \cdot 5}{4 \cdot 8} = \frac{15}{32}$ Multiply by the reciprocal of $\frac{8}{5}$.

(b) $\frac{3}{4} \div \frac{5}{8} = \frac{3}{4} \cdot \frac{8}{5} = \frac{3 \cdot 8}{4 \cdot 5} = \frac{3 \cdot 4 \cdot 2}{4 \cdot 5} = \frac{6}{5}$

OBJECTIVE 4 Add and subtract fractions. To find the **sum** of two fractions having the same denominator, add the numerators and keep the same denominator.

Adding Fractions

If $\frac{a}{b}$ and $\frac{c}{b}$ are fractions, then $\frac{a}{b} + \frac{c}{b} = \frac{a+c}{b}$.

EXAMPLE 5 Adding Fractions with the Same Denominator

Add.

(a) $\frac{3}{7} + \frac{2}{7} = \frac{3+2}{7} = \frac{5}{7}$ Add numerators and keep the same denominator.

(b) $\frac{2}{10} + \frac{3}{10} = \frac{2+3}{10} = \frac{5}{10} = \frac{1}{2}$

If the fractions to be added do not have the same denominators, the procedure above can still be used, but only *after* the fractions are rewritten with a common denominator. For example, to rewrite $\frac{3}{4}$ as a fraction with a denominator of 32,

$$\frac{3}{4} = \frac{?}{32},$$

find the number that can be multiplied by 4 to give 32. Since $4 \cdot 8 = 32$, use the number 8. By the basic principle, we can multiply the numerator and the denominator by 8.

$$\frac{3}{4} = \frac{3 \cdot 8}{4 \cdot 8} = \frac{24}{32}$$

Finding the Least Common Denominator

To add or subtract fractions with different denominators, find the **least common denominator (LCD)** as follows.

Step 1 Factor both denominators.

Step 2 For the LCD, use every factor that appears in any factored form. If a factor is repeated, use the largest number of repeats in the LCD.

The next example shows this procedure.

EXAMPLE 6 Adding Fractions with Different Denominators

Add the following fractions.

(a) $\frac{4}{15} + \frac{5}{9}$

To find the least common denominator, first factor both denominators.

$$15 = 5 \cdot 3 \qquad \text{and} \qquad 9 = 3 \cdot 3$$

Since 5 and 3 appear as factors, and 3 is a factor of 9 twice, the LCD is

$$5 \cdot 3 \cdot 3 \qquad \text{or} \qquad 45.$$

Write each fraction with 45 as denominator.

$$\frac{4}{15} = \frac{4 \cdot 3}{15 \cdot 3} = \frac{12}{45} \qquad \text{and} \qquad \frac{5}{9} = \frac{5 \cdot 5}{9 \cdot 5} = \frac{25}{45}$$

Now add the two equivalent fractions.

$$\frac{4}{15} + \frac{5}{9} = \frac{12}{45} + \frac{25}{45} = \frac{37}{45}$$

(b) $3\frac{1}{2} + 2\frac{3}{4}$

These numbers are called mixed numbers. A **mixed number** is understood to be the sum of a whole number and a fraction. We can add mixed numbers using either of two methods.

Method 1

Rewrite both numbers as follows.

$$3\frac{1}{2} = 3 + \frac{1}{2} = \frac{3}{1} + \frac{1}{2} = \frac{6}{2} + \frac{1}{2} = \frac{6+1}{2} = \frac{7}{2}$$

$$2\frac{3}{4} = 2 + \frac{3}{4} = \frac{8}{4} + \frac{3}{4} = \frac{8+3}{4} = \frac{11}{4}$$

Now add. The common denominator is 4.

$$3\frac{1}{2} + 2\frac{3}{4} = \frac{7}{2} + \frac{11}{4} = \frac{14}{4} + \frac{11}{4} = \frac{25}{4} \quad \text{or} \quad 6\frac{1}{4}$$

Method 2

Write $3\frac{1}{2}$ as $3\frac{2}{4}$. Then add vertically.

$$\begin{array}{r} 3\dfrac{1}{2} \\ +\,2\dfrac{3}{4} \\ \hline \end{array} \quad \rightarrow \quad \begin{array}{r} 3\dfrac{2}{4} \\ +\,2\dfrac{3}{4} \\ \hline 5\dfrac{5}{4} \end{array}$$

Since $\frac{5}{4} = 1\frac{1}{4}$,

$$5\frac{5}{4} = 5 + 1\frac{1}{4} = 6\frac{1}{4}, \quad \text{or} \quad \frac{25}{4}.$$

NOTE

To multiply and divide mixed numbers, follow the same general procedure shown in Example 6(b), Method 1. First change to fractions, then perform the operation, and then convert back to a mixed number if desired. For example,

$$3\frac{1}{2} \cdot 2\frac{3}{4} = \frac{7}{2} \cdot \frac{11}{4} = \frac{77}{8} \quad \text{or} \quad 9\frac{5}{8}$$

$$3\frac{1}{2} \div 2\frac{3}{4} = \frac{7}{2} \div \frac{11}{4} = \frac{7}{2} \cdot \frac{4}{11} = \frac{14}{11} \quad \text{or} \quad 1\frac{3}{11}.$$

The **difference** between two numbers is found by subtraction. For example, $9 - 5 = 4$ so the difference between 9 and 5 is 4. Subtraction of fractions is similar to addition. Just subtract the numerators instead of adding them; again, keep the same denominator.

Subtracting Fractions

$$\frac{a}{b} - \frac{c}{b} = \frac{a - c}{b}$$

EXAMPLE 7 Subtracting Fractions

Subtract. Write the differences in lowest terms.

(a) $\dfrac{15}{8} - \dfrac{3}{8} = \dfrac{15 - 3}{8}$ Subtract numerators; keep the same denominator.

$= \dfrac{12}{8} = \dfrac{3}{2}$ Lowest terms

(b) $\dfrac{7}{18} - \dfrac{4}{15}$

Here, $18 = 2 \cdot 3 \cdot 3$ and $15 = 3 \cdot 5$, so the LCD is $2 \cdot 3 \cdot 3 \cdot 5 = 90$.

$$\frac{7}{18} - \frac{4}{15} = \frac{7 \cdot 5}{2 \cdot 3 \cdot 3 \cdot 5} - \frac{4 \cdot 2 \cdot 3}{2 \cdot 3 \cdot 3 \cdot 5} = \frac{35}{90} - \frac{24}{90} = \frac{11}{90}$$

(c) $\dfrac{15}{32} - \dfrac{11}{45}$

Since $32 = 2 \cdot 2 \cdot 2 \cdot 2 \cdot 2$ and $45 = 3 \cdot 3 \cdot 5$, there are no common factors, and the LCD is $32 \cdot 45 = 1440$.

$$\begin{aligned} \frac{15}{32} - \frac{11}{45} &= \frac{15 \cdot 45}{32 \cdot 45} - \frac{11 \cdot 32}{45 \cdot 32} && \text{Get a common denominator.} \\ &= \frac{675}{1440} - \frac{352}{1440} \\ &= \frac{323}{1440} && \text{Subtract.} \end{aligned}$$

OBJECTIVE 5 Solve applied problems that involve fractions. Applied problems often require work with fractions. For example, when a carpenter reads diagrams and plans, he or she often must work with fractions whose denominators are 2, 4, 8, 16, or 32, as shown in the next example.

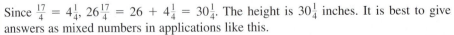

EXAMPLE 8 Adding Fractions to Solve a Manufacturing (Woodworking) Problem

The diagram in Figure 1 appears in the book *Woodworker's 39 Sure-Fire Projects.* It is the front view of a corner bookcase/desk. Add the fractions shown in the diagram to find the height of the bookcase/desk.

We must add the following measures (in inches):

$$\frac{3}{4},\quad 4\frac{1}{2},\quad 9\frac{1}{2},\quad \frac{3}{4},\quad 9\frac{1}{2},\quad \frac{3}{4},\quad 4\frac{1}{2}.$$

Begin by changing $4\frac{1}{2}$ to $4\frac{2}{4}$ and $9\frac{1}{2}$ to $9\frac{2}{4}$, since the common denominator is 4. Then, use Method 2 from Example 6(b).

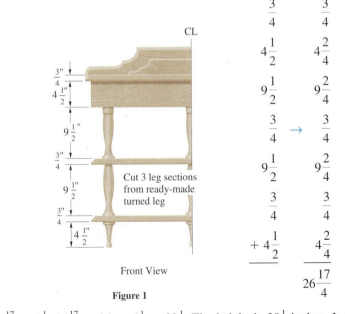

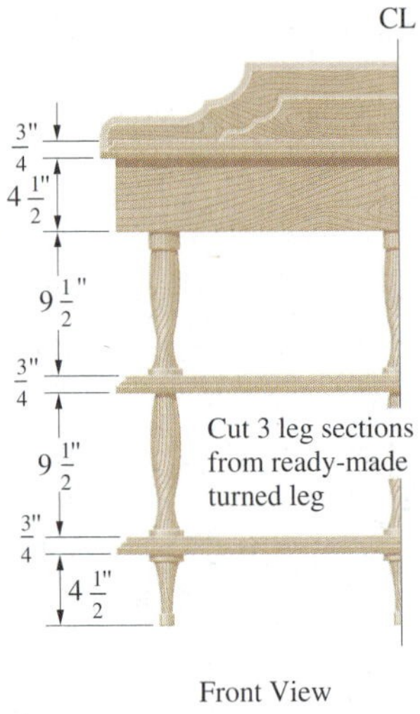

Figure 1

$$\begin{array}{rcr} \frac{3}{4} & & \frac{3}{4} \\ 4\frac{1}{2} & & 4\frac{2}{4} \\ 9\frac{1}{2} & & 9\frac{2}{4} \\ \frac{3}{4} & \rightarrow & \frac{3}{4} \\ 9\frac{1}{2} & & 9\frac{2}{4} \\ \frac{3}{4} & & \frac{3}{4} \\ +\,4\frac{1}{2} & & 4\frac{2}{4} \\ \hline & & 26\frac{17}{4} \end{array}$$

Since $\frac{17}{4} = 4\frac{1}{4}$, $26\frac{17}{4} = 26 + 4\frac{1}{4} = 30\frac{1}{4}$. The height is $30\frac{1}{4}$ inches. It is best to give answers as mixed numbers in applications like this.

OBJECTIVE 6 Interpret data in a circle graph. A **circle graph** or **pie chart** is often used to give a pictorial representation of data. A circle is used to indicate the total of all the categories represented. The circle is divided into sectors, or wedges (like pieces of pie) whose sizes show the relative magnitudes of the categories. The sum of all the fractional parts must be 1 (for 1 whole circle).

EXAMPLE 9 Using a Pie Chart to Interpret Information

The pie chart in Figure 2 shows the job categories of African Americans employed in 1993 (age 16 or older).

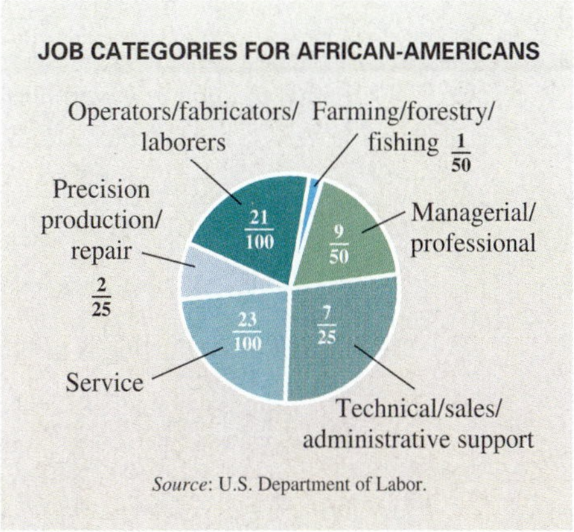

Figure 2

(a) In a group of 150,000 such employees, about how many would we expect to be employed in precision production/repair?

To find the answer, we multiply the fraction indicated in the chart for the category $\left(\frac{2}{25}\right)$ by the number of people in the group (150,000):

$$\frac{2}{25} \cdot 150{,}000 = \frac{300{,}000}{25} = 12{,}000.$$

About 12,000 people in the group are employed in precision production/repair.

(b) *Estimate* the number employed in service.

The fraction $\frac{23}{100}$ is approximately $\frac{1}{4}$. Therefore, a good estimate for this number is

$$\frac{1}{4} \cdot 150{,}000 = 37{,}500.$$

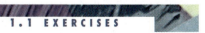

1.1 EXERCISES

Decide whether each statement is true or false. If it is false, say why.

1. In the fraction $\frac{3}{7}$, 3 is the numerator and 7 is the denominator.

2. The mixed number equivalent of $\frac{41}{5}$ is $8\frac{1}{5}$.

3. The fraction $\frac{17}{51}$ is in lowest terms.

4. The reciprocal of $\frac{8}{2}$ is $\frac{4}{1}$.

5. The product of 8 and 2 is 10.

6. The difference between 12 and 2 is 6.

Identify each number as prime, composite, or neither. If the number is composite, write it as the product of prime factors. See Example 1.

7. 19
8. 31
9. 64
10. 99
11. 3458
12. 1025
13. 1
14. 0
15. 30
16. 40
17. 500
18. 700
19. 124
20. 120
21. 29
22. 83

Write each fraction in lowest terms. See Example 2.

23. $\frac{8}{16}$
24. $\frac{4}{12}$
25. $\frac{15}{18}$
26. $\frac{16}{20}$
27. $\frac{15}{45}$
28. $\frac{16}{64}$
29. $\frac{144}{120}$
30. $\frac{132}{77}$

31. One of the following is the correct way to write $\frac{16}{24}$ in lowest terms. Which one is it?

(a) $\frac{16}{24} = \frac{8 + 8}{8 + 16} = \frac{8}{16} = \frac{1}{2}$
(b) $\frac{16}{24} = \frac{4 \cdot 4}{4 \cdot 6} = \frac{4}{6}$
(c) $\frac{16}{24} = \frac{8 \cdot 2}{8 \cdot 3} = \frac{2}{3}$
(d) $\frac{16}{24} = \frac{14 + 2}{21 + 3} = \frac{2}{3}$

32. For the fractions $\frac{p}{q}$ and $\frac{r}{s}$, which one of the following can serve as a common denominator?

(a) $q \cdot s$ (b) $q + s$ (c) $p \cdot r$ (d) $p + r$

Find each product or quotient, and write it in lowest terms. See Examples 3 and 4.

33. $\frac{4}{5} \cdot \frac{6}{7}$
34. $\frac{5}{9} \cdot \frac{10}{7}$
35. $\frac{1}{10} \cdot \frac{12}{5}$
36. $\frac{6}{11} \cdot \frac{2}{3}$
37. $\frac{15}{4} \cdot \frac{8}{25}$
38. $\frac{4}{7} \cdot \frac{21}{8}$
39. $2\frac{2}{3} \cdot 5\frac{4}{5}$
40. $3\frac{3}{5} \cdot 7\frac{1}{6}$
41. $\frac{5}{4} \div \frac{3}{8}$
42. $\frac{7}{6} \div \frac{9}{10}$
43. $\frac{32}{5} \div \frac{8}{15}$
44. $\frac{24}{7} \div \frac{6}{21}$
45. $\frac{3}{4} \div 12$
46. $\frac{2}{5} \div 30$
47. $2\frac{5}{8} \div 1\frac{15}{32}$
48. $2\frac{3}{10} \div 7\frac{4}{5}$

49. Write a summary explaining how to multiply and divide two fractions. Give examples.

50. Write a summary explaining how to add and subtract two fractions. Give examples.

Find each sum or difference, and write it in lowest terms. See Examples 5–7.

51. $\frac{7}{12} + \frac{1}{12}$
52. $\frac{3}{16} + \frac{5}{16}$
53. $\frac{5}{9} + \frac{1}{3}$
54. $\frac{4}{15} + \frac{1}{5}$
55. $3\frac{1}{8} + \frac{1}{4}$
56. $5\frac{3}{4} + \frac{2}{3}$
57. $\frac{7}{12} - \frac{1}{9}$
58. $\frac{11}{16} - \frac{1}{12}$
59. $6\frac{1}{4} - 5\frac{1}{3}$
60. $8\frac{4}{5} - 7\frac{4}{9}$
61. $\frac{5}{3} + \frac{1}{6} - \frac{1}{2}$
62. $\frac{7}{15} + \frac{1}{6} - \frac{1}{10}$

The following chart appears on a package of Quaker Quick Grits.

	Microwave	Stove Top		
Servings	1	1	4	6
Water	$\frac{3}{4}$ cup	1 cup	3 cups	4 cups
Grits	3 Tbsp	3 Tbsp	$\frac{3}{4}$ cup	1 cup
Salt (optional)	dash	dash	$\frac{1}{4}$ tsp	$\frac{1}{2}$ tsp

Use the chart to answer the questions in Exercises 63 and 64.

63. How many cups of water would be needed for 8 microwave servings?

64. How many tsp of salt would be needed for 5 stove top servings? (*Hint:* 5 is halfway between 4 and 6.)

Work each problem. See Example 8.

65. On Tuesday, February 10, 1998, Earthlink stock on the NASDAQ exchange closed the day at $4\frac{5}{8}$ (dollars) ahead of where it had opened. It closed at $38\frac{5}{8}$ (dollars). What was its opening price?

66. A report in *USA Today* on February 10, 1998, stated that Teva Pharmaceutical skidded $9\frac{9}{16}$ (dollars) to $37\frac{1}{2}$ (dollars) after the Israeli drugmaker said fourth-quarter net income was likely to be below expectations. What was its price before the skid?

67. A hardware store sells a 40-piece socket wrench set. The measure of the largest socket is $\frac{3}{4}$ inch, while the measure of the smallest socket is $\frac{3}{16}$ inch. What is the difference between these measures?

68. Two sockets in a socket wrench set have measures of $\frac{9}{16}$ inch and $\frac{3}{8}$ inch. What is the difference between these two measures?

69. A motel owner has decided to expand his business by buying a piece of property next to the motel. The property has an irregular shape, with five sides as shown in the figure. Find the total distance around the piece of property. (This is called the *perimeter* of the figure.)

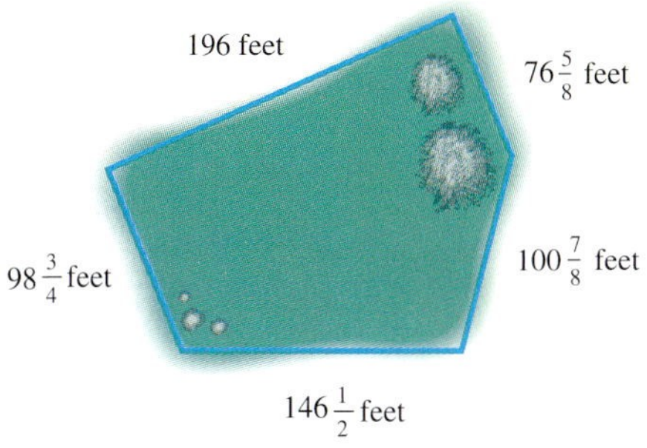

1.2 Exponents, Order of Operations, and Inequality

OBJECTIVES

1. Use exponents.
2. Use the order of operations rules.
3. Use more than one grouping symbol.
4. Know the meanings of $\neq$, $<$, $>$, $\leq$, and $\geq$.
5. Translate word statements to symbols.
6. Write statements that change the direction of inequality symbols.
7. Interpret data in a bar graph.

FOR EXTRA HELP

SSG Sec. 1.2
SSM Sec. 1.2
Pass the Test Software
InterAct Math Tutorial Software
Video 1

OBJECTIVE 1 Use exponents. In a multiplication problem, the same factor may appear several times. For example, in the product

$$3 \cdot 3 \cdot 3 \cdot 3 = 81,$$

the factor 3 appears four times. In algebra, repeated factors are written with an *exponent.* For example, in $3 \cdot 3 \cdot 3 \cdot 3$, the number 3 appears as a factor four times, so the product is written as 3^4, and is read "3 to the **fourth power**."

$$3 \cdot 3 \cdot 3 \cdot 3 = 3^4$$

The number 4 is the **exponent** or **power** and 3 is the **base** in the **exponential expression** 3^4. A natural number exponent, then, tells how many times the base is used as a factor. A number raised to the first power is simply that number. For example, $5^1 = 5$ and $\left(\frac{1}{2}\right)^1 = \frac{1}{2}$.

EXAMPLE 1 Evaluating an Exponential Expression

Find the values of the following.

(a) 5^2 $\quad 5 \cdot 5 = 25$ $\quad$ 5 is used as a factor 2 times.

Read 5^2 as "5 squared."

(b) 6^3 $\quad 6 \cdot 6 \cdot 6 = 216$ $\quad$ 6 is used as a factor 3 times.

Read 6^3 as "6 cubed."

(c) $2^5 = 2 \cdot 2 \cdot 2 \cdot 2 \cdot 2 = 32$ $\quad$ 2 is used as a factor 5 times.

Read 2^5 as "2 to the fifth power."

(d) $\left(\frac{2}{3}\right)^3 = \frac{2}{3} \cdot \frac{2}{3} \cdot \frac{2}{3} = \frac{8}{27}$ $\quad$ $\frac{2}{3}$ is used as a factor 3 times.

OBJECTIVE 2 Use the order of operations rules. Many problems involve more than one operation. To indicate the order in which the operations should be performed, we often use *grouping symbols.* If no grouping symbols are used, we apply the order of operations rules discussed below.

Consider the expression $5 + 2 \cdot 3$. To show that the multiplication should be performed before the addition, parentheses can be used to write

$$5 + (2 \cdot 3) = 5 + 6 = 11.$$

If addition is to be performed first, the parentheses should group $5 + 2$ as follows.

$$(5 + 2) \cdot 3 = 7 \cdot 3 = 21$$

Other grouping symbols used in more complicated expressions are brackets [ ], braces { }, and fraction bars. (For example, in $\frac{8-2}{3}$, the expression $8 - 2$ is considered to be grouped in the numerator.)

To work problems with more than one operation, use the following **order of operations.** This order is used by most calculators and computers.

Order of Operations

If grouping symbols are present, simplify within them, innermost first (and above and below fraction bars separately), in the following order.

Step 1 Apply all exponents.

Step 2 Do any multiplications or divisions in the order in which they occur, working from left to right.

Step 3 Do any additions or subtractions in the order in which they occur, working from left to right.

If no grouping symbols are present, start with Step 1.

A dot has been used to show multiplication; another way to show multiplication is with parentheses. For example, 3(7), (3)7, and (3)(7) each mean $3 \cdot 7$ or 21. The next example shows the use of parentheses for multiplication.

EXAMPLE 2 Using the Order of Operations

Find the values of the following.

(a) $9(6 + 11)$

Using the order of operations given above, work first inside the parentheses.

$$9(6 + 11) = 9(17) \quad \text{Work inside parentheses.}$$
$$= 153 \quad \text{Multiply.}$$

(b) $6 \cdot 8 + 5 \cdot 2$

Do any multiplications, working from left to right, and then add.

$$6 \cdot 8 + 5 \cdot 2 = 48 + 10 \quad \text{Multiply.}$$
$$= 58 \quad \text{Add.}$$

(c) $2(5 + 6) + 7 \cdot 3 = 2(11) + 7 \cdot 3$ Work inside parentheses.

$$= 22 + 21 \quad \text{Multiply.}$$
$$= 43 \quad \text{Add.}$$

(d) $9 - 2^3 + 5$

Find 2^3 first.

$$9 - 2^3 + 5 = 9 - 2 \cdot 2 \cdot 2 + 5 \quad \text{Use the exponent.}$$
$$= 9 - 8 + 5 \quad \text{Multiply.}$$
$$= 1 + 5 \quad \text{Subtract.}$$
$$= 6 \quad \text{Add.}$$

(e) $72 \div 2 \cdot 3 + 4 \cdot 2^3$

$$72 \div 2 \cdot 3 + 4 \cdot 2^3 = 72 \div 2 \cdot 3 + 4 \cdot 8 \quad \text{Use the exponent.}$$
$$= 36 \cdot 3 + 4 \cdot 8 \quad \text{Perform the division.}$$
$$= 108 + 32 \quad \text{Perform the multiplications.}$$
$$= 140 \quad \text{Add.}$$

Notice that the multiplications and divisions are performed from left to right *as they appear;* then the additions and subtractions should be done from left to right, *as they appear.*

OBJECTIVE 3 **Use more than one grouping symbol.** An expression with double (or *nested*) parentheses, such as 2(8 + 3(6 + 5)), can be confusing. For clarity, square brackets, [ ], often are used in place of one pair of parentheses. Fraction bars also act as grouping symbols. The next example explains these situations.

EXAMPLE 3 **Using Brackets and Fraction Bars as Grouping Symbols**

Simplify each expression.

(a) $2[8 + 3(6 + 5)]$

Work first within the parentheses, and then simplify inside the brackets until a single number remains.

$$\begin{aligned} 2[8 + 3(6 + 5)] &= 2[8 + 3(11)] \\ &= 2[8 + 33] \\ &= 2[41] \\ &= 82 \end{aligned}$$

(b) $\dfrac{4(5 + 3) + 3}{2(3) - 1}$

The expression can be written as the quotient

$$[4(5 + 3) + 3] \div [2(3) - 1],$$

which shows that the fraction bar groups the numerator and denominator separately. Simplify both numerator and denominator, then divide, if possible.

$$\begin{aligned} \frac{4(5 + 3) + 3}{2(3) - 1} &= \frac{4(8) + 3}{2(3) - 1} && \text{Work inside parentheses.} \\ &= \frac{32 + 3}{6 - 1} && \text{Multiply.} \\ &= \frac{35}{5} && \text{Add and subtract.} \\ &= 7 && \text{Divide.} \end{aligned}$$

NOTE Parentheses and fraction bars are used as grouping symbols to indicate an expression that represents a single number. That is why we must first simplify within parentheses and above and below fraction bars.

OBJECTIVE 4 **Know the meanings of ≠, <, >, ≤, and ≥.** So far, we have used the symbols for the operations of arithmetic and the symbol for equality (=). The equality symbol with a slash through it, ≠, means "is not equal to." For example,

$$7 \neq 8$$

indicates that 7 is not equal to 8.

If two numbers are not equal, then one of the numbers must be less than the other. The symbol $<$ represents "is less than," so that "7 is less than 8" is written

$$7 < 8.$$

Also, write "6 is less than 9" as $6 < 9$.

The symbol $>$ means "is greater than." Write "8 is greater than 2" as

$$8 > 2.$$

The statement "17 is greater than 11" becomes $17 > 11$.

Keep the meanings of the symbols $<$ and $>$ clear by remembering that the symbol always *points to the smaller number.* For example, write "8 is less than 15" by pointing the symbol toward the 8:

Smaller number → $8 < 15.$

Two other symbols, $\leq$ and $\geq$, also represent the idea of inequality. The symbol $\leq$ means "is less than or equal to," so that

$$5 \leq 9$$

means "5 is less than or equal to 9." This statement is true, since $5 < 9$ is true. **If either the $<$ part or the $=$ part is true, then the inequality $\leq$ is true.**

The symbol $\geq$ means "is greater than or equal to." Again,

$$9 \geq 5$$

is true because $9 > 5$ is true. Also, $8 \leq 8$ is true since $8 = 8$ is true. But it is not true that $13 \leq 9$ because neither $13 < 9$ nor $13 = 9$ is true.

EXAMPLE 4 Using Inequality Symbols

Determine whether each statement is true or false.

(a) $6 \neq 6$
The statement is false because 6 *is equal to* 6.

(b) $5 < 19$
Since 5 represents a number that is indeed less than 19, this statement is true.

(c) $15 \leq 20$
The statement $15 \leq 20$ is true, since $15 < 20$.

(d) $25 \geq 30$
Both $25 > 30$ and $25 = 30$ are false. Because of this, $25 \geq 30$ is false.

(e) $12 \geq 12$
Since $12 = 12$, this statement is true.

OBJECTIVE 5 Translate word statements to symbols. An important part of algebra deals with translating words into algebraic notation.

PROBLEM SOLVING

As we will see throughout this book, the ability to solve problems using mathematics is based on translating the words of the problem into symbols. The next example is the first of many that illustrate such translations.

EXAMPLE 5 Translating from Words to Symbols

Write each word statement in symbols.

(a) Twelve **equals** ten **plus** two.

$12 = 10 + 2$

(b) Nine **is less than** ten.

$9 < 10$

(c) Fifteen **is not equal to** eighteen.

$15 \neq 18$

(d) Seven **is greater than** four.

$7 > 4$

(e) Thirteen **is less than or equal to** forty.

$13 \leq 40$

(f) Eleven **is greater than or equal to** eleven.

$11 \geq 11$

OBJECTIVE 6 Write statements that change the direction of inequality symbols. Any statement with $<$ can be converted to one with $>$, and any statement with $>$ can be converted to one with $<$. We do this by reversing the order of the numbers and the direction of the symbol. For example, the statement $6 < 10$ can be written with $>$ as $10 > 6$. Similarly, the statement $4 \leq 10$ can be changed to $10 \geq 4$.

EXAMPLE 6 Converting between Inequality Symbols

The following examples show the same statement written in two equally correct ways.

(a) $9 < 16$ $\quad 16 > 9$

(b) $5 > 2$ $\quad 2 < 5$

(c) $3 \leq 8$ $\quad 8 \geq 3$

(d) $12 \geq 5$ $\quad 5 \leq 12$

Note that in each pair of inequalities, the point of the inequality symbol points toward the smaller number.

Here is a summary of the symbols discussed in this section.

Symbols of Equality and Inequality

$=$	is equal to	$\neq$	is not equal to
$<$	is less than	$>$	is greater than
$\leq$	is less than or equal to	$\geq$	is greater than or equal to

CAUTION The equality and inequality symbols are used to write mathematical *sentences* that describe the relationship between two numbers. On the other hand, the symbols for operations ($+$, $-$, $\times$, $\div$) are used to write mathematical *expressions* that represent a single number. For example, compare the sentence $4 < 10$ with the expression $4 + 10$, which represents the number 14.

OBJECTIVE 7 Interpret data in a bar graph. **Bar graphs** are often used to summarize data in a concise manner.

EXAMPLE 7 Interpreting Inequality Concepts Using a Bar Graph

Figure 3 shows a bar graph that depicts the percentage growth of the ten fastest-growing U.S. manufacturing industries from 1993 to 1994.

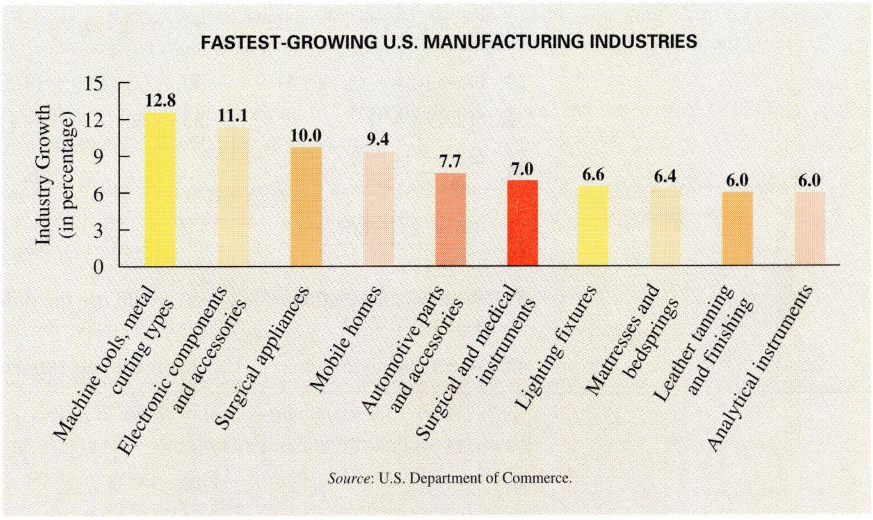

Figure 3

(a) Which industries showed growth greater than 9.4%?

Machine tools, metal cutting types (at 12.8%), electronic components and accessories (at 11.1%), and surgical appliances (at 10.0%) all showed growth greater than 9.4%.

(b) Which industries showed growth greater than or equal to 9.4%?

The three industries listed in part (a) and the mobile home industry, at 9.4%, showed growth greater than or equal to 9.4%. (These are industries that showed *at least* 9.4% growth.)

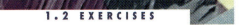

1.2 EXERCISES

Decide whether each statement is true or false. If it is false, explain why.

1. When evaluated, $4 + 3(8 - 2)$ is equal to 42.

2. $3^3 = 9$

3. The statement "4 is 12 less than 16" is interpreted $4 = 12 - 16$.

4. The statement "6 is 4 less than 10" is interpreted $6 < 10 - 4$.

Find the value of each exponential expression. See Example 1.

5. 7^2 **6.** 4^2 **7.** 12^2 **8.** 14^2

9. 4^3 **10.** 5^3 **11.** 10^3 **12.** 11^3

13. 3^4 **14.** 6^4 **15.** 4^5 **16.** 3^5

17. $\left(\frac{2}{3}\right)^4$ **18.** $\left(\frac{3}{4}\right)^3$ **19.** $(.04)^3$ **20.** $(.05)^4$

21. Explain in your own words how to evaluate a power of a number, such as 6^3.

22. Explain why any power of 1 must be equal to 1.

Find the value of each expression. See Examples 2 and 3.

23. $9 \cdot 5 - 13$ **24.** $7 \cdot 6 - 11$ **25.** $\frac{1}{4} \cdot \frac{2}{3} + \frac{2}{5} \cdot \frac{11}{3}$

4 Identify solutions of equations from a set of numbers.

5 Distinguish between an *expression* and an *equation.*

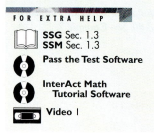

FOR EXTRA HELP

SSG Sec. 1.3
SSM Sec. 1.3
Pass the Test Software
InterAct Math Tutorial Software
Video 1

EXAMPLE 1 Evaluating Expressions

Find the numerical values of the following algebraic expressions when $m = 5$.

(a) $8m$
Replace m with 5, to get

$$8m = 8 \cdot 5 = 40.$$

(b) $3m^2$
For $m = 5$,

$$3m^2 = 3 \cdot 5^2 = 3 \cdot 25 = 75.$$

CAUTION In Example 1(b), notice that $3m^2$ means $3 \cdot m^2$; it *does not* mean $3m \cdot 3m$. The product $3m \cdot 3m$ is indicated by $(3m)^2$.

EXAMPLE 2 Evaluating Expressions

Find the value of each expression when $x = 5$ and $y = 3$.

(a) $2x + 7y$
Replace x with 5 and y with 3. Follow the order of operations; multiply first, then add.

$$\begin{aligned} 2x + 7y &= 2 \cdot 5 + 7 \cdot 3 && \text{Let } x = 5 \text{ and } y = 3. \\ &= 10 + 21 && \text{Multiply.} \\ &= 31 && \text{Add.} \end{aligned}$$

(b) $\dfrac{9x - 8y}{2x - y}$
Replace x with 5 and y with 3.

$$\begin{aligned} \frac{9x - 8y}{2x - y} &= \frac{9 \cdot 5 - 8 \cdot 3}{2 \cdot 5 - 3} && \text{Let } x = 5 \text{ and } y = 3. \\ &= \frac{45 - 24}{10 - 3} && \text{Multiply.} \\ &= \frac{21}{7} && \text{Subtract.} \\ &= 3 && \text{Divide.} \end{aligned}$$

(c)
$$\begin{aligned} x^2 - 2y^2 &= 5^2 - 2 \cdot 3^2 && \text{Let } x = 5 \text{ and } y = 3. \\ &= 25 - 2 \cdot 9 && \text{Use the exponents.} \\ &= 25 - 18 && \text{Multiply.} \\ &= 7 && \text{Subtract.} \end{aligned}$$

OBJECTIVE 2 Convert phrases from words to algebraic expressions. In Section 1.2 we saw how to translate from words to symbols.

PROBLEM SOLVING

Sometimes variables must be used to change word phrases into algebraic expressions. The next example illustrates this. Such translations are used in problem solving.

EXAMPLE 3 Changing Word Phrases to Algebraic Expressions

Change the following word phrases to algebraic expressions. Use x as the variable.

(a) The **sum** of a number and 9

"**Sum**" is the answer to an addition problem. This phrase translates as

$$x + 9 \qquad \text{or} \qquad 9 + x.$$

(b) 7 **minus** a number

"**Minus**" indicates subtraction, so the answer is $7 - x$.

CAUTION

Here $x - 7$ would *not* be correct; this statement translates as "a number minus 7," not "7 minus a number." The expressions $7 - x$ and $x - 7$ are rarely equal. For example, if $x = 10$, $10 - 7 \neq 7 - 10$. ($7 - 10$ is a *negative number,* discussed in Section 1.4.)

(c) 7 **less than** a number

Write 7 **less than** a number as $x - 7$. In this case $7 - x$ would not be correct, because "less than" means "subtracted from."

(d) The product of 11 and a number

$$11 \cdot x \qquad \text{or} \qquad 11x$$

As mentioned earlier, $11x$ means 11 times x. No symbol is needed to indicate the product of a number and a variable.

(e) 5 **divided by** a number

This translates as $\frac{5}{x}$. The expression $\frac{x}{5}$ would *not* be correct here.

(f) The **product** of 2, and the **sum** of a number and 8

We are multiplying 2 times another number. This number is the sum of x and 8, written $x + 8$. Using parentheses for this sum, the final expression is

$$2(x + 8).$$

OBJECTIVE 3 Identify solutions of equations. An **equation** is a statement that two algebraic expressions are equal. Therefore, an equation *always* includes the equality symbol, =. Examples of equations are

$$x + 4 = 11, \qquad 2y = 16, \qquad \text{and} \qquad 4p + 1 = 25 - p.$$

Solving an Equation

To **solve** an equation means to find the values of the variable that make the equation true. Such values of the variable are called the **solutions** of the equation.

EXAMPLE 4 Deciding whether a Number Is a Solution

Decide whether the given number is a solution of the equation.

(a) $5p + 1 = 36$; 7

Replace p with 7.

$$5p + 1 = 36$$
$$5 \cdot 7 + 1 = 36 \quad ? \qquad \text{Let } p = 7.$$
$$35 + 1 = 36 \quad ?$$
$$36 = 36 \qquad \text{True}$$

The number 7 is a solution of the equation.

(b) $9m - 6 = 32$; 4

$$9m - 6 = 32$$
$$9 \cdot 4 - 6 = 32 \quad ? \qquad \text{Let } m = 4.$$
$$36 - 6 = 32 \quad ?$$
$$30 = 32 \qquad \text{False}$$

The number 4 is not a solution of the equation.

OBJECTIVE 4 **Identify solutions of equations from a set of numbers.** A **set** is a collection of objects. In mathematics, these objects are most often numbers. The objects that belong to the set, called **elements** of the set, are written between **set braces.** For example, the set containing the numbers 1, 2, 3, 4, and 5 is written as

$$\{1, 2, 3, 4, 5\}.$$

For more information about sets, see Appendix B at the back of this book.

In some cases, the set of numbers from which the solutions of an equation must be chosen is specifically stated. One way of determining solutions is direct substitution of all possible replacements. The ones that lead to a true statement are solutions.

EXAMPLE 5 Finding a Solution from a Given Set

Change each word statement to an equation. Use x as the variable. Then find all solutions of the equation from the set

$$\{0, 2, 4, 6, 8, 10\}.$$

(a) The sum of a number and four is six.

The word "is" suggests "equals." If x represents the unknown number, then translate as follows.

The sum of a number and four	is	six.
↓	↓	↓
$x + 4$	$=$	6

Try each number from the given set $\{0, 2, 4, 6, 8, 10\}$, in turn, to see that 2 is the only solution of $x + 4 = 6$.

(b) 9 more than five times a number is 49.

Use x to represent the unknown number. We start with $5x$ and then add 9 to it. The word "is" translates as $=$.

$$5x + 9 = 49$$

Try each number from $\{0, 2, 4, 6, 8, 10\}$. The solution is **8**, since $5 \cdot 8 + 9 = 49$.

(c) The sum of a number and 12 is equal to four times the number.

If x represents the number, "the sum of a number and 12" is represented by $x + 12$. The translation is

$$x + 12 = 4x.$$

Trying each replacement leads to a true statement when $x = 4$, since $4 + 12 = 4(4) = 16$.

OBJECTIVE 5 Distinguish between an *expression* and an *equation*. Students often have trouble distinguishing between equations and expressions. Remember that an equation is a sentence; an expression is a phrase.

$4x + 5 = 9$	$4x + 5$
↑	↑
Equation (to solve)	Expression (to simplify or evaluate)

EXAMPLE 6 Distinguishing between Equations and Expressions

Decide whether each of the following is an equation or an expression.

(a) $2x - 5y$

There is no equals sign, so this is an expression.

(b) $2x = 5y$

Because of the equals sign, this is an equation.

1.3 EXERCISES

Fill in each blank with the correct response.

1. If $x = 3$, then the value of $x + 7$ is ______.

2. If $x = 1$ and $y = 2$, then the value of $4xy$ is ______.

3. The sum of 12 and x is represented by the expression ______. If $x = 9$, the value of that expression is ______.

4. If x can be chosen from the set $\{0, 1, 2, 3, 4, 5\}$, the only solution of $x + 5 = 9$ is ______.

5. Will the equation $x = x + 4$ ever have a solution? ______.

6. $2x + 3$ is a(n) __________________ (equation/expression), while $2x + 3 = 8$ is a(n) __________________ (equation/expression).

Exercises 7–12 cover some of the concepts introduced in this section. Give a short explanation for each.

7. Why is $2x^3$ not the same as $2x \cdot 2x \cdot 2x$? Explain.

8. Why are "5 less than a number" and "5 is less than a number" translated differently?

9. Explain in your own words why, when evaluating the expression $4x^2$ for $x = 3$, 3 must be squared *before* multiplying by 4.

10. What value of x would cause the expression $2x + 3$ to equal 9? Explain your reasoning.

11. There are many pairs of values of x and y for which $2x + y$ will equal 6. Name two such pairs and describe how you determined them.

12. Suppose that for the equation $3x - y = 9$, the value of x is given as 4. What would be the corresponding value of y? How do you know this?

*Find the numerical value (**a**) if $x = 4$ and (**b**) if $x = 6$. See Example 1.*

13. $x + 9$
14. $x - 1$
15. $5x$
16. $7x$
17. $4x^2$
18. $5x^2$
19. $\frac{x+1}{3}$
20. $\frac{x-2}{5}$
21. $\frac{3x-5}{2x}$
22. $\frac{4x-1}{3x}$
23. $3x^2 + x$
24. $2x + x^2$
25. $6.459x$
26. $.74x^2$

*Find the numerical value if (**a**) $x = 2$ and $y = 1$ and (**b**) $x = 1$ and $y = 5$. See Example 2.*

27. $8x + 3y + 5$
28. $4x + 2y + 7$
29. $3(x + 2y)$
30. $2(2x + y)$
31. $x + \frac{4}{y}$
32. $y + \frac{8}{x}$
33. $\frac{x}{2} + \frac{y}{3}$
34. $\frac{x}{5} + \frac{y}{4}$
35. $\frac{2x + 4y - 6}{5y + 2}$
36. $\frac{4x + 3y - 1}{x}$
37. $2y^2 + 5x$
38. $6x^2 + 4y$
39. $\frac{3x + y^2}{2x + 3y}$
40. $\frac{x^2 + 1}{4x + 5y}$
41. $.841x^2 + .32y^2$
42. $.941x^2 + .2y^2$

Change each word phrase to an algebraic expression. Use x as the variable to represent the number. See Example 3.

43. Twelve times a number
44. Nine times a number
45. Seven added to a number
46. Thirteen added to a number
47. Two subtracted from a number
48. Eight subtracted from a number
49. A number subtracted from seven
50. A number subtracted from fourteen
51. The difference between a number and 6
52. The difference between 6 and a number
53. 12 divided by a number
54. A number divided by 12
55. The product of 6 and four less than a number
56. The product of 9 and five more than a number
57. In the phrase "Four more than the product of a number and 6," does the word *and* signify the operation of addition? Explain.
58. Suppose that the directions on a test read "Solve the following expressions." How would you politely correct the person who wrote these directions? What alternative directions might you suggest?

Decide whether each given number is a solution of the equation. See Example 4.

59. $5m + 2 = 7;\ \ 1$
60. $3r + 5 = 8;\ \ 1$
61. $2y + 3(y - 2) = 14;\ \ 3$
62. $6a + 2(a + 3) = 14;\ \ 2$
63. $6p + 4p + 9 = 11;\ \ \frac{1}{5}$
64. $2x + 3x + 8 = 20;\ \ \frac{12}{5}$
65. $3r^2 - 2 = 46;\ \ 4$
66. $2x^2 + 1 = 19;\ \ 3$
67. $\frac{z+4}{2-z} = \frac{13}{5};\ \ \frac{1}{3}$
68. $\frac{x+6}{x-2} = \frac{37}{5};\ \ \frac{13}{4}$

Change each word statement to an equation. Use x as the variable. Find all solutions from the set {2, 4, 6, 8, 10}. See Example 5.

69. The sum of a number and 8 is 18.
70. A number minus three equals 1.
71. Sixteen minus three-fourths of a number is 13.
72. The sum of six-fifths of a number and 2 is 14.
73. One more than twice a number is 5.
74. The product of a number and 3 is 6.
75. Three times a number is equal to 8 more than twice the number.
76. Twelve divided by a number equals $\frac{1}{3}$ times that number.

Identify as an expression *or an* equation. *See Example 6.*

77. $3x + 2(x - 4)$ **78.** $5y - (3y + 6)$ **79.** $7t + 2(t + 1) = 4$

80. $9r + 3(r - 4) = 2$ **81.** $x + y = 3$ **82.** $x + y - 3$

Mathematicians who study statistics have developed methods of determining mathematical models. Loosely speaking, a **mathematical model** is an equation that can be used to determine unknown quantities. We cannot always expect a model to give us an accurate answer, but at least we can obtain a rough estimate. For example, based on data from the U.S. Bureau of Labor Statistics, average hourly earnings of production workers in manufacturing industries in the United States during the period from 1990 through 1996 can be approximated by the equation $y = .319x - 624.31$, where x represents the year and y represents the hourly earnings in dollars.

Use this equation to approximate the average hourly earnings during the following years. Do this by replacing x in the equation by the year, and then simplifying the right side of the equation to find y, the hourly earnings. Then compare your answer to the actual earnings (given in parentheses) and tell whether the approximation is greater than or less than the actual earnings, and by how much.

83. 1990 ($10.83)

84. 1994 ($12.07)

85. 1995 ($12.37)

86. 1996 ($12.78)

1.4 Real Numbers and the Number Line

OBJECTIVES

1. Set up number lines.
2. Classify numbers.
3. Tell which of two different real numbers is smaller.
4. Find additive inverses and absolute values of real numbers.
5. Interpret real number meanings from a table of data.

FOR EXTRA HELP

SSG Sec. 1.4
SSM Sec. 1.4
Pass the Test Software
InterAct Math Tutorial Software
Video 1

OBJECTIVE 1 Set up number lines. In Section 1.1 we introduced two important sets of numbers, the *natural numbers* and the *whole numbers.*

Natural Numbers

{1, 2, 3, 4, . . .} is the set of **natural numbers.**

Whole Numbers

{0, 1, 2, 3, . . .} is the set of **whole numbers.**

NOTE The three dots show that the list of numbers continues in the same way indefinitely.

These numbers, along with many others, can be represented on **number lines** like the one pictured in Figure 4. We draw a number line by choosing any point on the line and calling it 0. Choose any point to the right of 0 and call it 1. The distance between 0 and 1 gives a unit of measure used to locate other points, as shown in Figure 4. The points labeled in Figure 4 and those continuing in the same way to the right correspond to the set of whole numbers.

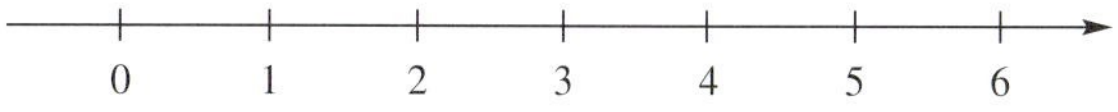

Figure 4

All the whole numbers starting with 1 are located to the right of 0 on the number line. But numbers may also be placed to the left of 0. These numbers, written -1, -2, -3, and so on, shown in Figure 5, are called **negative numbers.** (The minus sign is used to show that these numbers are located to the *left* of 0.) The numbers to the *right* of 0 are **positive numbers.** The number 0 itself is neither positive nor negative. Positive numbers and negative numbers are called **signed numbers.**

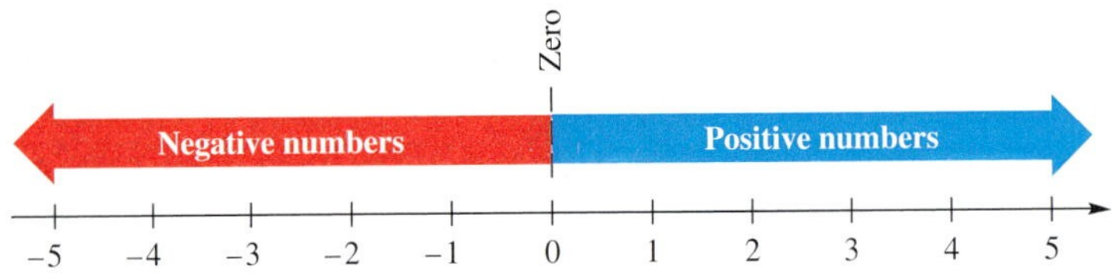

Figure 5

CONNECTIONS

There are many practical applications of negative numbers. For example, temperatures sometimes fall below zero. The lowest temperature ever recorded in meteorological records was -128.6°F at Vostok, Antarctica, on July 22, 1983. A business that spends more than it takes in has a negative "profit." Altitudes below sea level can be represented by negative numbers. The shore surrounding the Dead Sea is 1312 feet below sea level; this can be represented as -1312 feet. (*Source: The World Almanac and Book of Facts,* 1995.)

OBJECTIVE 2 **Classify numbers.** The set of numbers marked on the number line in Figure 5, including positive and negative numbers and zero, is part of the set of *integers.*

Integers

$\{\ldots, -3, -2, -1, 0, 1, 2, 3, \ldots\}$ is the set of **integers.**

Not all numbers are integers. For example, $\frac{1}{2}$ is not; it is a number halfway between the integers 0 and 1. Also, $3\frac{1}{4}$ is not an integer. Several numbers that are not integers are *graphed* in Figure 6. The **graph** of a number is a point on the number line. The number is called the **coordinate** of the point. Think of the graph of a set of numbers as a picture of the set. All the numbers in Figure 6 can be written as quotients of integers. These numbers are examples of *rational numbers.*

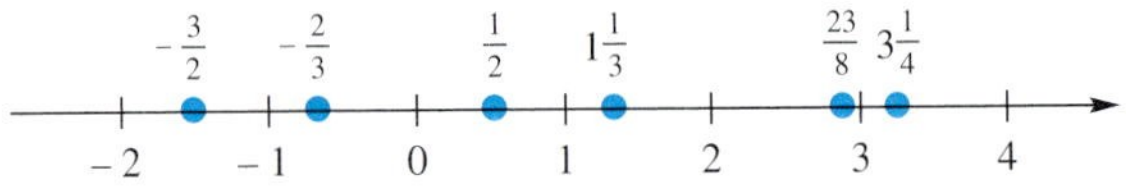

Figure 6

Rational Numbers

$\{x \mid x$ is a quotient of two integers, with denominator not $0\}$ is the set of **rational numbers.**

(Read the part in the braces as "the set of all numbers x such that x is a quotient of two integers, with denominator not 0.")

NOTE

The set symbolism used in the definition of rational numbers,

$$\{x \mid x \text{ has a certain property}\},$$

is called **set-builder notation.** This notation is convenient to use when it is not possible to list all the elements of the set.

Since any integer can be written as the quotient of itself and 1, all integers also are rational numbers.

Although many numbers are rational, not all are. For example, a square that measures one unit on a side has a diagonal whose length is the square root of 2, written $\sqrt{2}$. See Figure 7. It can be shown that $\sqrt{2}$ cannot be written as a quotient of integers. Because of this, $\sqrt{2}$ is not rational; it is *irrational*. Other examples of irrational numbers are $\sqrt{3}$, $\sqrt{7}$, $-\sqrt{10}$, and π (the ratio of the *circumference* of a circle to its diameter).

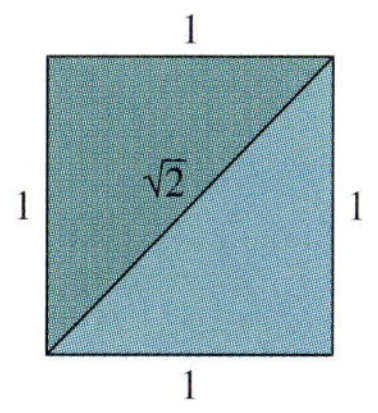

Figure 7

All numbers, both rational and irrational, can be represented by points on the number line and are called *real numbers*.

Real Numbers

$\{x \mid x$ is a rational or an irrational number$\}$ is the set of **real numbers.**

Real numbers can be written as decimals. Any rational number will have a decimal that either comes to an end (terminates) or repeats in a fixed "block" of digits. For example, $\frac{2}{5} = .4$ and $\frac{27}{100} = .27$ are rational numbers with terminating decimals; $\frac{1}{3} = .3333\ldots$ and $\frac{3}{11} = .27272727\ldots$ are repeating decimals. The decimal representation of an irrational number will neither terminate nor repeat. (A review of operations with decimals can be found in Appendix A.)

An example of a number that is not a real number is the square root of a negative number like $\sqrt{-5}$. These numbers are discussed in a later chapter.

Two ways to represent the relationships among the various types of numbers are shown in Figure 8. Part (a) also gives some examples. Notice that every real number is either a rational number or an irrational number.

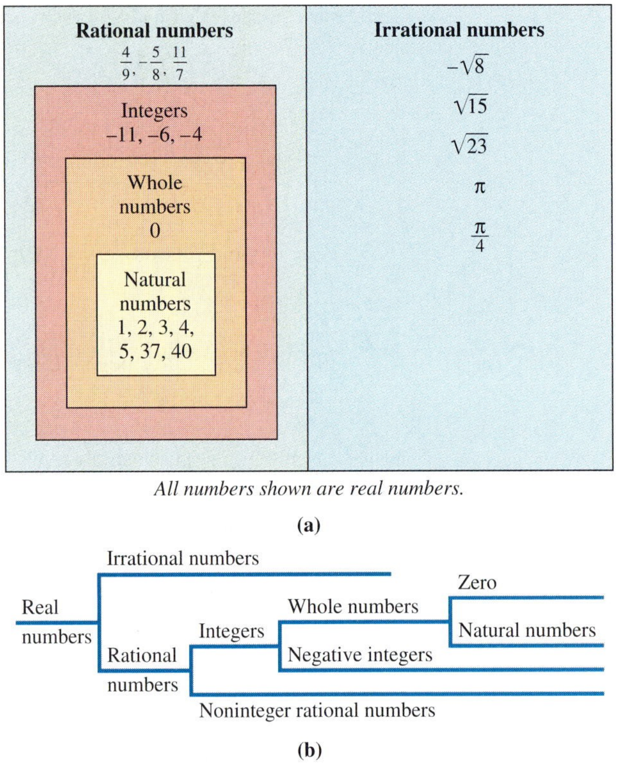

Figure 8

EXAMPLE 1 Determining whether a Number Belongs to a Set

List the numbers in the set

$$\left\{-5, \quad -\frac{2}{3}, \quad 0, \quad \sqrt{2}, \quad 3\frac{1}{4}, \quad 5, \quad 5.8\right\}$$

that belong to each of the following sets of numbers.

(a) Natural numbers

The only natural number in the set is 5.

(b) Whole numbers

The whole numbers consist of the natural numbers and 0. So the elements of the set that are whole numbers are 0 and 5.

(c) Integers

The integers in the set are -5, 0, and 5.

(d) Rational numbers

The rational numbers are $-5, -\frac{2}{3}, 0, 3\frac{1}{4}, 5, 5.8$, since each of these numbers *can* be written as the quotient of two integers. For example, $5.8 = \frac{58}{10}$.

(e) Irrational numbers

The only irrational number in the set is $\sqrt{2}$.

(f) Real numbers

All the numbers in the set are real numbers.

OBJECTIVE 3 **Tell which of two different real numbers is smaller.** Given any two whole numbers, you probably can tell which number is smaller. But what happens with negative numbers, as in the set of integers? Positive numbers decrease as the corresponding points on the number line go to the left. For example, $8 < 12$, and 8 is to the left of 12 on the number line. This ordering is extended to all real numbers by definition.

The Ordering of Real Numbers

For any two real numbers a and b, ***a* is less than *b*** if a is to the left of b on the number line.

This means that any negative number is smaller than 0, and any negative number is smaller than any positive number. Also, 0 is smaller than any positive number.

EXAMPLE 2 **Determining the Order of Real Numbers**

Is it true that $-3 < -1$?

To decide whether the statement is true, locate both numbers, -3 and -1, on a number line, as shown in Figure 9. Since -3 is to the left of -1 on the number line, -3 is smaller than -1. The statement $-3 < -1$ is true.

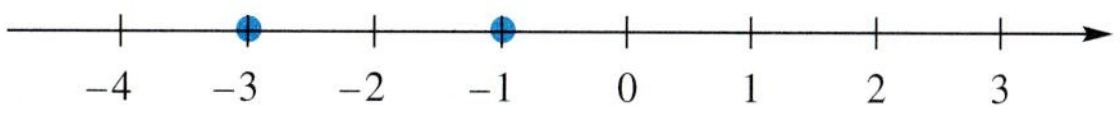

Figure 9

NOTE: In Section 1.2 we saw how it is possible to rewrite a statement involving $<$ as an equivalent statement involving $>$. The question in Example 2 can also be worded as follows: Is it true that $-1 > -3$? This is, of course, also a true statement.

We can say that for any two real numbers a and b, ***a* is greater than *b*** if a is to the right of b on the number line.

OBJECTIVE 4 **Find additive inverses and absolute values of real numbers.** By a property of the real numbers, for any real number x (except 0), there is exactly one number on the number line the same distance from 0 as x but on the opposite side of 0. For example, Figure 10 shows that the numbers 3 and -3 are each the same distance from 0 but are on opposite sides of 0. The numbers 3 and -3 are called **additive inverses,** or **opposites,** of each other.

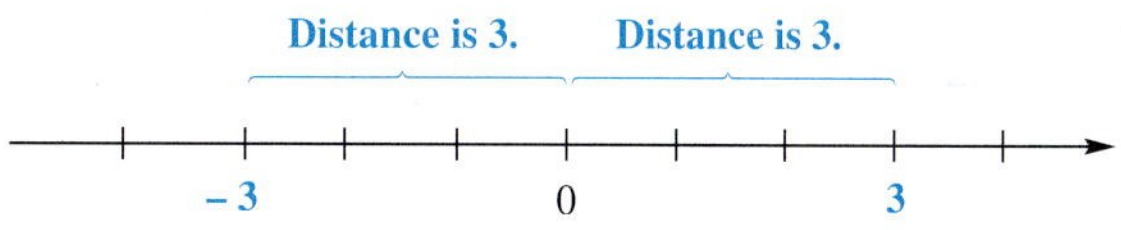

Figure 10

Additive Inverse

The **additive inverse** of a number x is the number that is the same distance from 0 on the number line as x, but on the opposite side of 0.

The additive inverse of the number 0 is 0 itself. In fact, 0 is the only real number that is its own additive inverse. Other additive inverses occur in pairs. For example, 4 and -4, and 5 and -5, are additive inverses of each other. Several pairs of additive inverses are shown in Figure 11.

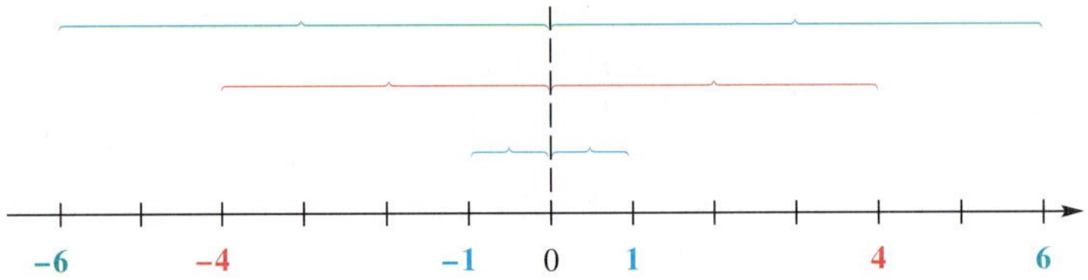

Figure 11

The additive inverse of a number can be indicated by writing the symbol $-$ in front of the number. With this symbol, the additive inverse of 7 is written -7. The additive inverse of -4 is written $-(-4)$, and can also be read "the opposite of -4" or "the negative of -4." Figure 11 suggests that 4 is an additive inverse of -4. Since a number can have only one additive inverse, the symbols 4 and $-(-4)$ must represent the same number, which means that

$$-(-4) = 4.$$

This idea can be generalized as follows.

Double Negative Rule

For any real number x,

$$-(-x) = x.$$

The following chart shows several numbers and their additive inverses.

Number	*Additive Inverse*
-4	$-(-4)$ or 4
0	0
19	-19
$-\frac{2}{3}$	$\frac{2}{3}$

NOTE The chart suggests that the additive inverse of a number is found by changing the sign of the number.

An important property of additive inverses will be studied later in this chapter: $a + (-a) = (-a) + a = 0$ for all real numbers a.

As mentioned above, additive inverses are numbers that are the same distance from 0 on the number line. See Figure 11. This idea can also be expressed by saying that a number and its additive inverse have the same absolute value. The **absolute value** of a real number can be defined as the distance between 0 and the number on the number line. The symbol for the absolute value of the number x is $|x|$, read "the absolute value of x." For example, the distance between 2 and 0 on the number line is 2 units, so that

$$|2| = 2.$$

Because the distance between -2 and 0 on the number line is also 2 units,

$$|-2| = 2.$$

Since distance is a physical measurement, which is never negative, **the absolute value of a number is never negative.** For example, $|12| = 12$ and $|-12| = 12$, since both 12 and -12 lie at a distance of 12 units from 0 on the number line. Also, since 0 is a distance of 0 units from 0, $|0| = 0$.

In symbols, the absolute value of x is defined as follows.

Formal Definition of Absolute Value

$$|x| = \begin{cases} x & \text{if } x \geq 0 \\ -x & \text{if } x < 0 \end{cases}$$

By this definition, if x is a positive number or 0, then its absolute value is x itself. For example, since 8 is a positive number, $|8| = 8$. However, if x is a negative number, then its absolute value is the additive inverse of x. This means that if $x = -9$, then $|-9| = -(-9) = 9$, since the additive inverse of -9 is 9.

CAUTION The formal definition of absolute value can be confusing if it is not read carefully. The "$-x$" in the second part of the definition *does not* represent a negative number. Since x is negative in the second part, $-x$ represents the opposite of a negative number, that is, a positive number. *The absolute value of a number is never negative.*

EXAMPLE 3 Finding Absolute Value

Simplify by finding the absolute value.

(a) $|5| = 5$

(b) $|-5| = -(-5) = 5$

(c) $-|5| = -(5) = -5$

(d) $-|-14| = -(14) = -14$

(e) $|8 - 2| = |6| = 6$

(f) $-|8 - 2| = -|6| = -6$

Part (e) of Example 3 shows that absolute value bars are also grouping symbols. You must perform any operations that appear inside absolute value symbols before finding the absolute value.

OBJECTIVE 5 Interpret real number meanings from a table of data. A table of data provides a concise way of relating information. The final example gives an illustration of this.

EXAMPLE 4 Interpreting Data

The table shows the annual percent change from the previous year in the Consumer Price Index for 1986 and 1987.

	Percent Change	
Category	*1986*	*1987*
Food	3.2	4.1
Shelter	5.5	4.7
Rent, residential	5.8	4.1
Fuel and other utilities	−2.3	−1.1
Apparel and upkeep	.9	4.4
Private transportation	−4.7	−3.0
New cars	4.2	3.6
Gasoline	−21.9	−4.0
Public transportation	5.9	3.5
Medical care	7.5	6.6
Entertainment	3.4	3.3
Commodities	−.9	3.2

Source: Bureau of Labor Statistics, U.S. Dept. of Labor.

(a) What category in which year represents the greatest drop?

To find the greatest drop, we must find the negative entry with the largest absolute value. The entry for gasoline in 1986 is -21.9, so this is the category and year with the greatest drop.

(b) Which percent change is represented by a larger drop: 1986 fuel and other utilities or 1987 gasoline?

The two entries are -2.3 and -4.0, respectively. The larger drop is -4.0, so 1987 gasoline is the answer.

(c) What is the smallest percent change (positive or negative) in these two years?

The smallest percent change (positive or negative) is .9 and $-.9$, both found in 1986.

1.4 EXERCISES

In Exercises 1–6, give a number that satisfies the given condition.

1. An integer between 3.5 and 4.5

2. A rational number between 3.8 and 3.9

3. A whole number that is not positive and is less than 1

4. A whole number greater than 4.5

5. An irrational number that is between $\sqrt{11}$ and $\sqrt{13}$

6. A real number that is neither negative nor positive

In Exercises 7–10, decide whether each statement is true or false.

7. Every natural number is positive.

8. Every whole number is positive.

9. Every integer is a rational number.

10. Every rational number is a real number.

For Exercises 11 and 12, see Example 1.

11. List all numbers from the set

$$\left\{-9, -\sqrt{7}, -1\frac{1}{4}, -\frac{3}{5}, 0, \sqrt{5}, 3, 5.9, 7\right\}$$

that are

(a) natural numbers; **(b)** whole numbers; **(c)** integers;
(d) rational numbers; **(e)** irrational numbers; **(f)** real numbers.

12. List all numbers from the set

$$\left\{-5.3, -5, -\sqrt{3}, -1, -\frac{1}{9}, 0, 1.2, 1.8, 3, \sqrt{11}\right\}$$

that are

(a) natural numbers; **(b)** whole numbers; **(c)** integers;
(d) rational numbers; **(e)** irrational numbers; **(f)** real numbers.

13. Explain in your own words the different sets of numbers introduced in this section, and give an example of each kind.

14. What two possible situations exist for the decimal representation of a rational number?

Use an integer to express each number representing a change in the following applications.

15. In February 1998, the number of housing starts in the United States increased from the previous month by 93,000 units. (*Source: The Wall Street Journal.*)

16. The Wolfsburg, Germany, Volkswagen plant turns out 1550 fewer cars per day than it did in 1991. (*Source:* Paul Klebnikov, "Bringing Back the Beetle," *Forbes,* April 7, 1997.)

17. The boiling point of chlorine is approximately 30° below 0° Fahrenheit.

18. The height of Mt. Arenal, an active volcano in Costa Rica, is 5436 feet above sea level. (*Source: The Universal Almanac,* 1997, John W. Wright, General Editor.)

19. Between 1980 and 1990, the population of the District of Columbia decreased by 31,532. (*Source:* U.S. Bureau of the Census.)

20. In 1994, the country of Taiwan produced 159,376 more passenger cars than commercial vehicles. (*Source:* American Automobile Manufacturers Association.)

21. The city of New Orleans lies 8 feet below sea level. (*Source:* U.S. Geological Survey, *Elevations and Distances in the United States,* 1990.)

22. When the wind speed is 20 miles per hour and the actual temperature is 25° Fahrenheit, the wind-chill factor is 3° below 0° Fahrenheit. (Give three responses.)

Graph each group of numbers on a number line. See Figure 6.

23. $0, 3, -5, -6$

24. $2, 6, -2, -1$

25. $-2, -6, -4, 3, 4$

26. $-5, -3, -2, 0, 4$

27. $\frac{1}{4}, 2\frac{1}{2}, -3\frac{4}{5}, -4, -1\frac{5}{8}$

28. $5\frac{1}{4}, 4\frac{5}{9}, -2\frac{1}{3}, 0, -3\frac{2}{5}$

29. Match each expression in Column I with its value in Column II. Some choices in Column II may not be used.

I	II
(a) $\|-7\|$	**A.** 7
(b) $-(-7)$	**B.** -7
(c) $-\|-7\|$	**C.** neither A nor B
(d) $-\|-(-7)\|$	**D.** both A and B

30. Fill in the blanks with the correct values: The opposite of -2 is ______, while the absolute value of -2 is ______. The additive inverse of -2 is ______, while the additive inverse of the absolute value of -2 is ______.

*Find **(a)** the opposite (or additive inverse) of each number and **(b)** the absolute value of each number.*

31. -2 **32.** -8 **33.** 6 **34.** 11

35. $7 - 4$ **36.** $8 - 3$ **37.** $7 - 7$ **38.** $3 - 3$

39. Look at Exercises 35 and 36 and use the results to complete the following: If $a - b > 0$, then the absolute value of $a - b$ in terms of a and b is ______.

40. Look at Exercises 37 and 38 and use the results to complete the following: If $a - b = 0$, then the absolute value of $a - b$ is ______.

Select the smaller of the two given numbers. See Examples 2 and 3.

41. $-12, -4$ **42.** $-9, -14$ **43.** $-8, -1$

44. $-15, -16$ **45.** $3, |-4|$ **46.** $5, |-2|$

47. $|-3|, |-4|$ **48.** $|-8|, |-9|$ **49.** $-|-6|, -|-4|$

50. $-|-2|, -|-3|$ **51.** $|5 - 3|, |6 - 2|$ **52.** $|7 - 2|, |8 - 1|$

Decide whether each statement is true or false. See Examples 2 and 3.

53. $6 > -(-2)$ **54.** $-8 > -(-2)$ **55.** $-4 \leq -(-5)$

56. $-6 \leq -(-3)$ **57.** $|-6| < |-9|$ **58.** $|-12| < |-20|$

59. $-|8| > |-9|$ **60.** $-|12| > |-15|$ **61.** $-|-5| \geq -|-9|$

62. $-|-12| \leq -|-15|$ **63.** $|6 - 5| \geq |6 - 2|$ **64.** $|13 - 8| \leq |7 - 4|$

The table shows the percent change in the Producer Price Index for selected commodities from 1994 to 1995 and from 1995 to 1996. Use the table to answer the questions in Exercises 65–68. See Example 4.

Commodity	*Change from 1994 to 1995*	*Change from 1995 to 1996**
Interior solvent-based paint	16.4	11.1
Plastic construction products	10.9	−3.1
Softwood plywood	11.3	−14.1
Bright nails	1.8	−.3
Gypsum products	18.4	−.3

*1996 data is preliminary.
Source: U.S. Bureau of Labor Statistics.

65. Which category in what year represents the greatest drop?

66. Which of these is the greater number: the change in plastic construction products from 1995 to 1996 or the change in bright nails from 1994 to 1995?

67. True or false? The absolute value of the data in the first column for softwood plywood is less than the absolute value of the data in the second column for the same commodity.

68. Which category represents the greatest increase?

For each statement give a pair of values for a and b that make it true, and then give a pair of values that make it false.

69. $|a + b| = |a - b|$

70. $|a - b| = |b - a|$

71. $|a + b| = -|a + b|$

72. $|-(a + b)| = -(a + b)$

Give three numbers between -6 and 6 that satisfy each given condition.

73. Positive real numbers but not integers

74. Real numbers but not positive numbers

75. Real numbers but not whole numbers

76. Rational numbers but not integers

77. Real numbers but not rational numbers

78. Rational numbers but not negative numbers

79. Students often say "Absolute value is always positive." Is this true? If not, explain why.

80. True or false: If a is negative, $|a| = -a$.

1.5 Addition and Subtraction of Real Numbers

OBJECTIVES

1. Add two numbers with the same sign.
2. Add positive and negative numbers.
3. Use the definition of subtraction.
4. Use the order of operations with real numbers.
5. Interpret words and phrases involving addition and subtraction.
6. Use signed numbers to interpret data.

FOR EXTRA HELP

SSG Sec. 1.5
SSM Sec. 1.5
Pass the Test Software
InterAct Math Tutorial Software
Video 2

In this section and the next section, we extend the rules for operations with positive numbers to the negative numbers, beginning with addition and subtraction.

OBJECTIVE 1 Add two numbers with the same sign. A number line can be used to explain addition of real numbers.

EXAMPLE 1 Adding Positive Numbers on a Number Line

Use a number line to find the sum $2 + 3$.

Add the positive numbers 2 and 3 on the number line by starting at 0 and drawing an arrow two units to the *right,* as shown in Figure 12. This arrow represents the number 2 in the sum $2 + 3$. Then, from the right end of this arrow draw another arrow three units to the right. The number below the end of this second arrow is 5, so $2 + 3 = 5$.

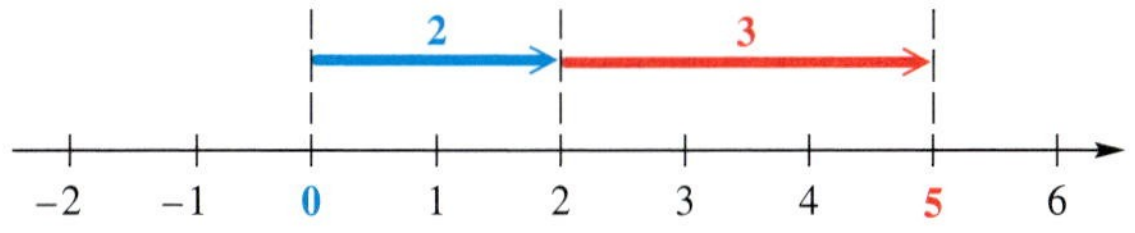

Figure 12

EXAMPLE 2 Adding Negative Numbers on a Number Line

Use a number line to find the sum $-2 + (-4)$. (Parentheses are placed around the -4 to avoid the confusing use of + and − next to each other.)

Add the negative numbers -2 and -4 on the number line by starting at 0 and drawing an arrow two units to the *left,* as shown in Figure 13. The arrow is drawn to the left to represent the addition of a *negative* number. From the left end of this first arrow, draw a second arrow four units to the left. The number below the end of this second arrow is -6, so $-2 + (-4) = -6$.

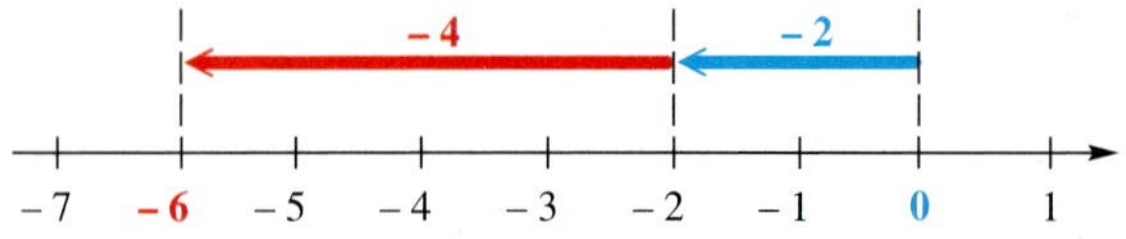

Figure 13

In Example 2, the sum of the two negative numbers -2 and -4 is a negative number whose distance from 0 is the sum of the distance of -2 from 0 and the distance of -4 from 0. That is, *the sum of two negative numbers is the negative of the sum of their absolute values.*

$$-2 + (-4) = -(|-2| + |-4|) = -(2 + 4) = -6$$

Adding Numbers with the Same Signs

To add two numbers with the *same* signs, add the absolute values of the numbers. The sum has the same sign as the numbers being added.

EXAMPLE 3 **Adding Two Negative Numbers**

Find the sums.

(a) $-2 + (-9) = -(|-2| + |-9|) = -(2 + 9) = -11$

(b) $-8 + (-12) = -20$

(c) $-15 + (-3) = -18$

OBJECTIVE 2 **Add positive and negative numbers.** We can use a number line to explain the sum of a positive number and a negative number.

EXAMPLE 4 **Adding Numbers with Different Signs**

Use a number line to find the sum $-2 + 5$.

Find the sum $-2 + 5$ on the number line by starting at 0 and drawing an arrow two units to the left. From the left end of this arrow, draw a second arrow five units to the right, as shown in Figure 14. The number below the end of the second arrow is 3, so $-2 + 5 = 3$.

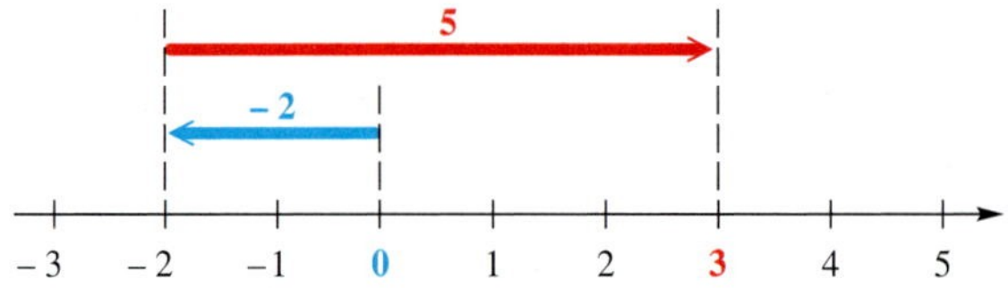

Figure 14

Adding Numbers with Different Signs

To add two numbers with *different* signs, subtract the smaller absolute value from the larger absolute value. The answer has the sign of the number with the larger absolute value.

For example, to add -12 and 5, find their absolute values: $|-12| = 12$ and $|5| = 5$. Then find the difference between these absolute values: $12 - 5 = 7$. Since $|-12| > |5|$, the sum will be negative, so that the final answer is $-12 + 5 = -7$.

While a number line is useful in showing the rules for addition, it is important to be able to do the problems quickly, "in your head."

EXAMPLE 5 Adding Mentally

Check each answer, trying to work the addition mentally. If you get stuck, use a number line.

(a) $7 + (-4) = 3$

(b) $-8 + 12 = 4$

(c) $-\frac{1}{2} + \frac{1}{8} = -\frac{4}{8} + \frac{1}{8} = -\frac{3}{8}$ Remember to get a common denominator first.

(d) $\frac{5}{6} + \left(-\frac{4}{3}\right) = -\frac{1}{2}$

(e) $-4.6 + 8.1 = 3.5$

(f) $-16 + 16 = 0$

(g) $42 + (-42) = 0$

Parts (f) and (g) in Example 5 suggest that the sum of a number and its additive inverse is 0. This is always true, and this property is discussed further in Section 1.7.

The rules for adding signed numbers are summarized as follows.

Adding Signed Numbers

Like signs Add the absolute values of the numbers. The sum has the same sign as the given numbers.

Unlike signs Find the difference between the larger absolute value and the smaller. The sum has the sign of the number with the larger absolute value.

OBJECTIVE 3 Use the definition of subtraction. We can illustrate subtraction of 4 from 7, written $7 - 4$, using a number line. As seen in Figure 15, we begin at 0 and draw an arrow seven units to the right. From the right end of this arrow, draw an arrow four units to the left. The number at the end of the second arrow shows that $7 - 4 = 3$.

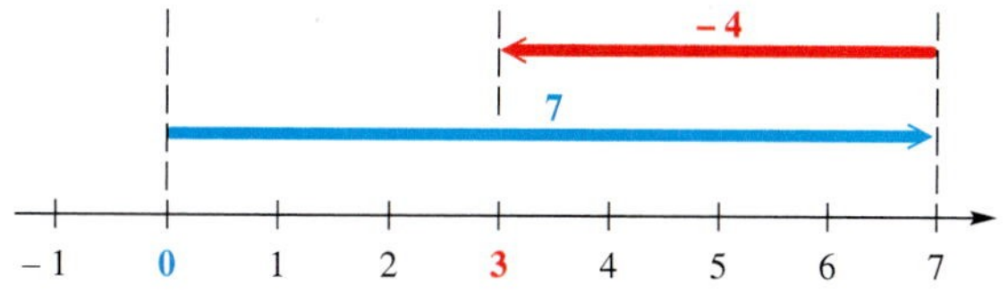

Figure 15

The procedure used to find $7 - 4$ is exactly the same procedure that would be used to find $7 + (-4)$, so that

$$7 - 4 = 7 + (-4).$$

This suggests that *subtraction* of a positive number from a larger positive number is the same as *adding* the additive inverse of the smaller number to the larger. This result leads to the definition of subtraction for all real numbers.

Definition of Subtraction

For any real numbers x and y,

$$x - y = x + (-y).$$

That is, to *subtract* y from x, *add the additive inverse* (or opposite) of y to x. This definition gives the following procedure for subtracting signed numbers.

Subtracting Signed Numbers

Step 1 Change the subtraction symbol to the addition symbol.

Step 2 Change the sign of the number being subtracted.

Step 3 Add.

EXAMPLE 6 Using the Definition of Subtraction

Subtract.

(Annotations: No change; Change − to +.; Additive inverse of 3)

(a) $12 - 3 = 12 + (-3) = 9$

(b) $5 - 7 = 5 + (-7) = -2$

(Annotations: No change; Change − to +.; Additive inverse of −5)

(c) $-3 - (-5) = -3 + (5) = 2$

(d) $-6 - 9 = -6 + (-9) = -15$

(e) $8 - (-4) = 8 + (4) = 12$

OBJECTIVE 4 Use the order of operations with real numbers. As before, with problems that have grouping symbols, first do any operations inside the parentheses and brackets. Work from the inside out.

EXAMPLE 7 Adding and Subtracting with Grouping Symbols

Perform the indicated operations.

(a)

$$\begin{aligned} -6 - [2 - (8 + 3)] &= -6 - [2 - 11] && \text{Add.} \\ &= -6 - [2 + (-11)] && \text{Use the definition of subtraction.} \\ &= -6 - (-9) && \text{Add.} \\ &= -6 + (9) && \text{Use the definition of subtraction.} \\ &= 3 && \text{Add.} \end{aligned}$$

(b)

$$\begin{aligned} 5 + [(-3 - 2) - (4 - 1)] &= 5 + [(-3 + (-2)) - 3] \\ &= 5 + [(-5) - 3] \\ &= 5 + [(-5) + (-3)] \\ &= 5 + (-8) \\ &= -3 \end{aligned}$$

(c) $\frac{2}{3} - \left[\frac{1}{12} - \left(-\frac{1}{4}\right)\right] = \frac{8}{12} - \left[\frac{1}{12} - \left(-\frac{3}{12}\right)\right]$ Get a common denominator.

$= \frac{8}{12} - \left[\frac{1}{12} + \frac{3}{12}\right]$ Use the definition of subtraction.

$= \frac{8}{12} - \frac{4}{12}$ Add.

$= \frac{4}{12}$ Subtract.

$= \frac{1}{3}$ Lowest terms

(d) $|4 - 7| + 2|6 - 3| = |-3| + 2|3|$ Work within absolute value symbols.

$= 3 + 2 \cdot 3$ Evaluate the absolute values.

$= 3 + 6$ Multiply.

$= 9$ Add.

OBJECTIVE 5 **Interpret words and phrases involving addition and subtraction.** We begin by working with addition words and phrases.

PROBLEM SOLVING

As we mentioned earlier, problem solving often requires translating words and phrases into symbols. The word *sum* is one of the words that indicates addition. The chart below lists some of the words and phrases that also signify addition.

Word or Phrase	*Example*	*Numerical Expression and Simplification*
Sum of	The ***sum of*** -3 and 4	$-3 + 4 = 1$
Added to	5 ***added to*** -8	$-8 + 5 = -3$
More than	12 ***more than*** -5	$-5 + 12 = 7$
Increased by	-6 ***increased by*** 13	$-6 + 13 = 7$
Plus	3 ***plus*** 14	$3 + 14 = 17$

EXAMPLE 8 Interpreting Words and Phrases Involving Addition

Write a numerical expression for each phrase, and simplify the expression.

(a) The **sum of** -8 and 4 and 6

$$-8 + 4 + 6 = -4 + 6 = 2$$

Add in order from left to right.

(b) 3 **more than** -5, **increased by** 12

$$-5 + 3 + 12 = -2 + 12 = 10$$

We now look at how we interpret words and phrases that involve subtraction.

PROBLEM SOLVING

In order to solve problems that involve subtraction, we must be able to interpret key words and phrases that indicate subtraction. *Difference* is one of them. Some of these are given in the chart below.

Word or Phrase	Example	Numerical Expression and Simplification
Difference between	The **difference between** −3 and −8	$-3 - (-8) = -3 + 8 = 5$
Subtracted from	12 **subtracted from** 18	$18 - 12 = 6$
Less	6 **less** 5	$6 - 5 = 1$
Less than	6 **less than** 5	$5 - 6 = 5 + (-6) = -1$
Decreased by	9 **decreased by** −4	$9 - (-4) = 9 + 4 = 13$
Minus	8 **minus** 5	$8 - 5 = 3$

CAUTION When you are subtracting two numbers, it is important that you write them in the correct order, because, in general, $a - b \neq b - a$. For example, $5 - 3 \neq 3 - 5$. For this reason, *think carefully before interpreting an expression involving subtraction.* (This difficulty did not arise for addition.)

EXAMPLE 9 Interpreting Words and Phrases Involving Subtraction

Write a numerical expression for each phrase, and simplify the expression.

(a) The **difference between** −8 and 5

It is conventional to write the numbers in the order they are given when "difference between" is used.

$$-8 - 5 = -8 + (-5) = -13$$

(b) 4 **subtracted from** the sum of 8 and −3

Here addition is also used, as indicated by the word *sum.* First, add 8 and −3. Next, subtract 4 *from* this sum.

$$[8 + (-3)] - 4 = 5 - 4 = 1$$

(c) 4 **less than** −6

Be careful with order here. 4 must be taken *from* −6.

$$-6 - 4 = -6 + (-4) = -10$$

Notice that "4 less than −6" differs from "4 *is less than* −6." The second of these is symbolized as $4 < -6$ (which is a false statement).

(d) 8, **decreased by** 5 **less than** 12

First, write "5 less than 12" as $12 - 5$. Next, subtract $12 - 5$ from 8.

$$8 - (12 - 5) = 8 - 7 = 1$$

OBJECTIVE 6 Use signed numbers to interpret data. The Producer Price Index is the oldest continuous statistical series published by the Bureau of Labor Statistics. It measures the average changes in prices received by producers of all commodities produced in the United States. The next example shows how signed numbers can be used to interpret such data.

EXAMPLE 10 Using a Signed Number in Data Interpretation

The bar graph in Figure 16 gives the Producer Price Index (PPI) for construction materials between 1985 and 1990.

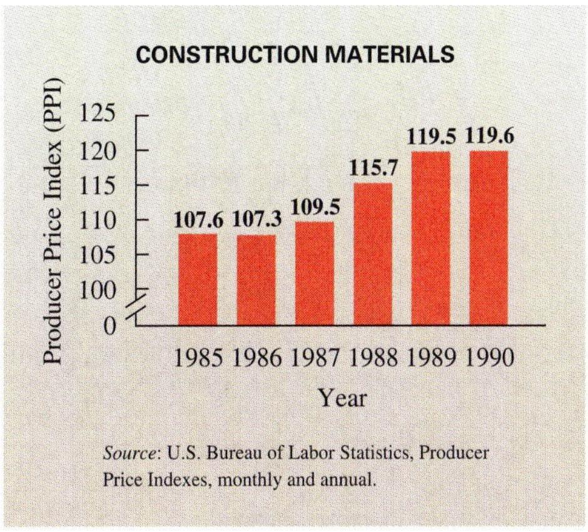

Figure 16

(a) Use a signed number to represent the change in the PPI from 1987 to 1988.

To find this change, we start with the index number from 1988 and subtract from it the index number from 1987.

$$\underbrace{115.7}_{\text{The 1988 index}} - \underbrace{109.5}_{\text{The 1987 index}} = \underbrace{6.2}_{\text{A positive number indicates an increase.}}$$

(b) Use a signed number to represent the change in the PPI from 1985 to 1986.

Use the same procedure as in part (a).

$$\underbrace{107.3}_{\text{The 1986 index}} - \underbrace{107.6}_{\text{The 1985 index}} = 107.3 + (-107.6) = \underbrace{-.3}_{\text{A negative number indicates a decrease.}}$$

EXAMPLE 11 Solving a Problem Involving Subtraction

The record high temperature in the United States was 134° Fahrenheit, recorded at Death Valley, California, in 1913. The record low was −80°F, at Prospect Creek, Alaska, in 1971. See Figure 17. What is the difference between these highest and lowest temperatures? (*Source: The World Almanac and Book of Facts,* 1998.)

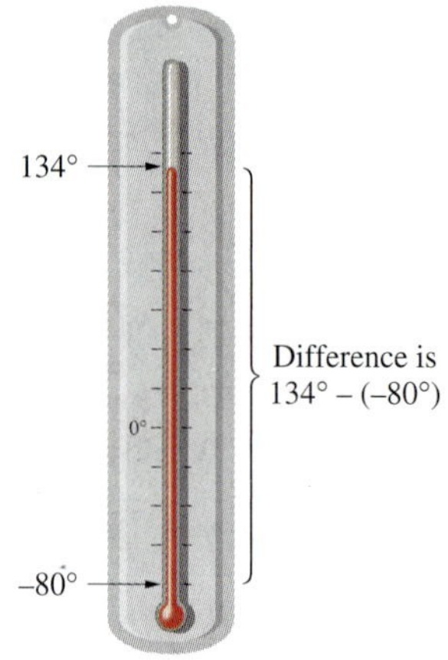

Figure 17

We must subtract the lowest temperature from the highest temperature.

$$134 - (-80) = 134 + 80 \quad \text{Use the definition of subtraction.}$$
$$= 214 \quad \text{Add.}$$

The difference between the two temperatures is 214°F.

1.5 EXERCISES

Fill in each blank with the correct response.

1. The sum of two negative numbers will always be a ________________ number. (positive/negative)
2. The sum of a number and its opposite will always be ________________.
3. To simplify the expression $8 + [-2 + (-3 + 5)]$, I should begin by adding ______ and ______, according to the rule for order of operations.
4. If I am adding a positive number and a negative number, and the negative number has the larger absolute value, the sum will be a ________________ number. (positive/negative)
5. Explain in words how to add signed numbers. Consider the various cases and give examples.
6. Explain in words how to subtract signed numbers.

Find each sum or difference. See Examples 1–7.

7. $6 + (-4)$
8. $12 + (-9)$
9. $7 + (-10)$
10. $4 + (-8)$
11. $-7 + (-3)$
12. $-11 + (-4)$
13. $-10 + (-3)$
14. $-16 + (-7)$
15. $-12.4 + (-3.5)$
16. $-21.3 + (-2.5)$
17. $-8 + 7$
18. $-12 + 10$
19. $5 + [14 + (-6)]$
20. $7 + [3 + (-14)]$
21. $4 - 7$
22. $8 - 13$

23. $6 - 10$

24. $9 - 14$

25. $-7 - 3$

26. $-12 - 5$

27. $-10 - 6$

28. $-13 - 16$

29. $7 - (-4)$

30. $9 - (-6)$

31. $6 - (-13)$

32. $13 - (-3)$

33. $-7 - (-3)$

34. $-8 - (-6)$

35. $3 - (4 - 6)$

36. $6 - (7 - 14)$

37. $-3 - (6 - 9)$

38. $-4 - (5 - 12)$

39. $\frac{1}{2} - \left(-\frac{1}{4}\right)$

40. $\frac{1}{3} - \left(-\frac{4}{3}\right)$

41. $-8 + [3 + (-1) + (-2)]$

42. $-7 + [5 + (-8) + 3]$

43. $\frac{5}{8} - \left(-\frac{1}{2} - \frac{3}{4}\right)$

44. $\frac{9}{10} - \left(\frac{1}{8} - \frac{3}{10}\right)$

45. $[(-3.1) - 4.5] - (.8 - 2.1)$

46. $[(-7.8) - 9.3] - (.6 - 3.5)$

47. $[-5 + (-7)] + [-4 + (-9)] + [13 + (-12)]$

48. $[-3 + (-11)] + [13 + (-3)] + [19 + (-7)]$

49. $-4 + [(-6 - 9) - (-7 + 4)]$

50. $-8 + [(-3 - 10) - (-4 + 1)]$

51. Is it possible to add a negative number to another negative number and get a positive number? If so, give an example.

52. Under what conditions will the sum of a positive number and a negative number be a number which is neither negative nor positive?

53. Make up a subtraction problem so that the difference between two negative numbers is a negative number.

54. Make up a subtraction problem so that the difference between two negative numbers is a positive number.

Simplify each expression involving absolute value. Remember that absolute value bars serve as grouping symbols. See Example 7(d).

55. $|3 - 8| - |-2 + 8|$

56. $|-2 - 7| - |9 - (-3)|$

57. $\left|\frac{2}{3} - \frac{7}{3}\right| + 2\left|-\frac{4}{9} + \frac{5}{9}\right|$

58. $\left|-\frac{1}{5} + \frac{3}{5}\right| + 3\left|\frac{3}{10} - \frac{8}{10}\right|$

Write a numerical expression for each phrase and simplify. See Example 8.

59. The sum of -5 and 12 and 6

60. The sum of -3 and 5 and -12

61. 14 added to the sum of -19 and -4

62. -2 added to the sum of -18 and 11

63. The sum of -4 and -10, increased by 12

64. The sum of -7 and -13, increased by 14

65. 4 more than the sum of 8 and -18

66. 10 more than the sum of -4 and -6

Write a numerical expression for each phrase and simplify. See Example 9.

67. The difference between 4 and -8

68. The difference between 7 and -14

69. 8 less than -2

70. 9 less than -13

71. The sum of 9 and -4, decreased by 7

72. The sum of 12 and -7, decreased by 14

73. 12 less than the difference between 8 and -5

74. 19 less than the difference between 9 and -2

The bar graph shows the federal budget outlays for national defense for the years 1985 through 1994. Use a signed number to represent the change in outlay for each of the following time periods. See Example 10.

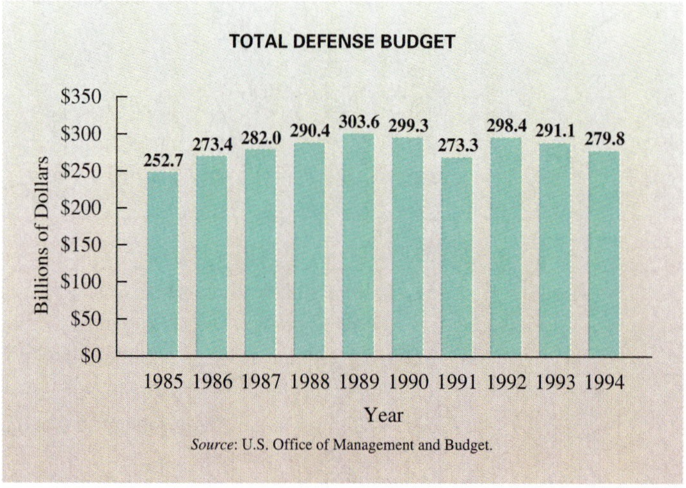

75. 1988 to 1989

76. 1986 to 1987

77. 1990 to 1991

78. 1993 to 1994

The two charts show the heights of some selected mountains and the depths of some selected trenches. Use the information given to find the answers in Exercises 79–82.

Mountain	*Height (in feet)*
Foraker	17,400
Wilson	14,246
Pikes Peak	14,110

Trench	*Depth (in feet, as a negative number)*
Philippine	−32,995
Cayman	−24,721
Java	−23,376

Source: The World Almanac and Book of Facts, 1998.

79. What is the difference between the height of Mt. Foraker and the depth of the Philippine Trench?

80. What is the difference between the height of Pikes Peak and the depth of the Java Trench?

81. How much deeper is the Cayman Trench than the Java Trench?

82. How much deeper is the Philippine Trench than the Cayman Trench?

Use this idea when working Exercises 83 and 84: If a player steals 65 *bases in 1990 and* 58 *in 1991, the change in stolen bases is represented by* $58 - 65 = -7$. *His number went* down *by* 7.

83. The bar graph shows the number of bases stolen by baseball player Rickey Henderson. Use a signed number to represent the change in the number of stolen bases from one year to the next. See Example 10. (*Source:* David S. Neft and Richard M. Cohen, *The Sports Encyclopedia: Baseball 1997.*)

(a) From 1991 to 1992 **(b)** From 1992 to 1993
(c) From 1989 to 1990 **(d)** From 1993 to 1994

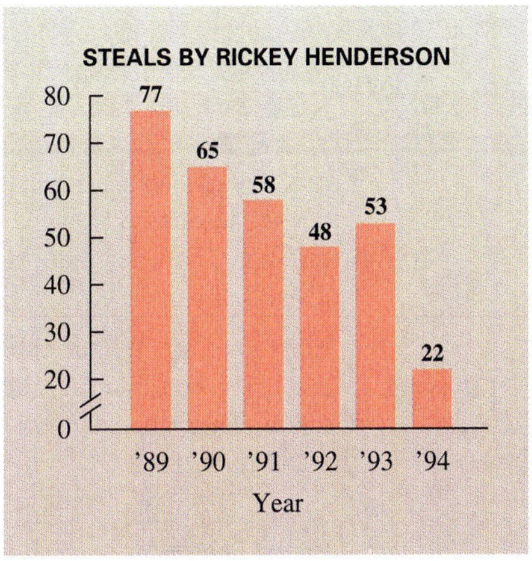

84. The bar graph shows the number of bases stolen by baseball player Kenny Lofton. Use a signed number to represent the change in the number of stolen bases from one year to the next. See Example 10. (*Source:* David S. Neft and Richard M. Cohen, *The Sports Encyclopedia: Baseball 1997.*)

(a) From 1993 to 1994 **(b)** From 1994 to 1995
(c) From 1995 to 1996 **(d)** From 1996 to 1997

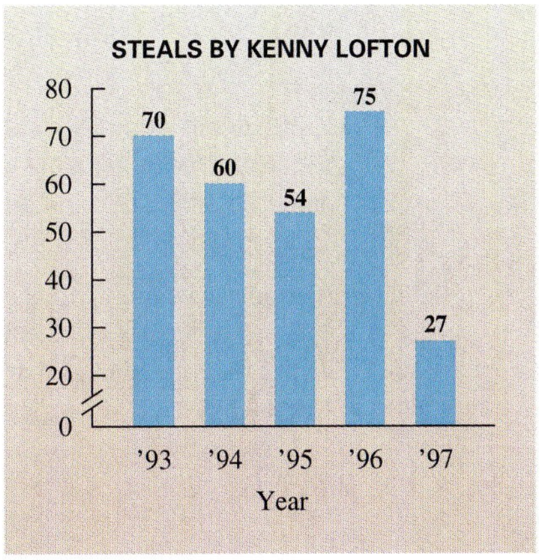

Solve each problem. (Refer to Figure 17 for Exercises 85–87.) See Example 11.

85. On January 23, 1943, the temperature rose 49°F in two minutes in Spearfish, South Dakota. If the starting temperature was -4°F, what was the temperature two minutes later? (*Source: The Guinness Book of World Records.*)

86. The lowest temperature ever recorded in Little Rock, Arkansas, was -5°F. The highest temperature ever recorded there was 117°F more than the lowest. What was this highest temperature? (*Source: The World Almanac and Book of Facts,* 1998.)

87. The coldest temperature recorded in Chicago, Illinois, was $-27°F$ in 1985. The record low in Huron, South Dakota, was set in 1994 and was 14°F lower than $-27°F$. What was the record low in Huron? (*Source: The World Almanac and Book of Facts,* 1998.)

88. No one knows just why humpback whales love to heave their 45-ton bodies out of the water, but leap they do. (This activity is called *breaching.*) Mark and Debbie, two researchers based on the island of Maui, noticed that one of their favorite whales, "Pineapple," leaped 15 feet above the surface of the ocean while her mate cruised 12 feet below the surface. What is the difference between these two distances?

89. The top of Mt. Whitney, visible from Death Valley, has an altitude of 14,494 feet above sea level. The bottom of Death Valley is 282 feet below sea level. Using zero as sea level, find the difference between these two elevations. (*Source: The World Almanac and Book of Facts,* 1998.)

90. The highest point in Louisiana is Driskill Mountain, at an altitude of 535 feet. The lowest point is at Spanish Fort, 8 feet below sea level. Using zero as sea level, find the difference between these two elevations. (*Source: The World Almanac and Book of Facts,* 1998.)

91. A female polar bear weighed 660 pounds when she entered her winter den. She lost 45 pounds during each of the first two months of hibernation, and another 205 pounds before leaving the den with her two cubs in March. How much did she weigh when she left the den?

92. On three consecutive passes, Troy Aikman of the Dallas Cowboys passed for a gain of 6 yards, was sacked for a loss of 12 yards, and passed for a gain of 43 yards. What positive or negative number represents the total net yardage for the plays?

93. Kim Falgout owes \$870.00 on her Master Card account. She returns two items costing \$35.90 and \$150.00 and receives credits for these on the account. Next, she makes a purchase of \$82.50, and then two more purchases of \$10.00 each. She makes a payment of \$500.00. She then incurs a finance charge of \$37.23. How much does she still owe?

94. A welder working with stainless steel must use precise measurements. Suppose a welder attaches two pieces of steel that are each 3.60 inches in length, and then attaches an additional three pieces that are each 9.10 inches long. She finally cuts off a piece that is 7.60 inches long. Find the length of the welded piece of steel.

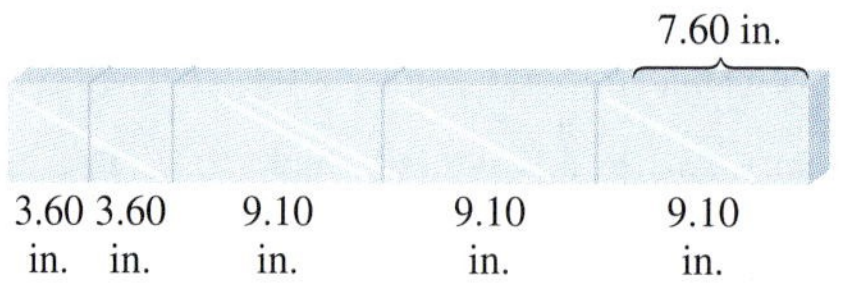

1.6 Multiplication and Division of Real Numbers

OBJECTIVES

1. Find the product of a positive number and a negative number.
2. Find the product of two negative numbers.
3. Identify factors of integers.
4. Use the reciprocal of a number to apply the definition of division.
5. Use the order of operations when multiplying and dividing signed numbers.
6. Evaluate expressions involving variables.
7. Interpret words and phrases involving multiplication and division.
8. Translate simple sentences into equations.

FOR EXTRA HELP

SSG Sec. 1.6
SSM Sec. 1.6
Pass the Test Software
InterAct Math Tutorial Software
Video 2

In this section we learn how to multiply with positive and negative numbers. We already know how to multiply positive numbers and that the product of two positive numbers is positive. We also know that the product of 0 and any positive number is 0, so we extend that property to all real numbers.

Multiplication by Zero

For any real number x, $\quad x \cdot 0 = 0.$

OBJECTIVE 1 **Find the product of a positive number and a negative number.** In order to define the product of a positive and a negative number so that the result is consistent with the multiplication of two positive numbers, look at the following pattern.

$$3 \cdot 5 = 15$$
$$3 \cdot 4 = 12$$
$$3 \cdot 3 = 9$$
$$3 \cdot 2 = 6$$
$$3 \cdot 1 = 3$$
$$3 \cdot 0 = 0$$
$$3 \cdot (-1) = ?$$

Numbers decrease by 3.

What should $3(-1)$ equal? The product $3(-1)$ represents the sum

$$-1 + (-1) + (-1) = -3,$$

so the product should be -3. Also,

$$3(-2) = -2 + (-2) + (-2) = -6$$

and

$$3(-3) = -3 + (-3) + (-3) = -9.$$

These results maintain the pattern in the list, which suggests the following rule.

Multiplying Numbers with Different Signs

For any positive real numbers x and y,

$$x(-y) = -(xy) \quad \text{and} \quad (-x)y = -(xy).$$

That is, the product of two numbers with opposite signs is negative.

EXAMPLE 1 **Multiplying a Positive Number and a Negative Number**

Find the products using the multiplication rule given above.

(a) $8(-5) = -(8 \cdot 5) = -40$

(b) $5(-4) = -(5 \cdot 4) = -20$

(c) $(-7)(2) = -(7 \cdot 2) = -14$

(d) $(-9)(3) = -(9 \cdot 3) = -27$

OBJECTIVE 2 Find the product of two negative numbers. The product of two positive numbers is positive, and the product of a positive and a negative number is negative. What about the product of two negative numbers? Look at another pattern.

$$\begin{aligned} (-5)(4) &= -20 \\ (-5)(3) &= -15 \\ (-5)(2) &= -10 \\ (-5)(1) &= -5 \\ (-5)(0) &= 0 \\ (-5)(-1) &= ? \end{aligned}$$

Numbers increase by 5.

The numbers on the left of the equals sign (in color) decrease by 1 for each step down the list. The products on the right increase by 5 for each step down the list. To maintain this pattern, $(-5)(-1)$ should be 5 more than $(-5)(0)$, or 5 more than 0, so

$$(-5)(-1) = 5.$$

The pattern continues with

$$\begin{aligned} (-5)(-2) &= 10 \\ (-5)(-3) &= 15 \\ (-5)(-4) &= 20 \\ (-5)(-5) &= 25, \end{aligned}$$

and so on. This pattern suggests the next rule.

Multiplying Two Negative Numbers

For any positive real numbers x and y,

$$(-x)(-y) = xy.$$

The product of two negative numbers is positive.

EXAMPLE 2 Multiplying Two Negative Numbers

Find the products using the multiplication rule given above.

(a) $(-9)(-2) = 9 \cdot 2 = 18$

(b) $(-6)(-12) = 6 \cdot 12 = 72$

(c) $(-8)(-1) = 8 \cdot 1 = 8$

(d) $(-15)(-2) = 15 \cdot 2 = 30$

A summary of multiplying signed numbers is given here.

Multiplying Signed Numbers

The product of two numbers having the *same* sign is *positive,* and the product of two numbers having *different* signs is *negative.*

OBJECTIVE 3 Identify factors of integers. In Section 1.1 the definition of a *factor* was given for whole numbers. (For example, since $9 \cdot 5 = 45$, both 9 and 5 are factors of 45.) The definition can now be extended to integers.

If the product of two integers is a third integer, then each of the two integers is a *factor* of the third. For example, $(-3)(-4) = 12$, so -3 and -4 are both factors of 12. The factors of 12 are $-12, -6, -4, -3, -2, -1, 1, 2, 3, 4, 6$, and 12.

The following chart shows several integers and the factors of those integers.

Integer	*Factors*
18	−18, −9, −6, −3, −2, −1, 1, 2, 3, 6, 9, 18
20	−20, −10, −5, −4, −2, −1, 1, 2, 4, 5, 10, 20
15	−15, −5, −3, −1, 1, 3, 5, 15
7	−7, −1, 1, 7
1	−1, 1

OBJECTIVE 4 Use the reciprocal of a number to apply the definition of division. In Section 1.5 we saw that the difference between two numbers is found by adding the additive inverse of the second number to the first. Similarly, the *quotient* of two numbers is found by *multiplying* by the *reciprocal,* or *multiplicative inverse.* By definition, since

$$8 \cdot \frac{1}{8} = \frac{8}{8} = 1 \qquad \text{and} \qquad \frac{5}{4} \cdot \frac{4}{5} = \frac{20}{20} = 1,$$

the reciprocal or multiplicative inverse of 8 is $\frac{1}{8}$, and of $\frac{5}{4}$ is $\frac{4}{5}$.

Reciprocal or Multiplicative Inverse

Pairs of numbers whose product is 1 are **reciprocals,** or **multiplicative inverses,** of each other.

The following chart shows several numbers and their multiplicative inverses.

Number	*Multiplicative Inverse (Reciprocal)*
4	$\frac{1}{4}$
$.3 = \frac{3}{10}$	$\frac{10}{3}$
-5	$\frac{1}{-5}$ or $-\frac{1}{5}$
$-\frac{5}{8}$	$-\frac{8}{5}$
0	None
1	1
−1	−1

Why is there no multiplicative inverse for the number 0? Suppose that k is to be the multiplicative inverse of 0. Then $k \cdot 0$ should equal 1. But $k \cdot 0 = 0$ for any number k. Since there is no value of k that is a solution of the equation $k \cdot 0 = 1$, the following statement can be made.

0 has no multiplicative inverse.

By definition, the quotient of x and y is the product of x and the multiplicative inverse of y.

Definition of Division

For any real numbers x and y, with $y \neq 0$, $\dfrac{x}{y} = x \cdot \dfrac{1}{y}$.

The definition of division indicates that y, the number to divide by, cannot be 0. The reason is that 0 has no multiplicative inverse, so that $\frac{1}{0}$ is not a number. **Because 0 has no multiplicative inverse, *division by 0 is undefined.* If a division problem turns out to involve division by 0, write "undefined."**

NOTE: While division by zero $\left(\frac{a}{0}\right)$ is undefined, we may divide 0 by any nonzero number. In fact, if $a \neq 0$,

$$\frac{0}{a} = 0.$$

Since division is defined in terms of multiplication, all the rules for multiplying signed numbers also apply to dividing them.

EXAMPLE 3 Using the Definition of Division

Find the quotients using the definition of division.

(a) $\dfrac{12}{3} = 12 \cdot \dfrac{1}{3} = 4$

(b) $\dfrac{-10}{2} = -10 \cdot \dfrac{1}{2} = -5$

(c) $\dfrac{-14}{-7} = -14\left(\dfrac{1}{-7}\right) = 2$

(d) $-\dfrac{2}{3} \div \left(-\dfrac{4}{5}\right) = -\dfrac{2}{3} \cdot \left(-\dfrac{5}{4}\right) = \dfrac{5}{6}$

(e) $\dfrac{0}{13} = 0\left(\dfrac{1}{13}\right) = 0$ $\quad \dfrac{0}{a} = 0 \quad (a \neq 0)$

(f) $\dfrac{-10}{0}$ Undefined

When dividing fractions, multiplying by the reciprocal works well. However, using the definition of division directly with integers is awkward. It is easier to divide in the usual way, then determine the sign of the answer. The following rule for division can be used instead of multiplying by the reciprocal.

Dividing Signed Numbers

The quotient of two numbers having the same sign is positive; the quotient of two numbers having *different* signs is *negative.*

Note that these are the same as the rules for multiplication.

EXAMPLE 4 Dividing Signed Numbers

Find the quotients.

(a) $\frac{8}{-2} = -4$

(b) $\frac{-4.5}{-.09} = 50$

(c) $-\frac{1}{8} \div \left(-\frac{3}{4}\right) = -\frac{1}{8} \cdot \left(-\frac{4}{3}\right) = \frac{1}{6}$

From the definitions of multiplication and division of real numbers,

$$\frac{-40}{8} = -40 \cdot \frac{1}{8} = -5, \text{ and } \frac{40}{-8} = 40\left(\frac{1}{-8}\right) = -5, \text{ so that}$$

$$\frac{-40}{8} = \frac{40}{-8}.$$

Based on this example, the quotient of a positive and a negative number can be expressed in any of the following three forms.

For any positive real numbers x and y, $\frac{-x}{y} = \frac{x}{-y} = -\frac{x}{y}$.

The quotient of two negative numbers can be expressed as a quotient of two positive numbers.

For any positive real numbers x and y, $\frac{-x}{-y} = \frac{x}{y}$.

OBJECTIVE 5 Use the order of operations when multiplying and dividing signed numbers.

EXAMPLE 5 Using the Order of Operations

Perform the indicated operations.

(a) $(-9)(2) - (-3)(2)$

First find all products, working from left to right.

$$(-9)(2) - (-3)(2) = -18 - (-6)$$

Now perform the subtraction.

$$-18 - (-6) = -18 + 6 = -12$$

(b) $(-6)(-2) - (3)(-4) = 12 - (-12) = 12 + 12 = 24$

(c) $-5(-2 - 3) = -5(-5) = 25$

(d) $\dfrac{5(-2) - (3)(4)}{2(1 - 6)}$

Simplify the numerator and denominator separately. Then write in lowest terms.

$$\frac{5(-2) - (3)(4)}{2(1 - 6)} = \frac{-10 - 12}{2(-5)} = \frac{-22}{-10} = \frac{11}{5}$$

We now summarize the rules for operations with signed numbers.

Operations with Signed Numbers

Addition

Like signs Add the absolute values of the numbers. The sum has the same sign as the numbers.

Unlike signs Subtract the number with the smaller absolute value from the one with the larger. Give the sum the sign of the number having the larger absolute value.

Subtraction

Add the additive inverse, or opposite, of the second number.

Multiplication and Division

Like signs The product or quotient of two numbers with like signs is positive.

Unlike signs The product or quotient of two numbers with unlike signs is negative.

Division by 0 is undefined.

0 divided by a nonzero number equals 0.

OBJECTIVE 6 Evaluate expressions involving variables. The next examples show numbers substituted for variables where the rules for multiplying and dividing signed numbers must be used.

EXAMPLE 6 Evaluating an Expression for Numerical Values

Evaluate each expression, given that $x = -1$, $y = -2$, and $m = -3$.

(a) $(3x + 4y)(-2m)$

First substitute the given values for the variables. Then use the order of operations to find the value of the expression.

$$\begin{aligned} (3x + 4y)(-2m) &= [3(-1) + 4(-2)][-2(-3)] && \text{Put parentheses around the number for each variable.} \\ &= [-3 + (-8)][6] && \text{Find the products.} \\ &= (-11)(6) && \text{Add inside the parentheses.} \\ &= -66 && \text{Multiply.} \end{aligned}$$

(b) $2x^2 - 3y^2$

Use parentheses as shown.

$$\begin{aligned} 2(-1)^2 - 3(-2)^2 &= 2(1) - 3(4) && \text{Apply the exponents.} \\ &= 2 - 12 && \text{Multiply.} \\ &= -10 && \text{Subtract.} \end{aligned}$$

(c) $\dfrac{4y^2 + x}{m}$

$$\frac{4(-2)^2+(-1)}{-3} = \frac{4(4) + (-1)}{-3} \quad \text{Apply the exponent.}$$

$$= \frac{16 + (-1)}{-3} \quad \text{Multiply.}$$

$$= \frac{15}{-3} \quad \text{Add.}$$

$$= -5 \quad \text{Divide.}$$

Notice how the fraction bar was used as a grouping symbol.

OBJECTIVE 7 **Interpret words and phrases involving multiplication and division.** Just as there are words and phrases that indicate addition and subtraction, certain ones also indicate multiplication and division.

PROBLEM SOLVING

The word *product* refers to multiplication. The chart below gives other key words and phrases that indicate multiplication.

Word or Phrase	*Example*	*Numerical Expression and Simplification*
Product of	The **product of** −5 and −2	$(-5)(-2) = 10$
Times	13 ***times*** −4	$13(-4) = -52$
Twice (meaning "2 times")	***Twice*** 6	$2(6) = 12$
Of (used with fractions)	$\frac{1}{2}$ ***of*** 10	$\frac{1}{2}(10) = 5$
Percent of	12**%** ***of*** −16	$.12(-16) = -1.92$

EXAMPLE 7 **Interpreting Words and Phrases Involving Multiplication**

Write a numerical expression for each phrase and simplify. Use the order of operations.

(a) The **product of** 12 and the sum of 3 and −6
Here 12 is multiplied by "the sum of 3 and −6."

$$12[3 + (-6)] = 12(-3) = -36$$

(b) **Twice** the difference between 8 and −4

$$2[8 - (-4)] = 2[8 + 4] = 2(12) = 24$$

(c) Two-thirds **of** the sum of −5 and −3

$$\frac{2}{3}[-5 + (-3)] = \frac{2}{3}[-8] = -\frac{16}{3}$$

(d) 15**%** **of** the difference between 14 and −2
Remember that 15% = .15.

$$.15[14 - (-2)] = .15(14 + 2) = .15(16) = 2.4$$

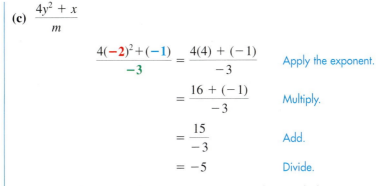

PROBLEM SOLVING

The word *quotient* refers to the answer in a division problem. In algebra, quotients are usually represented with a fraction bar; the symbol ÷ is seldom used. When translating applied problems involving division, use a fraction bar. The chart gives some key phrases associated with division.

Phrase	*Example*	*Numerical Expression and Simplification*
Quotient of	The *quotient of* −24 and 3	$\frac{-24}{3} = -8$
Divided by	−16 *divided by* −4	$\frac{-16}{-4} = 4$
Ratio of	The *ratio of* 2 to 3	$\frac{2}{3}$

It is customary to write the first number named as the numerator and the second as the denominator when interpreting a phrase involving division, as shown in the next example.

EXAMPLE 8 Interpreting Words and Phrases Involving Division

Write a numerical expression for each phrase, and simplify the expression.

(a) The **quotient of** 14 and the sum of −9 and 2

"Quotient" indicates division. The number 14 is the numerator and "the sum of −9 and 2" is the denominator.

$$\frac{14}{-9+2} = \frac{14}{-7} = -2$$

(b) The product of 5 and −6, **divided by** the difference between −7 and 8

The numerator of the fraction representing the division is obtained by multiplying 5 and −6. The denominator is found by subtracting −7 and 8.

$$\frac{5(-6)}{-7-8} = \frac{-30}{-15} = 2$$

OBJECTIVE 8 Translate simple sentences into equations. In this section and the previous one, important words and phrases involving the four operations of arithmetic have been introduced. We can use these words and phrases to interpret sentences that translate into equations.

EXAMPLE 9 Translating Words Into an Equation

Write the following in symbols, using x as the variable, and guess or use trial and error to find the solution. All solutions come from the list of integers between −12 and 12, inclusive.

(a) Three **times** a number **is** −18.

The word *times* indicates multiplication, and the word *is* translates as the equals sign (=).

$$3x = -18$$

Since the integer between -12 and 12, inclusive, that makes this statement true is -6, the solution of the equation is -6.

(b) The sum of a number and 9 is 12.

$$x + 9 = 12$$

Since $3 + 9 = 12$, the solution of this equation is 3.

(c) The difference between a number and 5 is 0.

$$x - 5 = 0$$

Since $5 - 5 = 0$, the solution of this equation is 5.

(d) The quotient of 24 and a number is -2.

$$\frac{24}{x} = -2$$

Here, x must be a negative number, since the numerator is positive and the quotient is negative. Since $\frac{24}{-12} = -2$, the solution is -12.

CAUTION

It is important to recognize the distinction between the types of problems found in Examples 7 and 8 and Example 9. In Examples 7 and 8, the phrases translate as *expressions,* while in Example 9, the sentences translate as *equations.* Remember that an equation is a sentence with an = sign, while an expression is a phrase.

$\frac{5(-6)}{-7-8}$ is an **expression,**

$3x = -18$ is an **equation.**

1.6 EXERCISES

Fill in each blank with one of the following: greater than 0, less than 0, equal to 0.

1. A positive number is ______________.

2. A negative number is ______________.

3. The product or the quotient of two numbers with the same sign is ______________.

4. The product or the quotient of two numbers with different signs is ______________.

5. If three negative numbers are multiplied together, the product is ______________.

6. If two negative numbers are multiplied and then their product is divided by a negative number, the result is ______________.

7. If a negative number is squared and the result is added to a positive number, the final answer is ______________.

8. The reciprocal of a negative number is ______________.

9. If three positive numbers, five negative numbers, and zero are multiplied, the product is ______________.

10. The fifth power of a negative number is ______________.

Find each product. See Examples 1 and 2.

11. $(-4)(-5)$ **12.** $(-4)(-6)$ **13.** $(-7)(4)$

14. $(-8)(5)$ **15.** $(-4)(-20)$ **16.** $(-8)(-30)$

17. $(-8)(0)$ **18.** $(-12)(0)$ **19.** $\left(-\frac{3}{8}\right)\left(-\frac{20}{9}\right)$

20. $\left(-\frac{5}{4}\right)\left(-\frac{6}{25}\right)$ **21.** $(-6)\left(-\frac{1}{4}\right)$ **22.** $(-8)\left(-\frac{1}{2}\right)$

Find all integer factors of each given number.

23. 32 **24.** 36 **25.** 40 **26.** 50 **27.** 31 **28.** 17

Find each quotient. See Examples 3 and 4.

29. $\frac{-15}{5}$ **30.** $\frac{-18}{6}$ **31.** $\frac{20}{-10}$ **32.** $\frac{28}{-4}$

33. $\frac{-160}{-10}$ **34.** $\frac{-260}{-20}$ **35.** $\frac{0}{-3}$ **36.** $\frac{0}{-6}$

37. $\frac{-10.252}{-.4}$ **38.** $\frac{-29.584}{-.8}$ **39.** $\left(-\frac{3}{4}\right) \div \left(-\frac{1}{2}\right)$ **40.** $\left(-\frac{3}{16}\right) \div \left(-\frac{5}{8}\right)$

41. Explain in words how to multiply signed numbers. Consider the various cases and give examples.

42. Explain in words how to divide signed numbers. Give examples.

Perform the indicated operations. See Examples 5(a), (b), and (c).

43. $7 - 3 \cdot 6$ **44.** $8 - 2 \cdot 5$

45. $-10 - (-4)(2)$ **46.** $-11 - (-3)(6)$

47. $-7(3 - 8)$ **48.** $-5(4 - 7)$

49. $(12 - 14)(1 - 4)$ **50.** $(8 - 9)(4 - 12)$

51. $(7 - 10)(10 - 4)$ **52.** $(5 - 12)(19 - 4)$

53. $(-2 - 8)(-6) + 7$ **54.** $(-9 - 4)(-2) + 10$

55. $3(-5) + |3 - 10|$ **56.** $4(-8) + |4 - 15|$

Perform each indicated operation. See Example 5(d).

57. $\frac{-5(-6)}{9 - (-1)}$ **58.** $\frac{-12(-5)}{7 - (-5)}$ **59.** $\frac{-21(3)}{-3 - 6}$

60. $\frac{-40(3)}{-2 - 3}$ **61.** $\frac{-10(2) + 6(2)}{-3 - (-1)}$ **62.** $\frac{8(-1) + 6(-2)}{-6 - (-1)}$

63. $\frac{-27(-2) - (-12)(-2)}{-2(3) - 2(2)}$ **64.** $\frac{-13(-4) - (-8)(-2)}{(-10)(2) - 4(-2)}$

65. Explain the method you would use to evaluate $3x + 2y$ if $x = -3$ and $y = 4$.

66. If x and y are both replaced by negative numbers, is the value of $4x + 8y$ positive or negative?

Evaluate each expression if $x = 6$, $y = -4$, and $a = 3$. See Example 6.

67. $5x - 2y + 3a$ **68.** $6x - 5y + 4a$ **69.** $(2x + y)(3a)$

70. $(5x - 2y)(-2a)$ **71.** $\left(\frac{1}{3}x - \frac{4}{5}y\right)\left(-\frac{1}{5}a\right)$ **72.** $\left(\frac{5}{6}x + \frac{3}{2}y\right)\left(-\frac{1}{3}a\right)$

73. $(-5 + x)(-3 + y)(3 - a)$ **74.** $(6 - x)(5 + y)(3 + a)$ **75.** $-2y^2 + 3a$

76. $5x - 4a^2$ **77.** $\frac{2y^2 - x}{a + 10}$ **78.** $\frac{xy + 8a}{x - y}$

Write a numerical expression for each phrase and simplify. See Examples 7 and 8.

79. The product of -9 and 2, added to 9

80. The product of 4 and -7, added to -12

81. Twice the product of -1 and 6, subtracted from -4

82. Twice the product of -8 and 2, subtracted from -1

83. Nine subtracted from the product of 1.5 and -3.2

84. Three subtracted from the product of 4.2 and -8.5

85. The product of 12 and the difference between 9 and -8

86. The product of -3 and the difference between 3 and -7

87. The quotient of -12 and the sum of -5 and -1

88. The quotient of -20 and the sum of -8 and -2

89. The sum of 15 and -3, divided by the product of 4 and -3

90. The sum of -18 and -6, divided by the product of 2 and -4

91. The product of $-\frac{1}{2}$ and $\frac{3}{4}$, divided by $-\frac{2}{3}$

92. The product of $-\frac{2}{3}$ and $-\frac{1}{5}$, divided by $\frac{1}{7}$

Write each statement in symbols, using x as the variable, and find the solution by guessing or by using trial and error. All solutions come from the set of integers between -12 *and* 12, *inclusive. See Example 9.*

93. Six times a number is -42.

94. Four times a number is -36.

95. The quotient of a number and 3 is -3.

96. The quotient of a number and 4 is -1.

97. 6 less than a number is 4.

98. 7 less than a number is 2.

99. When 5 is added to a number, the result is -5.

100. When 6 is added to a number, the result is -3.

To find the average of a group of numbers, we add the numbers and then divide the sum by the number of terms added. For example, to find the average of 14, 8, 3, 9, *and* 1, *we add them and then divide by* 5:

$$\frac{14 + 8 + 3 + 9 + 1}{5} = \frac{35}{5} = 7.$$

The average of these numbers is 7.

Exercises 101–106 involve finding the average of a group of numbers.

101. Find the average of 23, 18, 13, -4, and -8.

102. Find the average of 18, 12, 0, -4, and -10.

103. What is the average of all integers between -10 and 14, inclusive of both?

104. What is the average of all even integers between -18 and 4, inclusive of both?

105. The table shows average hourly earnings of various fields in private industry for 1995. Find the average of these amounts.

Private Industry Group	*1995 Hourly Earnings*
Mining	$15.30
Construction	15.08
Manufacturing	12.37
Transportation, public utilities	14.23
Wholesale trade	12.43
Retail trade	7.69
Finance, insurance, real estate	12.33
Service	11.39

Source: U.S. Bureau of Labor Statistics.

106. During the 1990s, the number of labor union or employee association members has remained fairly constant. The chart shows the number of members, in thousands, for the years 1990–1996.

Year	*Number (in thousands)*
1990	16,740
1991	16,568
1992	16,390
1993	16,598
1994	16,748
1995	16,360
1996	16,269

Source: U.S. Bureau of Labor Statistics.

What is the average number, in thousands, for this seven-year period?

107. If the average of a group of numbers is 0, what is the sum of all the numbers?

108. Suppose there is a group of numbers with some positive and some negative. Under what conditions will the average be a positive number? Under what conditions will the average be negative?

The operation of division is used in divisibility tests. *A divisibility test allows us to determine whether a given number is divisible (without remainder) by another number. For example, a number is divisible by 2 if its last digit is divisible by 2, and not otherwise.*

109. Tell why **(a)** 3,473,986 is divisible by 2 and **(b)** 4,336,879 is not divisible by 2.

110. An integer is divisible by 3 if the sum of its digits is divisible by 3, and not otherwise. Show that **(a)** 4,799,232 is divisible by 3 and **(b)** 2,443,871 is not divisible by 3.

111. An integer is divisible by 4 if its last two digits form a number divisible by 4, and not otherwise. Show that **(a)** 6,221,464 is divisible by 4 and **(b)** 2,876,335 is not divisible by 4.

112. An integer is divisible by 5 if its last digit is divisible by 5, and not otherwise. Show that **(a)** 3,774,595 is divisible by 5 and **(b)** 9,332,123 is not divisible by 5.

113. An integer is divisible by 6 if it is divisible by both 2 and 3, and not otherwise. Show that **(a)** 1,524,822 is divisible by 6 and **(b)** 2,873,590 is not divisible by 6.

114. An integer is divisible by 8 if its last three digits form a number divisible by 8, and not otherwise. Show that **(a)** 2,923,296 is divisible by 8 and **(b)** 7,291,623 is not divisible by 8.

115. An integer is divisible by 9 if the sum of its digits is divisible by 9, and not otherwise. Show that **(a)** 4,114,107 is divisible by 9 and **(b)** 2,287,321 is not divisible by 9.

116. An integer is divisible by 12 if it is divisible by both 3 and 4, and not otherwise. Show that **(a)** 4,253,520 is divisible by 12 and **(b)** 4,249,474 is not divisible by 12.

1.7 Properties of Real Numbers

OBJECTIVES

1. Use the commutative properties.
2. Use the associative properties.
3. Use the identity properties.
4. Use the inverse properties.
5. Use the distributive property.

FOR EXTRA HELP

SSG Sec. 1.7
SSM Sec. 1.7
Pass the Test Software
InterAct Math Tutorial Software
Video 2

If you were asked to find the sum $3 + 89 + 97$, you might mentally add $3 + 97$ to get 100, and then add $100 + 89$ to get 189. While the rule for order of operations says to add from left to right, we may change the order of the terms and group them in any way we choose without affecting the sum. These are examples of shortcuts that we use in everyday mathematics. These shortcuts are justified by the basic properties of addition and multiplication, discussed in this section. In these properties, a, b, and c represent real numbers.

OBJECTIVE 1 Use the commutative properties. The word *commute* means to go back and forth. Many people commute to work or to school. If you travel from home to work and follow the same route from work to home, you travel the same distance each time. The **commutative properties** say that if two numbers are added or multiplied in any order, the result is the same.

Commutative Properties

$$a + b = b + a$$
$$ab = ba$$

EXAMPLE 1 Using the Commutative Properties

Use a commutative property to complete each statement.

(a) $-8 + 5 = 5 + \underline{\quad\quad}$

By the commutative property for addition, the missing number is -8, since $-8 + 5 = 5 + (-8)$.

(b) $(-2)(7) = \underline{\quad\quad}(-2)$

By the commutative property for multiplication, the missing number is 7, since $(-2)(7) = (7)(-2)$.

OBJECTIVE 2 **Use the associative properties.** When we *associate* one object with another, we tend to think of those objects as being grouped together. The **associative properties** say that when we add or multiply three numbers, we can group the first two together or the last two together and get the same answer.

Associative Properties

$$(a + b) + c = a + (b + c)$$
$$(ab)c = a(bc)$$

EXAMPLE 2 **Using the Associative Properties**

Use an associative property to complete each statement.

(a) $8 + (-1 + 4) = (8 + \_\_\_\_) + 4$
The missing number is -1.

(b) $[2 \cdot (-7)] \cdot 6 = 2 \cdot \_\_\_\_$
The completed expression on the right should be $2 \cdot [(-7) \cdot 6]$.

By the associative property of addition, the sum of three numbers will be the same no matter how the numbers are "associated" in groups. For this reason, parentheses can be left out in many addition problems. For example, both

$$(-1 + 2) + 3 \qquad \text{and} \qquad -1 + (2 + 3)$$

can be written as

$$-1 + 2 + 3.$$

In the same way, parentheses also can be left out of many multiplication problems.

EXAMPLE 3 **Distinguishing between the Associative and Commutative Properties**

(a) Is $(2 + 4) + 5 = 2 + (4 + 5)$ an example of the associative property or the commutative property?

The order of the three numbers is the same on both sides of the equals sign. The only change is in the grouping, or association, of the numbers. Therefore, this is an example of the associative property.

(b) Is $6(3 \cdot 10) = 6(10 \cdot 3)$ an example of the associative property or the commutative property?

The same numbers, 3 and 10, are grouped on each side. On the left, the 3 appears first, but on the right, the 10 appears first. Since the only change involves the order of the numbers, this statement is an example of the commutative property.

(c) Is $(8 + 1) + 7 = 8 + (7 + 1)$ an example of the associative property or the commutative property?

In the statement, both the order and the grouping are changed. On the left the order of the three numbers is 8, 1, and 7. On the right it is 8, 7, and 1. On the left the 8 and 1 are grouped, and on the right the 7 and 1 are grouped. Therefore, both the associative and the commutative properties are used.

EXAMPLE 4 Using the Commutative and Associative Properties

Find the sum: $23 + 41 + 2 + 9 + 25$.

The commutative and associative properties make it possible to choose pairs of numbers whose sums are easy to add.

$$\begin{aligned} 23 + 41 + 2 + 9 + 25 &= (41 + 9) + (23 + 2) + 25 \\ &= 50 + 25 + 25 \\ &= 100 \end{aligned}$$

OBJECTIVE 3 Use the identity properties. If a child wears a costume on Halloween, the child's appearance is changed, but his or her *identity* is unchanged. The identity of a real number is left unchanged when identity properties are applied. The **identity properties** say that the sum of 0 and any number equals that number, and the product of 1 and any number equals that number.

Identity Properties

$$a + 0 = a \quad \text{and} \quad 0 + a = a$$
$$a \cdot 1 = a \quad \text{and} \quad 1 \cdot a = a$$

The number 0 leaves the identity, or value, of any real number unchanged by addition. For this reason, 0 is called the **identity element for addition.** Since multiplication by 1 leaves any real number unchanged, 1 is the **identity element for multiplication.**

EXAMPLE 5 Using the Identity Properties

These statements are examples of the identity properties.

(a) $-3 + 0 = -3$ **(b)** $1 \cdot \frac{1}{2} = \frac{1}{2}$

We use the identity property for multiplication to write fractions in lowest terms and to get common denominators.

EXAMPLE 6 Using the Identity Property to Simplify Expressions

Simplify the following expressions.

(a) $\frac{49}{35}$

$$\begin{aligned} \frac{49}{35} &= \frac{7 \cdot 7}{5 \cdot 7} && \text{Factor.} \\ &= \frac{7}{5} \cdot \frac{7}{7} && \text{Write as a product.} \\ &= \frac{7}{5} \cdot 1 && \text{Divide.} \\ &= \frac{7}{5} && \text{Identity property} \end{aligned}$$

(c) $-2(x + 3) = -2x + (-2)(3)$ Distributive property
$= -2x - 6$ Multiply.

(d) $3(k - 9) = 3k - 3 \cdot 9$ Distributive property
$= 3k - 27$ Multiply.

(e) $8(3r + 11t + 5z) = 8(3r) + 8(11t) + 8(5z)$ Distributive property
$= (8 \cdot 3)r + (8 \cdot 11)t + (8 \cdot 5)z$ Associative property
$= 24r + 88t + 40z$ Multiply.

(f) $6 \cdot 8 + 6 \cdot 2 = 6(8 + 2)$ Distributive property
$= 6(10) = 60$ Add, then multiply.

(g) $4x - 4m = 4(x - m)$ Distributive property

(h) $6x - 12 = 6 \cdot x - 6 \cdot 2 = 6(x - 2)$ Distributive property

The symbol $-a$ may be interpreted as $-1 \cdot a$. Similarly, when a negative sign precedes an expression within parentheses, it may also be interpreted as a factor of -1. The distributive property is used to remove parentheses from expressions such as $-(2y + 3)$. We do this by first writing $-(2y + 3)$ as $-1 \cdot (2y + 3)$.

$$-(2y + 3) = -1 \cdot (2y + 3)$$
$$= -1 \cdot (2y) + (-1) \cdot (3) \quad \text{Distributive property}$$
$$= -2y - 3 \quad \text{Multiply.}$$

EXAMPLE 10 Using the Distributive Property to Remove Parentheses

Write without parentheses.

(a) $-(7r - 8) = -1(7r) + (-1)(-8)$ Distributive property
$= -7r + 8$ Multiply.

(b) $-(-9w + 2) = 9w - 2$

The properties discussed here are the basic properties that justify how we do algebra. You should know them by name because we will be referring to them frequently. Here is a summary of these properties.

Properties of Addition and Multiplication

For any real numbers a, b, and c, the following properties hold.

Commutative Properties $a + b = b + a$ $\quad ab = ba$

Associative Properties $(a + b) + c = a + (b + c)$
$(ab)c = a(bc)$

Identity Properties There is a real number 0 such that

$a + 0 = a$ and $0 + a = a$.

There is a real number 1 such that

$a \cdot 1 = a$ and $1 \cdot a = a$.

Properties of Addition and Multiplication (continued)

Inverse Properties

For each real number a, there is a single real number $-a$ such that

$$a + (-a) = 0 \quad \text{and} \quad (-a) + a = 0.$$

For each nonzero real number a, there is a single real number $\frac{1}{a}$ such that

$$a \cdot \frac{1}{a} = 1 \quad \text{and} \quad \frac{1}{a} \cdot a = 1.$$

Distributive Property

$$a(b + c) = ab + ac \qquad (b + c)a = ba + ca$$

1.7 EXERCISES

Match each item in Column I with the correct choice(s) from Column II. Choices may be used once, more than once, or not at all.

I

1. Identity element for addition
2. Identity element for multiplication
3. Additive inverse of a
4. Multiplicative inverse, or reciprocal, of the nonzero number a
5. The number that is its own additive inverse
6. The two numbers that are their own multiplicative inverses
7. The only number that has no multiplicative inverse
8. An example of the associative property
9. An example of the commutative property
10. An example of the distributive property

II

A. $(5 \cdot 4) \cdot 3 = 5 \cdot (4 \cdot 3)$

B. 0

C. $-a$

D. -1

E. $5 \cdot 4 \cdot 3 = 60$

F. 1

G. $(5 \cdot 4) \cdot 3 = 3 \cdot (5 \cdot 4)$

H. $5(4 + 3) = 5 \cdot 4 + 5 \cdot 3$

I. $\frac{1}{a}$

Decide whether each statement is an example of the commutative, associative, identity, inverse, or distributive property. See Examples 1, 2, 3, 5, 6, 7, and 9.

11. $7 + 18 = 18 + 7$

12. $13 + 12 = 12 + 13$

13. $5(13 \cdot 7) = (5 \cdot 13) \cdot 7$

14. $-4(2 \cdot 6) = (-4 \cdot 2) \cdot 6$

15. $-6 + (12 + 7) = (-6 + 12) + 7$

16. $(-8 + 13) + 2 = -8 + (13 + 2)$

17. $-6 + 6 = 0$

18. $12 + (-12) = 0$

19. $\left(\frac{2}{3}\right)\left(\frac{3}{2}\right) = 1$

20. $\left(\frac{5}{8}\right)\left(\frac{8}{5}\right) = 1$

21. $2.34 + 0 = 2.34$

22. $-8.456 + 0 = -8.456$

23. $(4 + 17) + 3 = 3 + (4 + 17)$

24. $(-8 + 4) + (-12) = -12 + (-8 + 4)$

25. $6(x + y) = 6x + 6y$

26. $14(t + s) = 14t + 14s$

27. $-\frac{5}{9} = -\frac{5}{9} \cdot \frac{3}{3} = -\frac{15}{27}$

28. $\frac{13}{12} = \frac{13}{12} \cdot \frac{7}{7} = \frac{91}{84}$

29. $5(2x) + 5(3y) = 5(2x + 3y)$

30. $3(5t) - 3(7r) = 3(5t - 7r)$

31. The following conversation actually took place between one of the authors of this book and his son, Jack, when Jack was four years old:

DADDY: "Jack, what is $3 + 0$?"
JACK: "3."
DADDY: "Jack, what is $4 + 0$?"
JACK: "4. And Daddy, *string* plus zero equals *string*!"

What property of addition did Jack recognize?

32. The distributive property holds for multiplication with respect to addition. Is there a distributive property for addition with respect to multiplication? If not, give an example to show why.

33. Write a paragraph explaining in your own words the following properties of addition and multiplication: commutative, associative, identity, inverse.

34. Write a paragraph explaining in your own words the distributive property of multiplication with respect to addition. Give examples.

Use the indicated property to write a new expression that is equal to the given expression. Then simplify the new expression if possible. See Examples 1, 2, 5, 7, and 9.

35. $r + 7$; commutative
36. $t + 9$; commutative
37. $s + 0$; identity
38. $w + 0$; identity
39. $-6(x + 7)$; distributive
40. $-5(y + 2)$; distributive
41. $(w + 5) + (-3)$; associative
42. $(b + 8) + (-10)$; associative

Use the properties of this section to simplify each expression. See Examples 7 and 8.

43. $6t + 8 - 6t + 3$
44. $9r + 12 - 9r + 1$
45. $\frac{2}{3}x - 11 + 11 - \frac{2}{3}x$
46. $\frac{1}{5}y + 4 - 4 - \frac{1}{5}y$
47. $\left(\frac{9}{7}\right)(-.38)\left(\frac{7}{9}\right)$
48. $\left(\frac{4}{5}\right)(-.73)\left(\frac{5}{4}\right)$
49. $t + (-t) + \frac{1}{2}(2)$
50. $w + (-w) + \frac{1}{4}(4)$

51. Evaluate $25 - (6 - 2)$ and evaluate $(25 - 6) - 2$. Do you think subtraction is associative?

52. Evaluate $180 \div (15 \div 3)$ and evaluate $(180 \div 15) \div 3$. Do you think division is associative?

53. Suppose that a student shows you the following work.

$$-3(4 - 6) = -3(4) - 3(6) = -12 - 18 = -30$$

The student has made a very common error. Explain the student's mistake, and work the problem correctly.

54. Explain how the procedure of changing $\frac{3}{4}$ to $\frac{9}{12}$ requires the use of the multiplicative identity element, 1.

Use the distributive property to rewrite each expression. Simplify if possible. See Example 9.

55. $5x + x$
56. $6q + q$
57. $4(t + 3)$
58. $5(w + 4)$
59. $-8(r + 3)$
60. $-11(x + 4)$
61. $-5(y - 4)$
62. $-9(g - 4)$
63. $-\frac{4}{3}(12y + 15z)$
64. $-\frac{2}{5}(10b + 20a)$
65. $8 \cdot z + 8 \cdot w$
66. $4 \cdot s + 4 \cdot r$
67. $7(2v) + 7(5r)$
68. $13(5w) + 13(4p)$
69. $8(3r + 4s - 5y)$

70. $2(5u - 3v + 7w)$ **71.** $q + q + q$ **72.** $m + m + m + m$

73. $-5x + x$ **74.** $-9p + p$

Use the distributive property to write each expression without parentheses. See Example 10.

75. $-(4t + 3m)$ **76.** $-(9x + 12y)$ **77.** $-(-5c - 4d)$

78. $-(-13x - 15y)$ **79.** $-(-3q + 5r - 8s)$ **80.** $-(-4z + 5w - 9y)$

81. The operations of "getting out of bed" and "taking a shower" are not commutative. Give an example of another pair of everyday operations that are not commutative.

82. The phrase "dog biting man" has two different meanings, depending on how the words are associated:

(dog biting) man dog (biting man)

Give another example of a three-word phrase that has different meanings depending on how the words are associated.

RELATING CONCEPTS (EXERCISES 83–86)

In Section 1.6 we used a pattern to see that the product of two negative numbers is a positive number. In the group of exercises that follows, we show another justification for determining the sign of the product of two negative numbers.

Work Exercises 83–86 in order.

83. Evaluate the expression $-3[5 + (-5)]$ by using the rules for order of operations.

84. Write the expression in Exercise 83 using the distributive property. Do not simplify the products.

85. The product -3×5 should be one of the terms you wrote when answering Exercise 84. Based on the results in Section 1.6, what is this product?

86. In Exercise 83, you should have obtained 0 as an answer. Now, consider the following, using the results of Exercises 83 and 85.

$$-3[5 + (-5)] = -3(5) + (-3)(-5)$$
$$0 = -15 + ?$$

The question mark represents the product $(-3)(-5)$. When added to -15, it must give a sum of 0. Therefore, how must we interpret $(-3)(-5)$?

Did you make the connection that a rule can be obtained in more than one way, with consistent results from each method?

1.8 Simplifying Expressions

OBJECTIVES

1. Simplify expressions.
2. Identify terms and numerical coefficients.
3. Identify like terms.

OBJECTIVE 1 Simplify expressions. In this section we show how to simplify expressions using the properties of addition and multiplication introduced in the previous section.

EXAMPLE 1 Simplifying Expressions

Simplify the following expressions.

(a) $4x + 8 + 9$

Since $8 + 9 = 17$,

$$4x + 8 + 9 = 4x + 17.$$

4 Combine like terms.

5 Simplify expressions from word phrases.

FOR EXTRA HELP

SSG Sec. 1.8
SSM Sec. 1.8
Pass the Test Software
InterAct Math Tutorial Software
Video 2

(b) $4(3m - 2n)$
Use the distributive property first.

$$4(3m - 2n) = 4(3m) - 4(2n)$$ Arrows denote distributive property.

$$= (4 \cdot 3)m - (4 \cdot 2)n$$ Associative property

$$= 12m - 8n$$

(c) $6 + 3(4k + 5) = 6 + 3(4k) + 3(5)$ Distributive property

$$= 6 + (3 \cdot 4)k + 3(5)$$ Associative property

$$= 6 + 12k + 15$$

$$= 6 + 15 + 12k$$ Commutative property

$$= 21 + 12k$$

(d) $5 - (2y - 8) = 5 - 1 \cdot (2y - 8)$ Replace $-$ with -1.

$$= 5 - 2y + 8$$ Distributive property

$$= 5 + 8 - 2y$$ Commutative property

$$= 13 - 2y$$

NOTE In Example 1, parts (c) and (d), a different use of the commutative property would have resulted in answers of $12k + 21$ and $-2y + 13$. These answers also would be acceptable.

The steps using the commutative and associative properties will not be shown in the rest of the examples, but you should be aware that they are usually involved.

OBJECTIVE 2 Identify terms and numerical coefficients. A **term** is a number, a variable, or a product or quotient of numbers and variables raised to powers.* Examples of terms include

$$-9x^2, \quad 15y, \quad -3, \quad 8m^2n, \quad \frac{2}{p}, \quad \text{and} \quad k.$$

The **numerical coefficient** of the term $9m$ is 9, the numerical coefficient of $-15x^3y^2$ is -15, the numerical coefficient of x is 1, and the numerical coefficient of 8 is 8. In the expression $\frac{x}{3}$, the numerical coefficient of x is $\frac{1}{3}$. Do you see why?

CAUTION It is important to be able to distinguish between *terms* and *factors.* For example, in the expression $8x^3 + 12x^2$, there are two *terms,* $8x^3$ and $12x^2$. On the other hand, in the one-term expression $(8x^3)(12x^2)$, $8x^3$ and $12x^2$ are *factors.*

Several examples of terms and their numerical coefficients follow.

*Another name for certain terms, **monomial,** is introduced in Chapter 4.

Term	*Numerical Coefficient*
$-7y$	-7
$8p$	8
$34r^3$	34
$-26x^5yz^4$	-26
$-k$	-1
$\frac{x}{7}$	$\frac{1}{7}$

OBJECTIVE 3 **Identify like terms.** Terms with exactly the same variables that have the same exponents are **like terms.** For example, $9m$ and $4m$ have the same variable and are like terms. Also, $6x^3$ and $-5x^3$ are like terms. The terms $-4y^3$ and $4y^2$ have different exponents and are **unlike terms.**

Here are some additional examples.

$5x$ and $-12x$	$3x^2y$ and $5x^2y$	Like terms
$4xy^2$ and $5xy$	$-7w^3z^3$ and $2xz^3$	Unlike terms

OBJECTIVE 4 **Combine like terms.** Recall the distributive property:

$$x(y + z) = xy + xz.$$

As seen in the previous section, this statement can also be written "backward" as

$$xy + xz = x(y + z).$$

This form of the distributive property may be used to find the sum or difference of like terms. For example,

$$3x + 5x = (3 + 5)x = 8x.$$

This process is called **combining like terms.**

NOTE Remember that *only like terms may be combined.* For example, $5x^2 + 2x \neq 7x^3$.

EXAMPLE 2 Combining Like Terms

Combine like terms in the following expressions.

(a) $9m + 5m$

Use the distributive property as given above.

$$9m + 5m = (9 + 5)m = 14m$$

(b) $6r + 3r + 2r = (6 + 3 + 2)r = 11r$ Distributive property

(c) $4x + x = 4x + 1x = (4 + 1)x = 5x$ (Note: $x = 1x$.)

(d) $16y^2 - 9y^2 = (16 - 9)y^2 = 7y^2$

(e) $32y + 10y^2$ cannot be combined because $32y$ and $10y^2$ are unlike terms. The distributive property cannot be used here to combine coefficients.

Write a numerical expression for each phrase, and simplify the expression.

67. 19 added to the sum of -31 and 12

68. 13 more than the sum of -4 and -8

69. The difference between -4 and -6

70. Five less than the sum of 4 and -8

Find the solution of the equation from the set $\{-3, -2, -1, 0, 1, 2, 3\}$ *by guessing or by trial and error.*

71. $x + (-2) = -4$

72. $12 + x = 11$

Solve each problem.

73. Like many people, Kareem Dunlap neglects to keep up his checkbook balance. When he finally balanced his account, he found the balance was $-\$23.75$, so he deposited \$50.00. What is his new balance?

74. The low temperature in Yellowknife, in the Canadian Northwest Territories, one January day was $-26°$F. It rose 16° that day. What was the high temperature?

75. Eric owed his brother \$28. He repaid \$13 but then borrowed another \$14. What positive or negative amount represents his present financial status?

76. If the temperature drops 7° below its previous level of $-3°$, what is the new temperature?

77. A football team gained 3 yards on the first play from scrimmage, lost 12 yards on the second play, and then gained 13 yards on the third play. How many yards did the team gain or lose altogether?

78. In 1985, the construction industry had 4,480,000 employees. By 1990, this number had increased by 759,000 employees, but it experienced a decrease of 530,000 employees by 1994. How many employees were there in 1994? (*Source:* U.S. Bureau of the Census.)

[1.6] *Perform the indicated operations.*

79. $(-12)(-3)$

80. $15(-7)$

81. $\left(-\frac{4}{3}\right)\left(-\frac{3}{8}\right)$

82. $(-4.8)(-2.1)$

83. $5(8 - 12)$

84. $(5 - 7)(8 - 3)$

85. $2(-6) - (-4)(-3)$

86. $3(-10) - 5$

87. $\dfrac{-36}{-9}$

88. $\dfrac{220}{-11}$

89. $-\frac{1}{2} \div \frac{2}{3}$

90. $-33.9 \div (-3)$

91. $\dfrac{-5(3) - 1}{8 - 4(-2)}$

92. $\dfrac{5(-2) - 3(4)}{-2[3 - (-2)] - 1}$

93. $\dfrac{10^2 - 5^2}{8^2 + 3^2 - (-2)}$

94. $\dfrac{(.6)^2 + (.8)^2}{(-1.2)^2 - (-.56)}$

Evaluate each expression if $x = -5$, $y = 4$, *and* $z = -3$.

95. $6x - 4z$

96. $5x + y - z$

97. $5x^2$

98. $z^2(3x - 8y)$

Write a numerical expression for each phrase, and simplify the expression.

99. Nine less than the product of -4 and 5

100. Five-sixths of the sum of 12 and -6

101. The quotient of 12 and the sum of 8 and -4

102. The product of -20 and 12, divided by the difference between 15 and -15

Write each sentence in symbols, using x as the variable, and find the solution by guessing or by trial and error. All solutions come from the list of integers between -12 *and* 12.

103. 8 times a number is -24.

104. The quotient of a number and 3 is -2.

105. The payrolls and average salaries of 5 of the 28 major league baseball teams on opening day of 1998 are listed here.

Team	*Payroll*	*Average Salary*
Baltimore Orioles	\$68,988,134	\$2,555,116
New York Yankees	63,460,567	2,440,791
Cleveland Indians	59,583,500	2,127,982
Atlanta Braves	59,536,000	2,126,286
Texas Rangers	55,304,595	1,975,164

Source: The Associated Press.

(a) What is the average of the five payrolls? Round to the nearest dollar.
(b) What is the average of the five average salaries? Round to the nearest dollar.

106. The bar graph shows the 1993 sales in millions of dollars of four of the largest brands in the United States. What was the average of these sales?

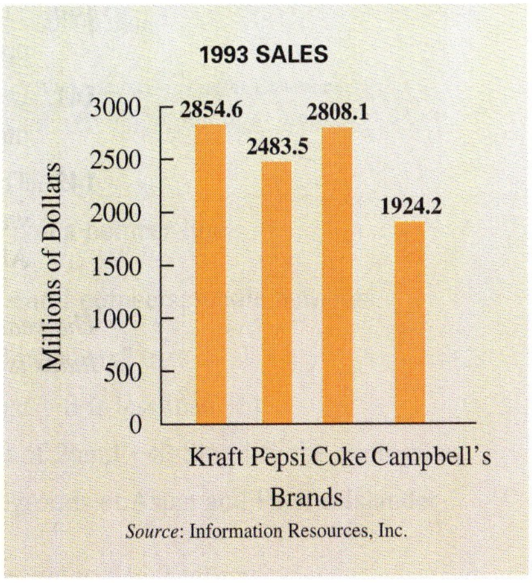

[1.7] *Decide whether each statement is an example of the commutative, associative, identity, inverse, or distributive property.*

107. $6 + 0 = 6$

108. $5 \cdot 1 = 5$

109. $-\frac{2}{3}\left(-\frac{3}{2}\right) = 1$

110. $17 + (-17) = 0$

111. $5 + (-9 + 2) = [5 + (-9)] + 2$

112. $w(xy) = (wx)y$

113. $3x + 3y = 3(x + y)$

114. $(1 + 2) + 3 = 3 + (1 + 2)$

Use the distributive property to rewrite each expression. Simplify if possible.

115. $7y + y$

116. $-12(4 - t)$

117. $3(2s) + 3(5y)$

118. $-(-4r + 5s)$

119. Evaluate $25 - (5 - 2)$ and $(25 - 5) - 2$. Use this example to explain why subtraction is not associative.

120. Evaluate $180 \div (15 \div 5)$ and $(180 \div 15) \div 5$. Use this example to explain why division is not associative.

[1.8] *Combine terms whenever possible.*

121. $2m + 9m$

122. $15p^2 - 7p^2 + 8p^2$

123. $5p^2 - 4p + 6p + 11p^2$

124. $-2(3k - 5) + 2(k + 1)$

125. $7(2m + 3) - 2(8m - 4)$

126. $-(2k + 8) - (3k - 7)$

13. $-6 - [-7 + (2 - 3)]$

14. $4^2 + (-8) - (2^3 - 6)$

15. $(-5)(-12) + 4(-4) + (-8)^2$

16. $\dfrac{-7 - (-6 + 2)}{-5 - (-4)}$

17. $\dfrac{30(-1 - 2)}{-9[3 - (-2)] - 12(-2)}$

Find the solution for each equation from the set $\{-6, -4, -2, 0, 2, 4, 6\}$ *by guessing or by trial and error.*

18. $-x + 3 = -3$

19. $-3x = -12$

Evaluate each expression, given $x = -2$ *and* $y = 4$.

20. $3x - 4y^2$

21. $\dfrac{5x + 7y}{3(x + y)}$

Solve each problem.

22. The average sales prices of new single-family homes in the United States for the years 1990 through 1995 are shown in the chart. Determine the change from one year to the next by subtraction.

	Year	Average Sales Price	Change from Previous Year
	1990	\$149,800	
	1991	\$147,200	−\$2600
(a)	1992	\$144,100	________
(b)	1993	\$147,700	________
(c)	1994	\$154,500	________
(d)	1995	\$158,700	________

Source: U.S. Bureau of the Census.

23. For a certain system of rating major league baseball relief pitchers, 3 points are awarded for a save, 3 points are awarded for a win, 2 points are subtracted for a loss, and 2 points are subtracted for a blown save. If Dennis Eckersley of the Boston Red Sox has 4 saves, 3 wins, 2 losses, and 1 blown save, how many points does he have?

24. The bar graph shows the number of corporate name changes that occurred during the years 1990–1996. These changes were mostly brought on by mergers and acquisitions. Use a signed number to represent each of the following.
(a) the change from 1990 to 1991
(b) the change from 1991 to 1996
(c) the change from 1992 to 1996

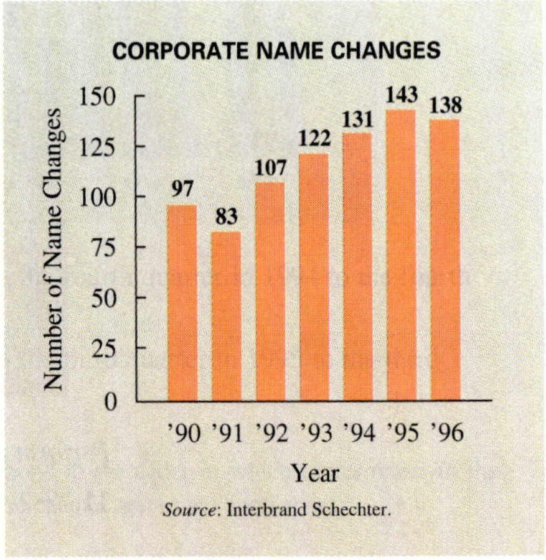

Source: Interbrand Schechter.

Match each property in Column I with the example of it in Column II.

I	II
25. Commutative	**A.** $3x + 0 = 3x$
26. Associative	**B.** $(5 + 2) + 8 = 8 + (5 + 2)$
27. Inverse	**C.** $-3(x + y) = -3x + (-3y)$
28. Identity	**D.** $-5 + (3 + 2) = (-5 + 3) + 2$
29. Distributive	**E.** $-\frac{5}{3}\left(-\frac{3}{5}\right) = 1$

30. What property is used to show that $3(x + 1) = 3x + 3$?

31. Consider the expression $-6[5 + (-2)]$.
(a) Evaluate it by first working within the brackets.
(b) Evaluate it by using the distributive property.
(c) Why must the answers in items (a) and (b) be the same?

Simplify by combining like terms.

32. $8x + 4x - 6x + x + 14x$

33. $5(2x - 1) - (x - 12) + 2(3x - 5)$

2 Linear Equations and Inequalities in One Variable

Sports

The first organized worldwide sporting event was the Summer Olympics I, held in 1896 in Athens, Greece. Three hundred eleven men from 13 nations gathered to compete. One hundred years later, the 1996 Summer Olympics, held in Atlanta, attracted 10,341 competitors—6562 men and 3779 women, from 197 nations. The results of an Olympic women's swimming event from 1968 to 1996 are given in the table. (Winning times are given in minutes and seconds to the nearest hundredth.)

200-Meter Freestyle

1968 Deborah Meyer, U.S.	2:10.50
1972 Shane Gould, Australia	2:03.56
1976 Kornelia Ender, East Germany	1:59.26
1980 Barbara Krause, East Germany	1:58.33
1984 Mary Wayte, U.S.	1:59.23
1988 Heike Friederich, East Germany	1:57.65
1992 Nicole Haislett, U.S.	1:57.90
1996 Claudia Poll, Costa Rica	1:58.16

Which year had the fastest winning time? In Section 2.6, we'll see how to determine the winning speed from this information.

In the United States, the first organized professional sport was baseball. The first baseball league was founded in 1876, and the first World Series playoff was held in 1882.* Both football and basketball worked their way from college campus to professional league status.

And, as you'll discover in this chapter, sports involves mathematics; the vast quantity of statistics kept by professional sports leagues fills numerous books.

*All data from *The Universal Almanac,* 1997, John W. Wright, General Editor.

Visit our Web site at www.LialAlgebra.com

2.1 The Addition and Multiplication Properties of Equality

OBJECTIVES

1. Identify linear equations.
2. Use the addition property of equality.
3. Use the multiplication property of equality.
4. Simplify equations, and then use the properties of equality.

FOR EXTRA HELP

SSG Sec. 2.1
SSM Sec. 2.1
Pass the Test Software
InterAct Math Tutorial Software
Video 3

Quite possibly the most important topic in the study of beginning algebra is the solution of equations. We will investigate many types of equations in this book, and the properties introduced in this section are essential in solving them.

OBJECTIVE 1 Identify linear equations. The simplest type of equation is a *linear equation.* Before we can solve a linear equation we must be able to recognize one.

Linear Equation

A **linear equation in one variable** can be written in the form

$$Ax + B = 0$$

for real numbers A and B, with $A \neq 0$.

For example,

$$4x + 9 = 0, \qquad 2x - 3 = 5, \qquad \text{and} \qquad x = 7$$

are linear equations; the last two can be written in the specified form using the properties to be developed in this section. However,

$$x^2 + 2x = 5, \qquad \frac{1}{x} = 6 \qquad \text{and} \qquad |2x + 6| = 0$$

are *not* linear equations.

Methods of solving linear equations are introduced in this section and in Section 2.2. As discussed in Section 1.3, a *solution* of an equation is a number that when substituted for the variable makes the equation true; that is, *satisfies* the equation. An equation is solved by finding its **solution set,** the set of all solutions. Equations that have exactly the same solution sets are **equivalent equations.** Linear equations are solved by using a series of steps to produce equivalent equations until an equation of the form

$$x = \text{a number}$$

is obtained.

OBJECTIVE 2 Use the addition property of equality. Consider the equation

$$x = 4.$$

Suppose we add 7 to both sides of the equation, getting

$$x + 7 = 11.$$

Now, adding $2x$ to both sides gives

$$3x + 7 = 2x + 11.$$

Undo what was done to B by first subtracting bh on both sides. Then divide both sides by h.

$$2A - bh = Bh \qquad \text{Subtract } bh.$$

$$\frac{2A - bh}{h} = B \quad \text{or} \quad B = \frac{2A - bh}{h} \qquad \text{Divide by } h.$$

The result can be written in a different form as follows.

$$B = \frac{2A - bh}{h} = \frac{2A}{h} - \frac{bh}{h} = \frac{2A}{h} - b$$

Either form is correct.

2.4 EXERCISES

1. In your own words, explain what is meant by the *perimeter* of a geometric figure.
2. In your own words, explain what is meant by the *area* of a geometric figure.

Decide whether perimeter or area would be used to solve a problem concerning the measure of the quantity.

3. Carpeting for a bedroom
4. Sod for a lawn
5. Fencing for a yard
6. Baseboards for a living room
7. Tile for a bathroom
8. Fertilizer for a garden
9. Determining the cost for replacing a linoleum floor with a wood floor
10. Determining the cost for planting rye grass in a lawn for the winter

In the following exercises a formula is given along with the values of all but one of the variables in the formula. Find the value of the variable that is not given. See Example 1.

11. $P = 2L + 2W$ (perimeter of a rectangle); $L = 6$, $W = 4$
12. $P = 2L + 2W$; $L = 8$, $W = 5$
13. $P = 4s$ (perimeter of a square); $s = 6$
14. $P = 4s$; $s = 12$
15. $A = \frac{1}{2}bh$ (area of a triangle); $b = 10$, $h = 14$
16. $A = \frac{1}{2}bh$; $b = 8$, $h = 16$
17. $d = rt$ (distance formula); $d = 100$, $t = 2.5$
18. $d = rt$; $d = 252$, $r = 45$
19. $I = prt$ (simple interest); $p = 5000$, $r = .025$, $t = 7$
20. $I = prt$; $p = 7500$, $r = .035$, $t = 6$
21. $A = \frac{1}{2}h(b + B)$ (area of a trapezoid); $h = 7$, $b = 12$, $B = 14$
22. $A = \frac{1}{2}h(b + B)$; $h = 3$, $b = 19$, $B = 31$
23. $C = 2\pi r$ (circumference of a circle); $C = 8.164$, $\pi = 3.14$
24. $C = 2\pi r$; $C = 16.328$, $\pi = 3.14$
25. $A = \pi r^2$ (area of a circle); $r = 12$, $\pi = 3.14$

26. $A = \pi r^2$; $r = 4$, $\pi = 3.14$

27. If a formula contains exactly five variables, how many values would you need to be given in order to find the value of any one variable?

28. The formula for changing Celsius to Fahrenheit is given in Example 5 as $F = \frac{9}{5}C + 32$. Sometimes it is seen as $F = \frac{9C}{5} + 32$. These are both correct. Why is it true that $\frac{9}{5}C$ is equal to $\frac{9C}{5}$?

*The **volume** of a three-dimensional object is a measure of the space occupied by the object. For example, we would need to know the volume of a gasoline tank in order to know how many gallons of gasoline it would take to completely fill the tank. In each of the following exercises, a formula for the volume, V, of a three-dimensional object is given, along with values for the other variables. Solve for V. See Example 1.*

29. $V = LWH$ (volume of a rectangular-sided box); $L = 12$, $W = 8$, $H = 4$

30. $V = LWH$; $L = 10$, $W = 5$, $H = 3$

31. $V = \frac{1}{3}Bh$ (volume of a pyramid); $B = 36$, $h = 4$

32. $V = \frac{1}{3}Bh$; $B = 12$, $h = 13$

33. $V = \frac{4}{3}\pi r^3$ (volume of a sphere); $r = 6$, $\pi = 3.14$

34. $V = \frac{4}{3}\pi r^3$; $r = 12$, $\pi = 3.14$

Use a formula to write an equation for each application, and use the six-step problem-solving method of Section 2.3 to solve it. Formulas may be found on the inside covers of this book. See Examples 2 and 3. Use 3.14 for π, if applicable.

35. Recently, a prehistoric ceremonial site dating to about 3000 B.C. was discovered at Stanton Drew in southwestern England. The site, which is larger than Stonehenge, is a near perfect circle, consisting of nine concentric rings that probably held upright wooden posts. Around this timber temple is a wide, encircling ditch enclosing an area with a diameter of 443 feet. Find this enclosed area. (*Source: Archaeology,* vol. 51, no. 1, Jan./Feb. 1998.)

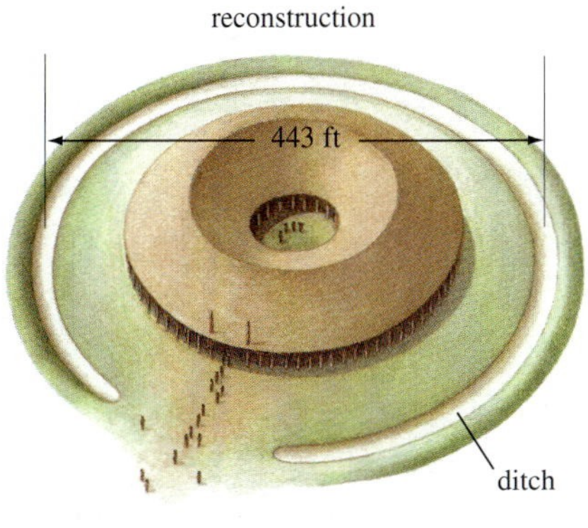

36. The U.S. Postal Service requires that any box sent through the mail have length plus girth (distance around) totaling no more than 108 inches. The maximum volume that meets this condition is contained by a box with a square end 18 inches on each side. What is the length of the box? What is the maximum volume?

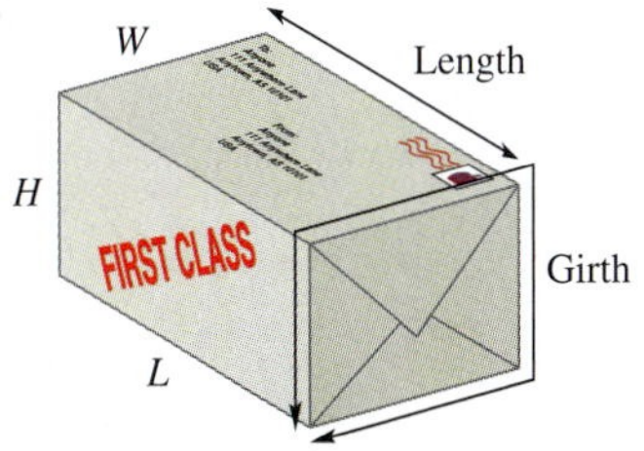

37. The Skydome in Toronto, Canada, is the first stadium with a hard-shell, retractable roof. The steel dome is 630 feet in diameter. To the nearest foot, what is the circumference of this dome?

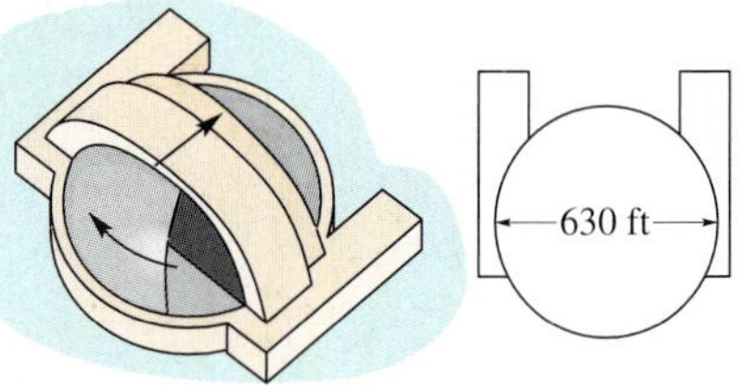

38. The largest drum ever constructed was played at the Royal Festival Hall in London in 1987. It had a diameter of 13 feet. What was the area of the circular face of the drum? (*Source: The Guinness Book of World Records.*)

39. The newspaper, *The Constellation,* printed in 1859 in New York City as part of the Fourth of July celebration, had length 51 inches and width 35 inches. What was the perimeter? What was the area? (*Source: The Guinness Book of World Records.*)

40. The *Daily Banner,* published in Roseberg, Oregon, in the nineteenth century, had page size 3 inches by 3.5 inches. What was the perimeter? What was the area? (*Source: The Guinness Book of World Records.*)

41. A color television set with a liquid crystal display was manufactured by Epson in 1985 and had dimensions 3 inches by $6\frac{3}{4}$ inches by $1\frac{1}{8}$ inches. What was the volume of this set? (*Source: The Guinness Book of World Records.*)

42. A sea lock at Zeebrugge, Belgium, measures 1640 feet by 187 feet by 75.4 feet. What is the volume of the lock? (It is a rectangular solid, so use the formula for the volume of such a figure.) (*Source: The Guinness Book of World Records.*)

43. The survey plat (a two-dimensional map plotted from a plane survey) in the figure shows two lots that form a trapezoid. The measures of the parallel sides are 115.80 feet and 171.00 feet. The height of the trapezoid is 165.97 feet. Find the combined area of the two lots. Round your answer to the nearest hundredth of a square foot.

44. Lot A in the figure is in the shape of a trapezoid. The parallel sides measure 26.84 feet and 82.05 feet. The height of the trapezoid is 165.97 feet. Find the area of Lot A. Round your answer to the nearest hundredth of a square foot.

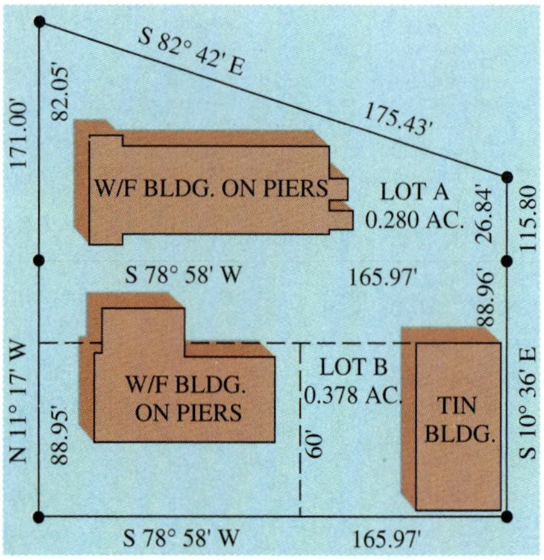

Property survey of lot in New Roads, Louisiana.

Find the measure of each marked angle. See Example 4.

45.

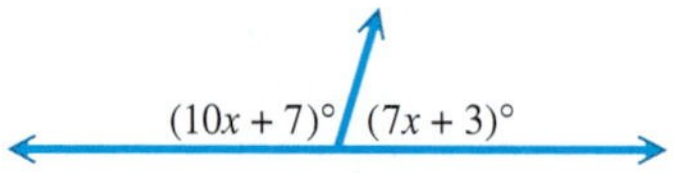

46.

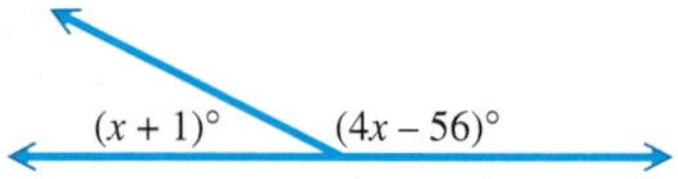

47.

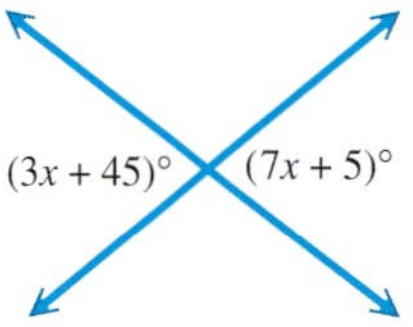

48.

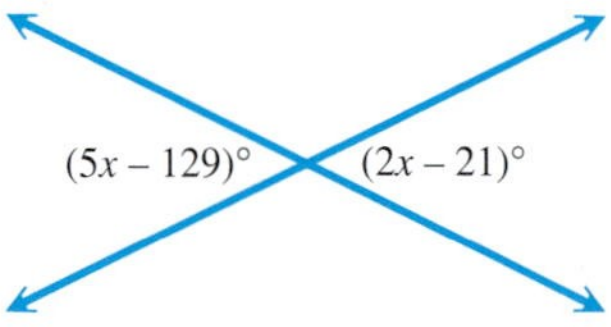

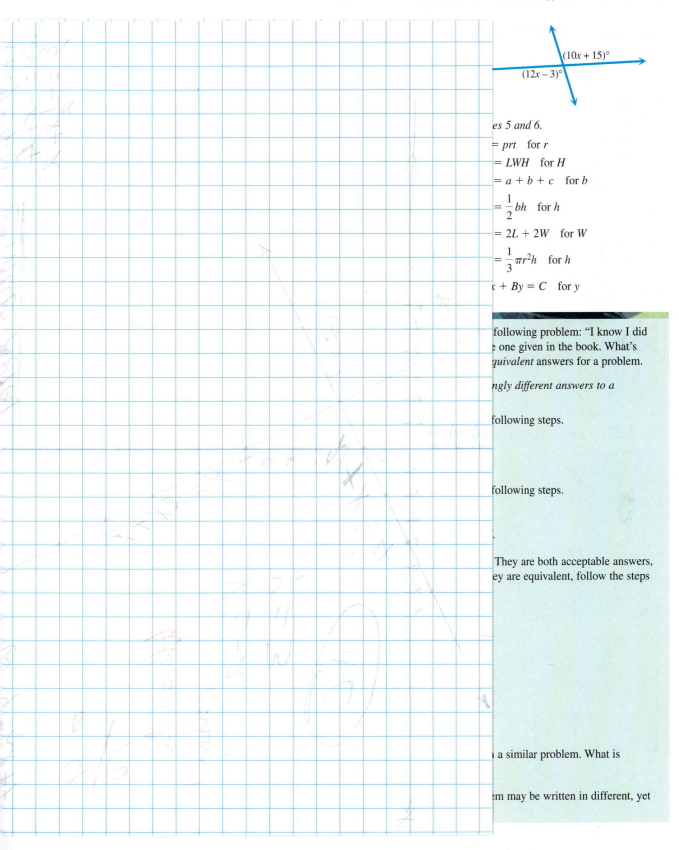

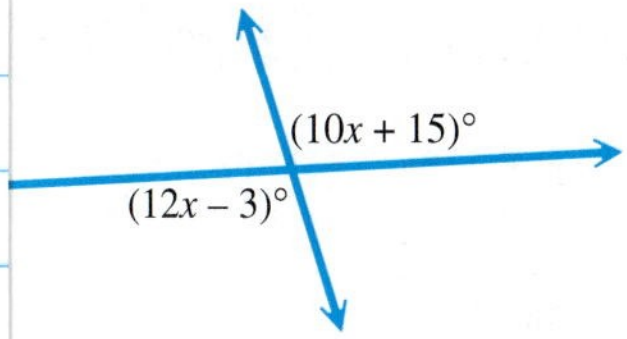

es 5 and 6.

= prt for r

= LWH for H

= $a + b + c$ for b

= $\frac{1}{2}bh$ for h

= $2L + 2W$ for W

= $\frac{1}{3}\pi r^2h$ for h

$\ldots x + By = C$ for y

following problem: "I know I did
e one given in the book. What's
quivalent answers for a problem.

ngly different answers to a

following steps.

following steps.

.

They are both acceptable answers,
ey are equivalent, follow the steps

a similar problem. What is

em may be written in different, yet

2.5 Ratios and Proportions

OBJECTIVES

1. Write ratios.
2. Decide whether proportions are true.
3. Solve proportions.
4. Solve applied problems using proportions.
5. Solve direct variation problems.

FOR EXTRA HELP

SSG Sec. 2.5
SSM Sec. 2.5
Pass the Test Software
InterAct Math Tutorial Software
Video 3

OBJECTIVE 1 **Write ratios.** Ratios provide a way of comparing two numbers or quantities using division. A **ratio** is a quotient of two quantities with the same units.

The ratio of the number a to the number b is written

$$a \text{ to } b, \qquad a : b, \qquad \text{or} \qquad \frac{a}{b}.$$

This last way of writing a ratio is most common in algebra. Note that the order must be $\frac{a}{b}$, not $\frac{b}{a}$.

Percents are ratios where the second number is always 100. For example, 50% represents the ratio of 50 to 100, 27% represents the ratio of 27 to 100, and so on.

When ratios are used in comparing units of measure, the units should be the same. This is shown in Example 1.

EXAMPLE 1 **Writing a Ratio**

Write a ratio for each word phrase.

(a) The ratio of 5 hours to 3 hours
This ratio can be written as $\frac{5}{3}$.

(b) The ratio of 5 hours to 3 days
First convert 3 days to hours: 3 days $= \mathbf{3 \cdot 24} = 72$ hours. The ratio of 5 hours to 3 days is thus

$$\frac{5 \text{ hours}}{72 \text{ hours}} \qquad \text{or} \qquad \frac{5}{72}.$$

NOTE Because the units of the two quantities in a ratio must be the same, ratios do not have units in their final forms.

OBJECTIVE 2 **Decide whether proportions are true.** A ratio is used to compare two numbers or amounts. A **proportion** is a statement that two ratios are equal. For example,

$$\frac{3}{4} = \frac{15}{20}$$

is a proportion that says that the ratios $\frac{3}{4}$ and $\frac{15}{20}$ are equal. In the proportion

$$\frac{a}{b} = \frac{c}{d},$$

$a, b, c,$ and d are the **terms** of the proportion. Beginning with the proportion

$$\frac{a}{b} = \frac{c}{d}$$

and multiplying both sides by the common denominator, bd, gives

$$\mathbf{bd} \cdot \frac{a}{b} = \mathbf{bd} \cdot \frac{c}{d}$$

$$ad = bc.$$

The same products *ad* and *bc* can be found by multiplying diagonally.

$$\frac{a}{b} = \frac{c}{d} \qquad \begin{matrix} bc \\ ad \end{matrix}$$

This is called **cross multiplication,** and *ad* and *bc* are called **cross products.**

Cross Products

If $\frac{a}{b} = \frac{c}{d}$, then the cross products *ad* and *bc* are equal.

Also, if $ad = bc$, then $\frac{a}{b} = \frac{c}{d}$ $\quad (b, d \neq 0)$.

From the rule given above,

$$\text{if} \quad \frac{a}{b} = \frac{c}{d} \quad \text{then} \quad ad = bc.$$

However, if $\frac{a}{c} = \frac{b}{d}$, then $ad = cb$, or $ad = bc$. This means that the two proportions are equivalent and

$$\text{the proportion } \frac{a}{b} = \frac{c}{d} \text{ can also be written as } \frac{a}{c} = \frac{b}{d}.$$

Sometimes one form is more convenient to work with than the other.

EXAMPLE 2 **Deciding Whether a Proportion Is True**

Decide whether the following proportions are true or false.

(a) $\frac{3}{4} = \frac{15}{20}$

Check to see whether the cross products are equal.

$$\frac{3}{4} = \frac{15}{20} \qquad \begin{matrix} 4 \cdot 15 = 60 \\ 3 \cdot 20 = 60 \end{matrix}$$

The cross products are equal, so the proportion is true.

(b) $\frac{6}{7} = \frac{30}{32}$

The cross products are $6 \cdot 32 = 192$ and $7 \cdot 30 = 210$. The cross products are different, so the proportion is false.

CAUTION

The cross product method cannot be used directly if there is more than one term on either side of the equals sign. For example, you cannot use the method directly to solve the equation

$$\frac{4}{x} + 3 = \frac{1}{9},$$

because there are two terms on the left side.

OBJECTIVE 3 **Solve proportions.** Four numbers are used in a proportion. If any three of these numbers are known, the fourth can be found.

EXAMPLE 3 Solving an Equation Using Cross Products

(a) Find x in the proportion

$$\frac{63}{x} = \frac{9}{5}.$$

The cross products must be equal, so

$$63 \cdot 5 = 9x$$
$$315 = 9x.$$

Divide both sides by 9 to get

$$35 = x.$$

The solution set is $\{35\}$.

(b) Solve the equation

$$\frac{m-2}{5} = \frac{m+1}{3}.$$

Find the cross products, and set them equal to each other.

$$3(m-2) = 5(m+1) \quad \text{Be sure to use parentheses.}$$
$$3m - 6 = 5m + 5 \quad \text{Distributive property}$$
$$3m = 5m + 11 \quad \text{Add 6.}$$
$$-2m = 11 \quad \text{Subtract } 5m.$$
$$m = -\frac{11}{2} \quad \text{Divide by } -2.$$

The solution set is $\{-\frac{11}{2}\}$.

OBJECTIVE 4 **Solve applied problems using proportions.** Proportions occur in many practical applications.

EXAMPLE 4 Applying Proportions

A local store is offering 3 packs of toothpicks for \$.87. How much would it charge for 10 packs?

Let x = the cost of 10 packs of toothpicks.

Set up a proportion. One ratio in the proportion can involve the number of packs, and the other can involve the costs. Make sure that the corresponding numbers appear in the numerator and the denominator.

$$\frac{\text{Cost of 3}}{\text{Cost of 10}} = \frac{3}{10}$$

$$\frac{.87}{x} = \frac{3}{10}$$

$$3x = .87(10) \qquad \text{Cross products}$$

$$3x = 8.7$$

$$x = 2.90 \qquad \text{Divide by 3.}$$

The 10 packs should cost $2.90. As shown earlier, the proportion could also be written as $\frac{3}{.87} = \frac{10}{x}$, which would give the same cross products.

NOTE Many people would solve the problem in Example 4 mentally as follows: Three packs cost $.87, so one pack costs $.87/3 = $.29. Then ten packs will cost 10($.29) = $2.90. If you do the problem this way, you are using proportions and probably not even realizing it!

An important application that uses proportions is *unit pricing*—deciding which size of an item offered in different sizes produces the best price per unit. For example, suppose you can buy 36 ounces of pancake syrup for $3.89. To find the price per unit, set up the proportion

$$\frac{\text{36 ounces}}{\text{1 ounce}} = \frac{3.89}{x}$$

and solve for x.

$$36x = 3.89 \qquad \text{Cross products}$$

$$x = \frac{3.89}{36} \qquad \text{Divide by 36.}$$

$$x \approx .108 \qquad \text{Use a calculator.}$$

Thus, the price for 1 ounce is $.108, or about 11 cents. Notice that the unit price is the ratio of the cost for 36 ounces, $3.89, to the number of ounces, 36, which means that the unit price for an item is found by dividing the cost by the number of units.

EXAMPLE 5 Determining Unit Price to Obtain the Best Buy

Besides the 36-ounce size discussed above, the local supermarket carries two other sizes of a popular brand of pancake syrup, priced as follows.

Size	*Price*
36-ounce	$3.89
24-ounce	$2.79
12-ounce	$1.89

Which size is the best buy? That is, which size has the lowest unit price?

To find the best buy, divide the price by the number of units to get the price per ounce. Each result in the following table was found by using a calculator and rounding the answer to three decimal places.

Size	Unit Cost (dollars per ounce)	
36-ounce	$\frac{\$3.89}{36} = \$.108$	← The best buy
24-ounce	$\frac{\$2.79}{24} = \$.116$	
12-ounce	$\frac{\$1.89}{12} = \$.158$	

Since the 36-ounce size produces the lowest price per unit, it would be the best buy. (*Be careful:* Sometimes the largest container *does not* produce the lowest price per unit.)

OBJECTIVE 5 Solve direct variation problems. Suppose that gasoline costs \$1.50 per gallon. Then 1 gallon costs \$1.50, 2 gallons cost 2(\$1.50) = \$3.00, 3 gallons cost 3(\$1.50) = \$4.50, and so on. Using a proportion,

$$\frac{\text{Cost of 1 gallon}}{\text{Total cost}} = \frac{1 \text{ gallon}}{x \text{ gallons}}$$

$$x \cdot \text{Cost of 1 gallon} = 1 \cdot \text{Total cost}.$$

Thus, the total cost is obtained by multiplying the number of gallons by the price per gallon. In general, if k equals the price per gallon and x equals the number of gallons, then the total cost y is equal to kx. Notice that as number of gallons increases, total cost increases.

The preceding discussion is an example of variation. As in the gasoline example, two variables *vary directly* if one is a multiple of the other.

Direct Variation

y **varies directly** as x if there exists a number k such that

$$y = kx.$$

Another way to say this is that y **is directly proportional to** x.

EXAMPLE 6 Using Direct Variation to Find the Cost of Gasoline

If 6 gallons of gasoline cost \$8.52, find the cost of 20 gallons.

Here, the total cost of the gasoline, y, varies directly as (or is directly proportional to) the number of gallons purchased, x. This means there is a number k such that $y = kx$. To find y when $x = 20$, use the fact that 6 gallons cost \$8.52.

$$y = kx$$

$$8.52 = k(6) \quad \text{Let } x = 6, y = 8.52.$$

$$1.42 = k \quad \text{Divide by 6.}$$

Since $k = 1.42$,

$$y = 1.42x.$$

When $x = 20$,

$$y = 1.42(20) = 28.4,$$

so the cost to purchase 20 gallons of gasoline is \$28.40.

2.5 EXERCISES

Determine the ratio and write it in lowest terms. See Example 1.

1. 25 feet to 40 feet

2. 16 miles to 48 miles

3. 18 dollars to 72 dollars

4. 300 people to 250 people

5. 144 inches to 6 feet

6. 60 inches to 2 yards

7. 5 days to 40 hours

8. 75 minutes to 2 hours

9. Which one of the following ratios is not the same as the ratio 2 to 5?
(a) .4 **(b)** 4 to 10 **(c)** 20 to 50 **(d)** 5 to 2

10. Give three ratios that are equivalent to the ratio 4 to 3.

11. Explain the distinction between *ratio* and *proportion.* Give examples.

12. Suppose that someone told you to use cross products in order to multiply fractions. How would you explain to the person what is wrong with his or her thinking?

Decide whether each proportion is true or false. See Example 2.

13. $\frac{5}{35} = \frac{8}{56}$

14. $\frac{4}{12} = \frac{7}{21}$

15. $\frac{120}{82} = \frac{7}{10}$

16. $\frac{27}{160} = \frac{18}{110}$

17. $\frac{\frac{1}{2}}{5} = \frac{1}{10}$

18. $\frac{\frac{1}{3}}{6} = \frac{1}{18}$

Solve each equation. See Example 3.

19. $\frac{k}{4} = \frac{175}{20}$

20. $\frac{49}{56} = \frac{z}{8}$

21. $\frac{x}{6} = \frac{18}{4}$

22. $\frac{z}{80} = \frac{20}{100}$

23. $\frac{3y - 2}{5} = \frac{6y - 5}{11}$

24. $\frac{2p + 7}{3} = \frac{p - 1}{4}$

Solve each problem by setting up and solving a proportion. See Example 4.

25. A chain saw requires a mixture of 2-cycle engine oil and gasoline. According to the directions on a bottle of Oregon 2-cycle Engine Oil, for a 50 to 1 ratio requirement, approximately 2.5 fluid ounces of oil are required for 1 gallon of gasoline. For 2.75 gallons, how many fluid ounces of oil are required?

26. The directions on the bottle mentioned in Exercise 25 indicate that if the ratio requirement is 24 to 1, approximately 5.5 ounces of oil are required for 1 gallon of gasoline. If gasoline is to be mixed with 22 ounces of oil, how much gasoline is to be used?

27. In 1998, the average exchange rate between U.S. dollars and United Kingdom pounds was 1 pound to \$1.6762. Margaret went to London and exchanged her U.S. currency for U.K. pounds, and received 400 pounds. How much in U.S. money did Margaret exchange?

28. If 3 U.S. dollars can be exchanged for 4.5204 Swiss francs, how many Swiss francs can be obtained for \$49.20? (Round to the nearest hundredth.)

29. If 6 gallons of premium unleaded gasoline cost \$3.72, how much would it cost to completely fill a 15-gallon tank?

30. If sales tax on a \$16.00 compact disc is \$1.32, how much would the sales tax be on a \$120.00 compact disc player?

31. The distance between Kansas City, Missouri, and Denver is 600 miles. On a certain wall map, this is represented by a length of 2.4 feet. On the map, how many feet would there be between Memphis and Philadelphia, two cities that are actually 1000 miles apart?

32. The distance between Singapore and Tokyo is 3300 miles. On a certain wall map, this distance is represented by 11 inches. The actual distance between Mexico City and Cairo is 7700 miles. How far apart are they on the same map?

33. Biologists tagged 250 fish in Willow Lake on October 5. On a later date they found 7 tagged fish in a sample of 350. Estimate the total number of fish in Willow Lake to the nearest hundred.

34. On May 13 researchers at Argyle Lake tagged 420 fish. When they returned a few weeks later, their sample of 500 fish contained 9 that were tagged. Give an approximation of the fish population in Argyle Lake to the nearest hundred.

The Olympic Committee has come to rely more and more on television rights and major corporate sponsors to finance the games. The pie charts show the funding plans for the first Olympics in Athens and the 1996 Olympics in Atlanta. Use proportions and the figures to answer the questions in Exercises 35 and 36.

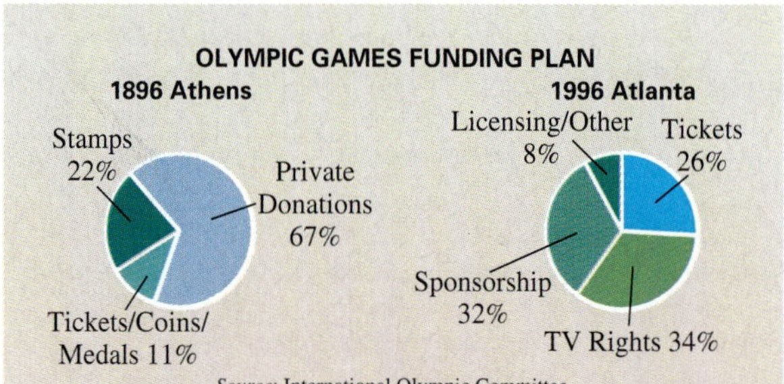

Source: International Olympic Committee.

35. In the 1996 Olympics, total revenue of \$350 million was raised. There were 10 major sponsors.
 (a) Write a proportion to find the amount of revenue provided by tickets. Solve it.
 (b) What amount was provided by sponsors? Assuming the sponsors contributed equally, how much was provided per sponsor?
 (c) What amount was raised by TV rights?

36. Suppose the amount of revenue raised in the 1896 Olympics was equivalent to the \$350 million in 1996.
 (a) Write a proportion for the amount of revenue provided by stamps and solve it.
 (b) What amount (in dollars) would have been provided by private donations?
 (c) In the 1988 Olympics, there were 9 major sponsors, and the total revenue was \$95 million. What is the ratio of major sponsors in 1988 to those in 1996? What is the ratio of revenue in 1988 to revenue in 1996?

A supermarket was surveyed to find the prices charged for items in various sizes. Find the best buy (based on price per unit) for each particular item. See Example 5.

37. Trash bags
20 count: \$3.09
30 count: \$4.59

38. Black pepper
1-ounce size: \$ .99
2-ounce size: \$1.65
4-ounce size: \$4.39

39. Breakfast cereal

15-ounce size:	\$2.99
25-ounce size:	\$4.49
31-ounce size:	\$5.49

40. Cocoa mix

8-ounce size:	\$1.39
16-ounce size:	\$2.19
32-ounce size:	\$2.99

41. Tomato ketchup

14-ounce size:	\$ .89
32-ounce size:	\$1.19
64-ounce size:	\$2.95

42. Cut green beans

8-ounce size:	\$ .45
16-ounce size:	\$ .49
50-ounce size:	\$1.59

Two triangles are **similar** if they have the same shape (but not necessarily the same size). Similar triangles have sides that are proportional. The figure shows two similar triangles. Notice that the ratios of the corresponding sides are all equal to $\frac{3}{2}$:

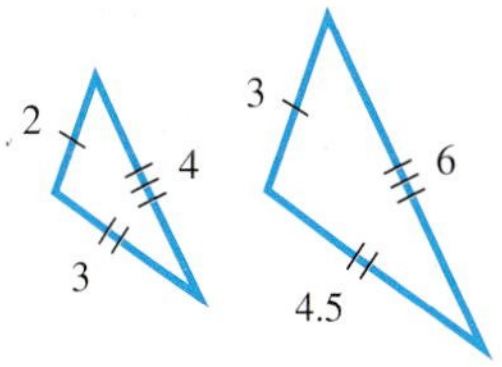

$$\frac{3}{2} = \frac{3}{2} \qquad \frac{4.5}{3} = \frac{3}{2} \qquad \frac{6}{4} = \frac{3}{2}.$$

If we know that two triangles are similar, we can set up a proportion to solve for the length of an unknown side.

Use a proportion to find the length x, given that the pair of triangles are similar.

43.

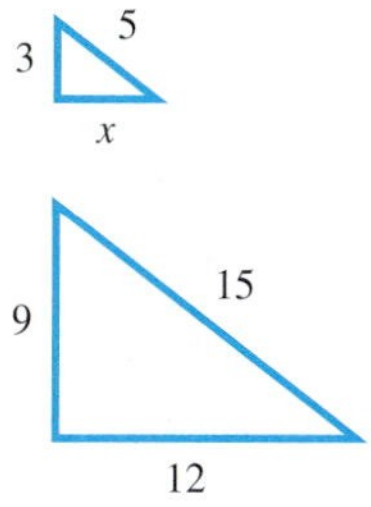

44.

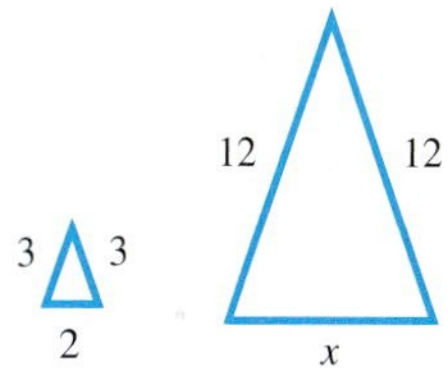

45.

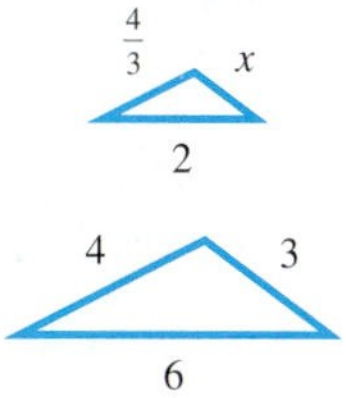

46.

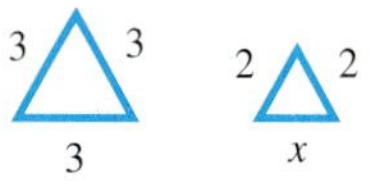

For the problems in Exercises 47 and 48, ***(a)*** *draw a sketch consisting of two right triangles, depicting the situation described, and* ***(b)*** *solve the problem. (*Source: The Guinness Book of World Records.*)*

47. An enlarged version of the chair used by George Washington at the Constitutional Convention casts a shadow 18 feet long at the same time a vertical pole 12 feet high casts a shadow 4 feet long. How tall is the chair?

48. One of the tallest candles ever constructed was exhibited at the 1897 Stockholm Exhibition. If it cast a shadow 5 feet long at the same time a vertical pole 32 feet high cast a shadow 2 feet long, how tall was the candle?

The Consumer Price Index, issued by the U.S. Bureau of Labor Statistics, provides a means of determining the purchasing power of the U.S. dollar from one year to the next. Using the period from 1982 to 1984 as a measure of 100.0, the Consumer Price Index from 1990 to 1995 is shown here.

Year	*Consumer Price Index*
1990	130.7
1991	136.2
1992	140.3
1993	144.5
1994	148.2
1995	152.4

Source: Bureau of Labor Statistics.

To use the Consumer Price Index to predict a price in a particular year, we can set up a proportion and compare it with a known price in another year, as follows:

$$\frac{\text{Price in year } A}{\text{Index in year } A} = \frac{\text{Price in year } B}{\text{Index in year } B}.$$

Use the Consumer Price Index figures above to find the amount that would be charged for the use of the same amount of electricity that cost \$225 in 1990. Give your answer to the nearest dollar.

49. in 1992 **50.** in 1993 **51.** in 1994 **52.** in 1995

53. The Consumer Price Index figures for shelter for the years 1981 and 1991 are 90.5 and 146.3. If shelter for a particular family cost \$3000 in 1981, what would be the comparable cost in 1991? Give your answer to the nearest dollar.

54. Due to a volatile fuel oil market in the early 1980s, the price of fuel decreased during the first three quarters of the decade. The Consumer Price Index figures for 1982 and 1986 were 105.0 and 74.1. If it cost you \$21.50 to fill your tank with fuel oil in 1982, how much would it have cost to fill the same tank in 1986? Give your answer to the nearest cent.

RELATING CONCEPTS (EXERCISES 55–58)

In Section 2.2 we learned that to make the solution process easier, if an equation involves fractions, we can multiply both sides of the equation by the least common denominator of all the fractions in the equation. A proportion consists of two fractions equal to each other, so a proportion is a special case of this kind of equation.

Work Exercises 55–58 in order *to see how the process of solving by cross products is justified.*

55. In the equation $\frac{x}{6} = \frac{2}{5}$, what is the least common denominator of the two fractions?

56. Solve the equation in Exercise 55 as follows:

(a) Multiply both sides by the LCD. What is the equation that you obtain?

(b) Solve for x by dividing both sides by the coefficient of x. What is the solution?

57. Solve the equation in Exercise 55 as follows:

(a) Set the cross products equal. What is the equation you obtain?

(b) Repeat part (b) of Exercise 56.

58. Compare your results from Exercises 56(b) and 57(b). What do you notice?

Did you make the connection that solving a proportion by cross products is justified by the general method of solving an equation with fractions?

Solve each variation problem. See Example 6.

59. The interest on an investment varies directly as the rate of simple interest. If the interest is \$48 when the interest rate is 5%, find the interest when the rate is 4.2%.

60. For a given base, the area of a triangle varies directly as its height. Find the area of a triangle with a height of 6 inches, if the area is 10 square inches when the height is 4 inches.

61. The distance a spring stretches is directly proportional to the force applied. If a force of 30 pounds stretches a spring 16 inches, how far will a force of 50 pounds stretch the spring?

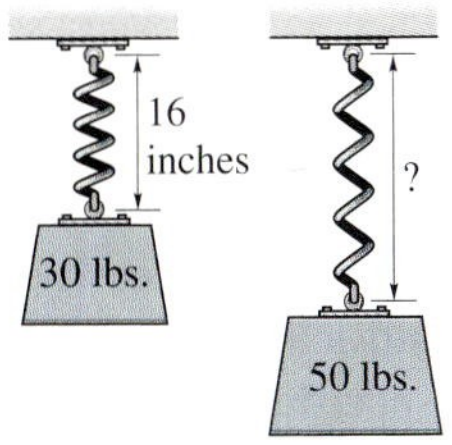

62. The perimeter of a square is directly proportional to the length of its side. What is the perimeter of a square with a 5-centimeter side, if a square with a 12-centimeter side has a perimeter of 48 centimeters?

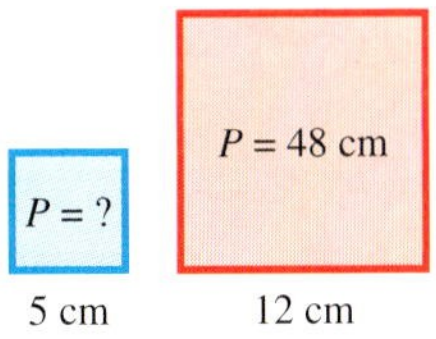

63. According to *The Guinness Book of World Records,* the longest recorded voyage in a paddleboat is 2226 miles in 103 days; the boat was propelled down the Mississippi River by the foot power of two boaters. Assuming a constant rate, how far would they have gone in 120 days? (*Hint:* Distance varies directly as time.)

2.6 More about Problem Solving

OBJECTIVES

1. Use percent in problems involving rates.
2. Solve problems involving mixtures.
3. Solve problems involving simple interest.
4. Solve problems involving denominations of money.
5. Solve problems involving distance, rate, and time.

FOR EXTRA HELP

SSG Sec. 2.6
SSM Sec. 2.6
Pass the Test Software
InterAct Math Tutorial Software
Video 4

OBJECTIVE 1 Use percent in problems involving rates. Recall that percent means "per hundred." Thus, percents are ratios where the second number is always 100. For example, 50% represents the ratio of 50 to 100 and 27% represents the ratio of 27 to 100.

PROBLEM SOLVING

Percents are often used in problems involving mixing different concentrations of a substance or different interest rates. In each case, to get the amount of pure substance or the interest, we multiply.

Mixture Problems	**Interest Problems (annual)**
base × rate (%) = percentage	principal × rate (%) = interest
$b \times r = p$	$p \times r = I$

In an equation, the percent always appears as a decimal. For example, 35% is written as .35, not 35.

30. LaShondra Williams inherited some money from her uncle. She deposited part of the money in a savings account paying 2%, and $3000 more than that amount in a different account paying 3%. Her annual interest income was $690. How much did she deposit at each rate?

31. With income earned by selling the rights to his life story, an actor invests some of the money at 3% and $30,000 more than twice as much at 4%. The total annual interest earned from the investments is $5600. How much is invested at each rate?

32. An artist invests her earnings in two ways. Some goes into a tax-free bond paying 6%, and $6000 more than three times as much goes into mutual funds paying 5%. Her total annual interest income from the investments is $825. How much does she invest at each rate?

Work Exercises 33–38 involving different monetary rates. See Example 4.

33. A bank teller has some five-dollar bills and some twenty-dollar bills. The teller has 5 more twenties than fives. The total value of the money is $725. Find the number of five-dollar bills that the teller has.

Number of Bills	Denomination	Value
x	5	
$x + 5$	20	

34. A coin collector has $1.70 in dimes and nickels. She has 2 more dimes than nickels. How many nickels does she have?

Number of Coins	Denomination	Value
x	.05	$.05x$
	.10	

35. A cashier has a total of 126 bills, made up of fives and tens. The total value of the money is $840. How many of each kind does he have?

36. A convention manager finds that she has $1290, made up of twenties and fifties. She has a total of 42 bills. How many of each kind does she have?

37. A merchant wishes to mix candy worth $5 per pound with 40 pounds of candy worth $2 per pound to get a mixture that can be sold for $3 per pound. How many pounds of $5 candy should be used?

38. At Vern's Grill, hamburgers cost 90 cents each, and a bag of french fries costs 40 cents. How many hamburgers and how many bags of french fries can a customer buy with $8.80 if he wants twice as many hamburgers as bags of french fries?

Vern's Grill
Hamburgers........90¢ each
French fries........40¢ a bag
Soda pop........25¢ each
Shakes........90¢ each
Cherry pie........90¢ a slice
Ice cream........40¢ a scoop

39. Read Example 2. Can a problem of this type have a fraction as an answer? Now read Example 4. Can a problem of this type have a fraction as an answer? Explain.

RELATING CONCEPTS (EXERCISES 40–43)

Sometimes applied problems that may seem different actually apply the same concepts.

Work Exercises 40–43 in order, to see how this happens.

40. Consider the following problem: A hoard of coins consists of only nickels and dimes. There are 3400 coins, and the value of the money is \$290. How many of each denomination are there?

(a) Write an equation you would use to solve the problem. Let x represent the number of nickels in the hoard.

(b) Solve the problem.

41. Consider the following problem: An investor deposits \$3400 in two accounts. One account is a passbook savings account that pays 5% interest, and the other is a money market account that pays 10% interest. After one year, the total interest earned is \$290. How much did she invest at each rate?

(a) Write an equation you would use to solve the problem. Let x represent the amount invested in the passbook savings account.

(b) Solve the problem.

42. Compare the equations you wrote in Exercises 40(a) and 41(a). What do you notice about them?

43. If, in either of the problems in Exercises 40 and 41, you let x represent the other unknown quantity, will you get the same *solution* to the equation? Will you get the same *answers* to the problem?

Did you make the connection that applications that appear different may be solved in the same way?

44. How do you think the pie charts in Exercises 15 and 16 were constructed so that the sections represented the appropriate percentages accurately?

Solve each problem. See Example 6.

45. A driver averaged 53 miles per hour and took 10 hours to travel from Memphis to Chicago. What is the distance between Memphis and Chicago?

46. A small plane traveled from Warsaw to Rome, averaging 164 miles per hour. The trip took 2 hours. What is the distance from Warsaw to Rome?

47. Suppose that an automobile averages 45 miles per hour, and travels for 30 minutes. Is the distance traveled $45 \times 30 = 1350$ miles? If not, explain why not, and give the correct distance.

48. Which of the following choices is the best *estimate* for the average speed of a trip of 405 miles that lasted 8.2 hours?

(a) 50 miles per hour **(b)** 30 miles per hour
(c) 60 miles per hour **(d)** 40 miles per hour

Solve each problem. See Examples 7 and 8.

49. St. Louis and Portland are 2060 miles apart. A small plane leaves Portland, traveling toward St. Louis at an average speed of 90 miles per hour. Another plane leaves St. Louis at the same time, traveling toward Portland, averaging 116 miles per hour. How long will it take them to meet?

	r	t	d
Plane leaving Portland	90	t	$90t$
Plane leaving St. Louis	116	t	$116t$

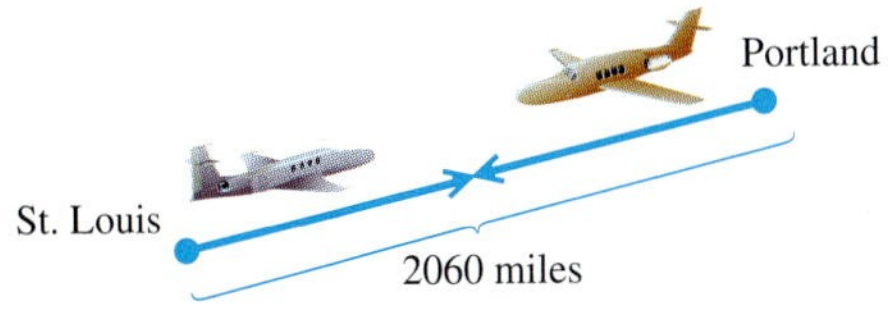

50. Atlanta and Cincinnati are 440 miles apart. John leaves Cincinnati, driving toward Atlanta at an average speed of 60 miles per hour. Pat leaves Atlanta at the same time, driving toward Cincinnati in her antique auto, averaging 28 miles per hour. How long will it take them to meet?

	r	t	d
John	60	t	$60t$
Pat	28	t	$28t$

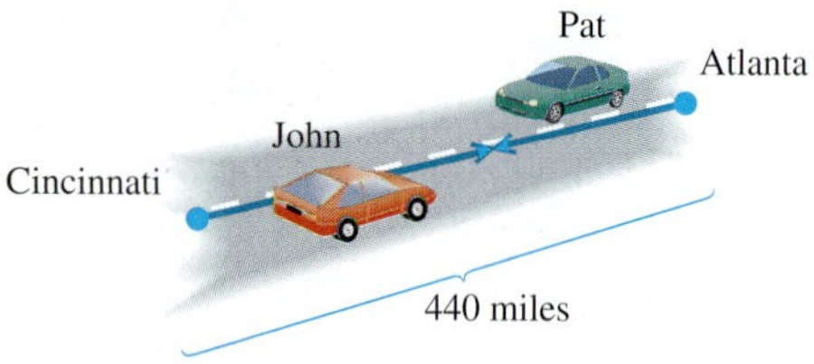

51. Two trains leave a city at the same time. One travels north and the other travels south at 20 miles per hour faster. In 2 hours the trains are 280 miles apart. Find their speeds.

	r	t	d
Northbound	x	2	
Southbound	$x + 20$	2	

52. Two planes leave an airport at the same time, one flying east, the other flying west. The eastbound plane travels 150 miles per hour slower. They are 2250 miles apart after 3 hours. Find the speed of each plane.

	r	t	d
Eastbound	$x - 150$	3	
Westbound	x	3	

53. At a given hour two steamboats leave a city in the same direction on a straight canal. One travels at 18 miles per hour and the other travels at 25 miles per hour. In how many hours will the boats be 35 miles apart?

54. From a point on a straight road, Lupe and Maria ride bicycles in the same direction. Lupe rides 10 miles per hour and Maria rides 12 miles per hour. In how many hours will they be 5 miles apart?

The remaining applications in this exercise set are not organized by type of problem. Use the problem-solving techniques described in this chapter to solve them.

55. According to a survey by Information Resources, Inc., single men spent an average of \$10.87 more than single women when buying frozen dinners during the 52 weeks that ended August 22, 1993. Together, a single man and a single woman would spend a total of \$92.29, according to the survey. What were the average expenditures for a single man and for a single woman?

56. Team Marketing Report, a sports-business newsletter, computes a Fan Cost Index for a trip to a major league baseball park. The index consists of the cost of four average-priced tickets, two small beers, four small sodas, four hot dogs, parking for one car, two game programs, and two twill baseball caps. For the 1994 season, the New York Yankees had the highest Fan Cost Index, while the Cincinnati Reds had the lowest. The Yankees' index was \$43.37 less than twice that of the Reds. What was the Fan Cost Index for each of these teams if the *total* of the two was \$194.56?

57. The pie chart shows the percents for the different categories of Macintosh computer units in use at the end of 1997. At that time there were approximately 21,782,000 units in use. Find the approximate number of units used in each category. (*Source:* Apple Computer, Inc., Cupertino, California.)

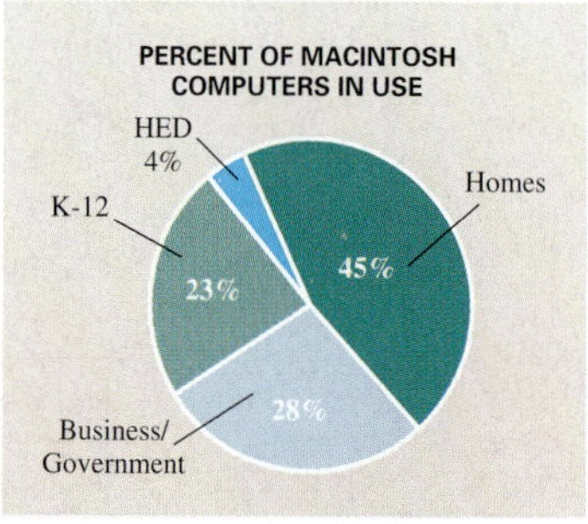

58. At the start of play on September 22, 1997, the standings of the Central Division of the American League were as shown. "Winning percentage" is commonly expressed as a decimal rounded to the nearest thousandth. To find the winning percentage of a team, divide the number of wins by the total number of games played. Find the winning percentage of each of the following teams.

(a) Cleveland **(b)** Chicago

(c) Milwaukee

	Won	*Lost*
Cleveland	83	71
Chicago	77	78
Milwaukee	76	78
Kansas City	64	90
Minnesota	63	91

59. In an automobile race, a driver was 240 miles from the finish line after 3 hours. Another driver, who was in a later race, traveled at the same speed as the first driver. After 2.5 hours the second driver was 300 miles from the finish. Find the speed of each driver.

60. Fran Liberto sold a painting that she found at a garage sale to an art dealer. With the \$12,000 she got, she invested part at 4% in a certificate of deposit and the rest at 5% in a municipal bond. Her total annual income from the two investments was \$515. How much did she invest at each rate? (*Hint:* Let x represent the amount invested at 4%. Then $12{,}000 - x$ represents the amount invested at 5%.)

2.7 The Addition and Multiplication Properties of Inequality

OBJECTIVES

1. Graph intervals on a number line.
2. Use the addition property of inequality.
3. Use the multiplication property of inequality.
4. Solve linear inequalities.
5. Solve applied problems by using inequalities.
6. Solve three-part inequalities.

FOR EXTRA HELP

SSG Sec. 2.7
SSM Sec. 2.7
Pass the Test Software
InterAct Math Tutorial Software
Video 4

Solving inequalities is closely related to the methods of solving equations. In this section we introduce properties that are essential for solving inequalities.

Inequalities are algebraic expressions related by

$<$ "is less than,"

$\leq$ "is less than or equal to,"

$>$ "is greater than,"

$\geq$ "is greater than or equal to."

We solve an inequality by finding all real number solutions for it. For example, the solution set of $x \leq 2$ includes all real numbers that are less than or equal to 2, and not just the integers less than or equal to 2. For example, -2.5, -1.7, -1, $\frac{7}{4}$, $\frac{1}{2}$, $\sqrt{2}$, and 2 are all real numbers less than or equal to 2, and are therefore solutions of $x \leq 2$.

OBJECTIVE 1 Graph intervals on a number line. A good way to show the solution set of an inequality is by graphing. We graph all the real numbers satisfying $x \leq 2$ by placing a square bracket at 2 on a number line and drawing an arrow extending from the bracket to the left (to represent the fact that all numbers less than 2 are also part of the graph). The graph is shown in Figure 12.

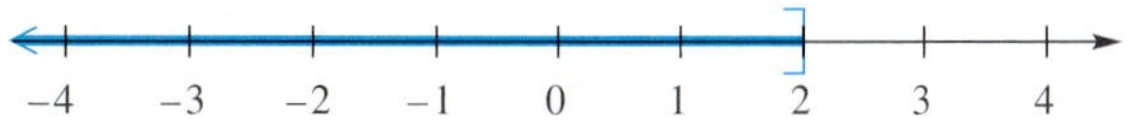

Figure 12

The set of numbers less than or equal to 2 is an example of an **interval** on the number line. To write intervals, we use **interval notation.** For example, using this notation, the interval of all numbers less than or equal to 2 is written as $(-\infty, 2]$. The **negative infinity** symbol $-\infty$ does not indicate a number. It is used to show that the interval includes all real numbers less than 2. As on the number line, the square bracket indicates that 2 is part of the solution. A parenthesis is always used next to the infinity symbol. The set of real numbers is written in interval notation as $(-\infty, \infty)$.

EXAMPLE 1 Graphing an Interval Written in Interval Notation on a Number Line

Write each inequality in interval notation and graph it.

(a) $x > -5$

The statement $x > -5$ says that x can represent any number greater than -5, but x cannot equal -5. The interval is written as $(-5, \infty)$. We show this on a graph by placing a parenthesis at -5 and drawing an arrow to the right, as in Figure 13. The parenthesis at -5 shows that -5 is not part of the graph.

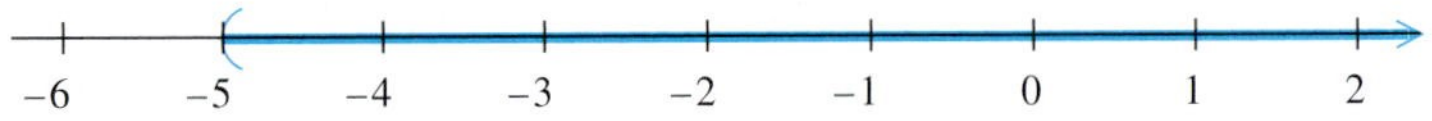

Figure 13

(b) $-1 \le x < 3$

This statement is read "-1 is less than or equal to x *and* x is less than 3." Thus, we want the set of numbers that are *between* -1 and 3, with -1 included and 3 excluded. In interval notation, we write this as $[-1, 3)$, using a square bracket at -1 because it is part of the graph, and a parenthesis at 3, because 3 is not part of the graph. The graph is shown in Figure 14.

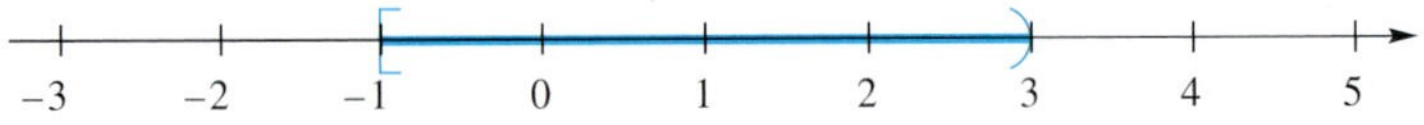

Figure 14

OBJECTIVE 2 **Use the addition property of inequality.** Inequalities such as $x + 4 \le 9$ can be solved in much the same way as equations. Consider the inequality $2 < 5$. If 4 is added to both sides of this inequality, the result is

$$2 + 4 < 5 + 4$$
$$6 < 9,$$

a true sentence. Now subtract 8 from both sides:

$$2 - 8 < 5 - 8$$
$$-6 < -3.$$

The result is again a true sentence. These examples suggest the **addition property of inequality,** which states that the same real number can be added to both sides of an inequality without changing the solutions.

Addition Property of Inequality

For any algebraic expressions A, B, and C that represent real numbers, the inequalities

$$A < B \quad \text{and} \quad A + C < B + C$$

have exactly the same solutions. In words, the same expression may be added to both sides of an inequality without changing the solutions.

The addition property of inequality also works with $>$, $\leq$, or $\geq$. Just as with the addition property of equality, the same expression may also be *subtracted* from both sides of an inequality.

The following examples show how the addition property is used to solve inequalities. We will write solution sets in interval notation.

EXAMPLE 2 Using the Addition Property of Inequality

Solve the inequality $7 + 3k > 2k - 5$.

Use the addition property of inequality twice, once to get the terms containing k alone on one side of the inequality and a second time to get the integers together on the other side. (These steps can be done in either order.)

$$7 + 3k > 2k - 5$$

$$7 + 3k - 2k > 2k - 5 - 2k \qquad \text{Subtract } 2k.$$

$$7 + k > -5 \qquad \text{Combine terms.}$$

$$7 + k - 7 > -5 - 7 \qquad \text{Subtract 7.}$$

$$k > -12$$

The solution set is $(-12, \infty)$. Its graph is shown in Figure 15.

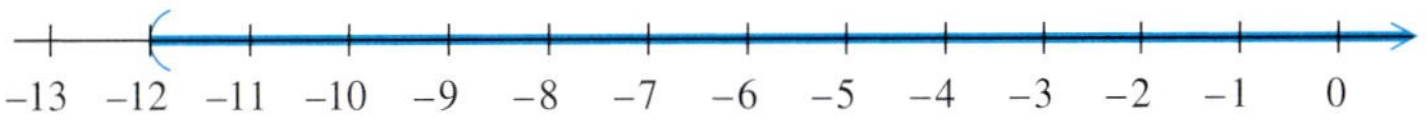

Figure 15

OBJECTIVE 3 **Use the multiplication property of inequality.** The addition property of inequality cannot be used to solve inequalities such as $4y \geq 28$. These inequalities require the multiplication property of inequality. To see how this property works, we look at some examples.

First, start with the inequality $3 < 7$ and multiply both sides by the positive number 2.

$$3 < 7$$

$$2(3) < 2(7) \qquad \text{Multiply both sides by 2.}$$

$$6 < 14 \qquad \text{True}$$

Now multiply both sides of $3 < 7$ by the negative number -5.

$$3 < 7$$

$$-5(3) < -5(7) \qquad \text{Multiply both sides by } -5.$$

$$-15 < -35 \qquad \text{False}$$

To get a true statement when multiplying both sides by -5 requires reversing the direction of the inequality symbol.

$$3 < 7$$
$$-5(3) > -5(7) \qquad \text{Multiply by } -5\text{; reverse the symbol.}$$
$$-15 > -35 \qquad \text{True}$$

Take the inequality $-6 < 2$ as another example. Multiply both sides by the positive number 4.

$$-6 < 2$$
$$4(-6) < 4(2) \qquad \text{Multiply by 4.}$$
$$-24 < 8 \qquad \text{True}$$

Multiplying both sides of $-6 < 2$ by -5 and at the same time reversing the direction of the inequality symbol gives

$$-6 < 2$$
$$(-5)(-6) > (-5)(2) \qquad \text{Multiply by } -5\text{; change } < \text{ to } >.$$
$$30 > -10. \qquad \text{True}$$

The two parts of the **multiplication property of inequality** are stated below.

Multiplication Property of Inequality

For any algebraic expressions A, B, and C that represent real numbers, with $C \neq 0$,

1. if C is *positive,* then the inequalities

$$A < B \quad \text{and} \quad AC < BC$$

have exactly the same solutions;

2. if C is *negative,* then the inequalities

$$A < B \quad \text{and} \quad AC > BC$$

have exactly the same solutions.

In other words, both sides of an inequality may be multiplied by the same expression representing a positive number without changing the solutions. If the expression represents a negative number, we must reverse the direction of the inequality symbol.

The multiplication property of inequality works with $>$, $\leq$, or $\geq$, as well. The multiplication property of inequality also permits *division* of both sides of an inequality by the same nonzero expression.

It is important to remember the differences in the multiplication property for positive and negative numbers.

1. When both sides of an inequality are multiplied or divided by a positive number, the direction of the inequality symbol *does not change.* Adding or subtracting terms on both sides also does not change the symbol.
2. When both sides of an inequality are multiplied or divided by a negative number, the direction of the symbol *does change. Reverse the symbol of inequality only when you multiply or divide both sides by a negative number.*

EXAMPLE 3 Using the Multiplication Property of Inequality

(a) Solve the inequality $3r < -18$.

Simplify this inequality by using the multiplication property of inequality and dividing both sides by 3. Since 3 is a positive number, the direction of the inequality symbol does not change.

$$3r < -18$$

$$\frac{3r}{3} < \frac{-18}{3} \qquad \text{Divide by 3.}$$

$$r < -6$$

The solution set is $(-\infty, -6)$. The graph is shown in Figure 16.

Figure 16

(b) Solve the inequality $-4t \geq 8$.

Here both sides of the inequality must be divided by -4, a negative number, which *does* change the direction of the inequality symbol.

$$-4t \geq 8$$

$$\frac{-4t}{-4} \leq \frac{8}{-4} \qquad \text{Divide by } -4\text{; symbol is reversed.}$$

$$t \leq -2$$

The solution set is $(-\infty, -2]$. The solutions are graphed in Figure 17.

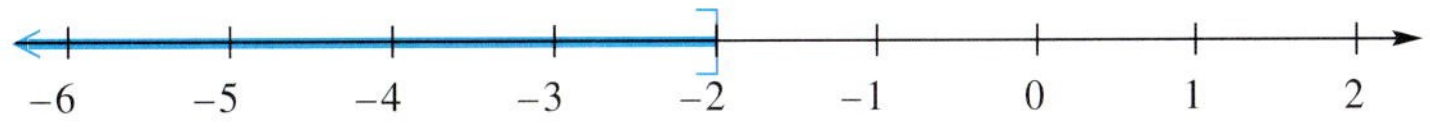

Figure 17

CAUTION Even though the number on the right side of the inequality in Example 3(a) is negative (-18), *do not reverse the direction of the inequality symbol.* Reverse the symbol only when multiplying or dividing by a negative number, as shown in Example 3(b).

OBJECTIVE 4 Solve linear inequalities. A **linear inequality** is an inequality that can be written in the form $ax + b < 0$, for real numbers a and b, with $a \neq 0$. ($<$ may be replaced with $>$, $\leq$, or $\geq$ in this definition.) To solve a linear inequality, follow these steps.

Solving a Linear Inequality

Step 1 **Simplify each side separately.** Use the properties to clear parentheses and combine terms.

Step 2 **Isolate the variable terms on one side.** Use the addition property of inequality to simplify the inequality to the form $ax < b$ or $ax > b$, where a and b are real numbers.

Step 3 **Isolate the variable.** Use the multiplication property of inequality to simplify the inequality to the form $x < d$ or $x > d$, where d is a real number.

Notice how these steps are used in the next example.

EXAMPLE 4 Solving a Linear Inequality

Solve the inequality $3z + 2 - 5 > -z + 7 + 2z$.

Step 1 Combine like terms and simplify.

$$3z + 2 - 5 > -z + 7 + 2z$$
$$3z - 3 > z + 7$$

Step 2 Use the addition property of inequality.

$$3z - 3 + 3 > z + 7 + 3 \qquad \text{Add 3.}$$
$$3z > z + 10$$
$$3z - z > z + 10 - z \qquad \text{Subtract } z.$$
$$2z > 10$$

Step 3 Use the multiplication property of inequality.

$$\frac{2z}{2} > \frac{10}{2} \qquad \text{Divide by 2.}$$
$$z > 5$$

Since 2 is positive, the direction of the inequality symbol was not changed in the third step. The solution set is $(5, \infty)$. Its graph is shown in Figure 18.

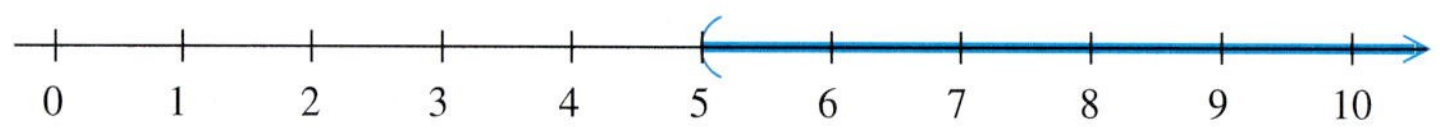

Figure 18

EXAMPLE 5 Solving a Linear Inequality

Solve $5(k - 3) - 7k \geq 4(k - 3) + 9$.

Step 1 Simplify and combine like terms.

$$5(k - 3) - 7k \geq 4(k - 3) + 9$$
$$5k - 15 - 7k \geq 4k - 12 + 9 \qquad \text{Distributive property}$$
$$-2k - 15 \geq 4k - 3 \qquad \text{Combine like terms.}$$

Step 2 Use the addition property.

$$-2k - 15 - 4k \geq 4k - 3 - 4k \quad \text{Subtract } 4k.$$
$$-6k - 15 \geq -3$$
$$-6k - 15 + 15 \geq -3 + 15 \quad \text{Add 15.}$$
$$-6k \geq 12$$

Step 3 Divide both sides by -6, a negative number. Change the direction of the inequality symbol.

$$\frac{-6k}{-6} \leq \frac{12}{-6} \quad \text{Divide by } -6\text{; symbol is reversed.}$$
$$k \leq -2$$

The solution set is $(-\infty, -2]$. Its graph is shown in Figure 19.

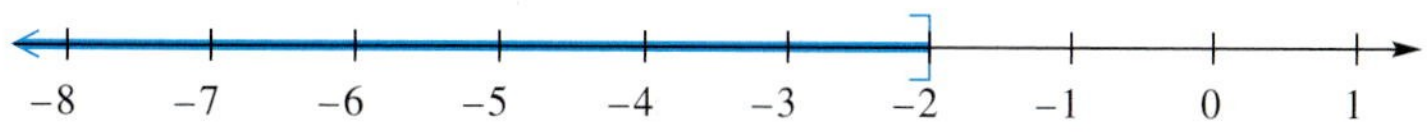

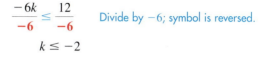

Figure 19

OBJECTIVE 5 **Solve applied problems by using inequalities.** Until now, the applied problems that we have studied have all led to equations.

PROBLEM SOLVING

Inequalities can be used to solve applied problems involving phrases that suggest inequality. The following chart gives some of the more common such phrases along with examples and translations.

Phrase	*Example*	*Inequality*
Is more than	A number ***is more than*** 4	$x > 4$
Is less than	A number ***is less than*** −12	$x < -12$
Is at least	A number ***is at least*** 6	$x \geq 6$
Is at most	A number ***is at most*** 8	$x \leq 8$

CAUTION Do not confuse statements like "5 is more than a number" with the phrase "5 more than a number." The first of these is expressed as "$5 > x$" while the second is expressed with addition, as "$x + 5$."

The next example shows an application of algebra that is important to anyone who has ever asked himself or herself "What score can I make on my next test and have a (particular grade) in this course?" It uses the idea of finding the average of a number of grades. In general, to find the average of n numbers, add the numbers, and divide by n.

EXAMPLE 6 Finding an Average Test Score

Brent has test grades of 86, 88, and 78 on his first three tests in geometry. If he wants an average of at least 80 after his fourth test, what are the possible scores he can make on his fourth test?

Let $x =$ Brent's score on his fourth test. To find his average after 4 tests, add the test scores and divide by 4.

Average ↓ is at least ↓ 80. ↓

$$\frac{86 + 88 + 78 + x}{4} \geq 80$$

$$\frac{252 + x}{4} \geq 80 \qquad \text{Add the known scores.}$$

$$4\left(\frac{252 + x}{4}\right) \geq 4(80) \qquad \text{Multiply by 4.}$$

$$252 + x \geq 320$$

$$252 - 252 + x \geq 320 - 252 \qquad \text{Subtract 252.}$$

$$x \geq 68 \qquad \text{Combine terms.}$$

He must score 68 or more on the fourth test to have an average of *at least* 80.

CAUTION

Errors often occur when the phrases "at least" and "at most" appear in applied problems. Remember that

at least translates as **greater than or equal to**

and

at most translates as **less than or equal to.**

OBJECTIVE 6 Solve three-part inequalities. Inequalities that say that one number is *between* two other numbers are *three-part inequalities.* For example,

$$-3 < 5 < 7$$

says that 5 is between -3 and 7. Three-part inequalities can also be solved by using the addition and multiplication properties of inequality. The idea is to get the inequality in the form

$$\textbf{a number} < x < \textbf{another number,}$$

using "is less than." The solution set can then easily be graphed.

EXAMPLE 7 Solving Three-Part Inequalities

(a) Solve $4 \leq 3x - 5 < 6$ and graph the solutions.

This inequality is equivalent to the statement $4 \leq 3x - 5$ and $3x - 5 < 6$. Using the inequality properties, we could solve each of these simple inequalities separately, and then combine the solutions with "and." We get the same result if we simply work with all three parts at once. Working separately, the first step would be to add 5 on each side. Do this to all three parts.

$$4 \le 3x - 5 < 6$$

$$4 + 5 \le 3x - 5 + 5 < 6 + 5 \qquad \text{Add 5.}$$

$$9 \le 3 < 11$$

Now divide each part by the positive number 3.

$$\frac{9}{3} \le \frac{3x}{3} < \frac{11}{3} \qquad \text{Divide by 3.}$$

$$3 \le x < \frac{11}{3}$$

The solution set is $[3, \frac{11}{3})$. Its graph is shown in Figure 20.

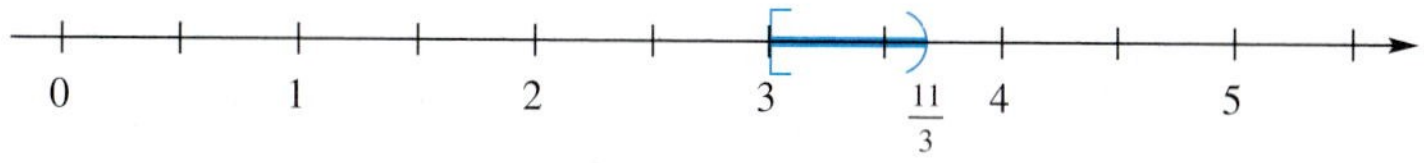

Figure 20

(b) Solve $-4 \le \frac{2}{3}m - 1 < 8$ and graph the solutions.

Recall from Section 2.2 that fractions as coefficients in equations can be eliminated by multiplying both sides by the least common denominator of the fractions. The same is true for inequalities. One way to begin is to multiply all three parts by 3.

$$-4 \le \frac{2}{3}m - 1 < 8$$

$$3(-4) \le 3\left(\frac{2}{3}m - 1\right) < 3(8) \qquad \text{Multiply by 3.}$$

$$-12 \le 2m - 3 < 24 \qquad \text{Distributive property}$$

Now add 3 to each part.

$$-12 + 3 \le 2m - 3 + 3 < 24 + 3 \qquad \text{Add 3.}$$

$$-9 \le 2m < 27$$

Finally, divide by 2 to get

$$-\frac{9}{2} \le m < \frac{27}{2}.$$

The solution set is $\left[-\frac{9}{2}, \frac{27}{2}\right)$. Its graph is shown in Figure 21.

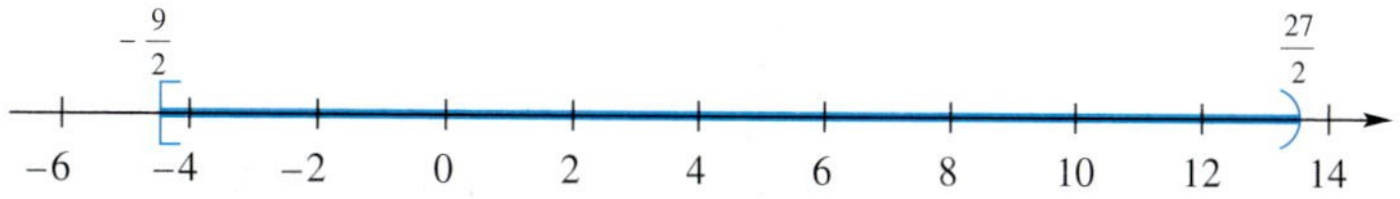

Figure 21

This inequality could also have been solved by first adding 1 to each part, and then multiplying each part by $\frac{3}{2}$.

CONNECTIONS

Many mathematical models involve inequalities rather than equations. This is often the case in economics. For example, a company that produces videocassettes has found that revenue from the sales of the cassettes is \$5 per cassette less sales costs of \$100. Production costs are \$125 plus \$4 per cassette. Profit (P) is given by revenue (R) less cost (C), so the company must find the production level x that makes

$$P = R - C > 0.$$

FOR DISCUSSION OR WRITING

Write an expression for revenue letting x represent the production level (number of cassettes to be produced). Write an expression for production costs using x. Write an expression for profit and solve the inequality shown above. Describe the solution in terms of the problem.

2.7 EXERCISES

1. Explain how to determine whether to use a parenthesis or a square bracket at the endpoint when graphing an inequality on a number line.
2. How does the graph of $t \geq -7$ differ from the graph of $t > -7$?

Write an inequality involving the variable x that describes each set of numbers graphed. See Example 1.

3. 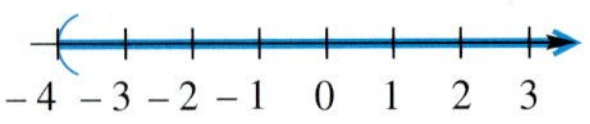

4. 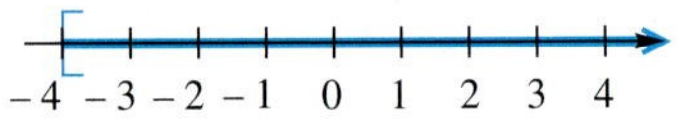

5. 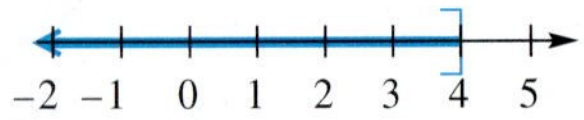

6. 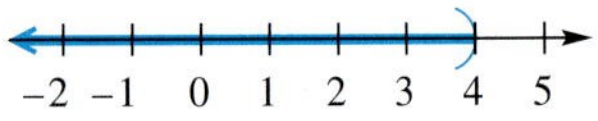

Write each inequality in interval notation and graph the interval on a number line. See Example 1.

7. $k \leq 4$
8. $r \leq -11$
9. $x < -3$
10. $y < 3$
11. $t > 4$
12. $m > 5$
13. $8 \leq x \leq 10$
14. $3 \leq x \leq 5$
15. $0 < y \leq 10$
16. $-3 \leq x < 5$
17. Why is it *wrong* to write $3 < x < -2$ to indicate that x is between -2 and 3?
18. If $p < q$ and $r < 0$, which one of the following statements is *false*?
    (a) $pr < qr$ **(b)** $pr > qr$ **(c)** $p + r < q + r$ **(d)** $p - r < q - r$

Solve each inequality and graph the solution set. See Example 2.

19. $z - 8 \geq -7$
20. $p - 3 \geq -11$
21. $2k + 3 \geq k + 8$
22. $3x + 7 \geq 2x + 11$
23. $3n + 5 < 2n - 6$
24. $5x - 2 < 4x - 5$
25. Under what conditions must the inequality symbol be reversed when solving an inequality?
26. Explain the steps you would use to solve the inequality $-5x > 20$.
27. Your friend tells you that when solving the inequality $6x < -42$ he reversed the direction of the inequality because of the presence of -42. How would you respond?
28. By what number must you *multiply* both sides of $.2x > 6$ to get just x on the left side?

Solve each inequality. Write the solution set in interval notation and graph it. See Example 3.

29. $3x < 18$ **30.** $5x < 35$ **31.** $2y \geq -20$ **32.** $6m \geq -24$

33. $-8t > 24$ **34.** $-7x > 49$ **35.** $-x \geq 0$ **36.** $-k < 0$

37. $-\frac{3}{4}r < -15$ **38.** $-\frac{7}{8}t < -14$

39. $-.02x \leq .06$ **40.** $-.03v \geq -.12$

Solve each inequality. Write the solution set in interval notation and graph it. See Examples 4 and 5.

41. $5r + 1 \geq 3r - 9$ **42.** $6t + 3 < 3t + 12$

43. $6x + 3 + x < 2 + 4x + 4$ **44.** $-4w + 12 + 9w \geq w + 9 + w$

45. $-x + 4 + 7x \leq -2 + 3x + 6$ **46.** $14y - 6 + 7y > 4 + 10y - 10$

47. $5(x + 3) - 6x \leq 3(2x + 1) - 4x$ **48.** $2(x - 5) + 3x < 4(x - 6) + 1$

49. $\frac{2}{3}(p + 3) > \frac{5}{6}(p - 4)$ **50.** $\frac{7}{9}(y - 4) \leq \frac{4}{3}(y + 5)$

51. $4x - (6x + 1) \leq 8x + 2(x - 3)$ **52.** $2y - (4y + 3) > 6y + 3(y + 4)$

53. $5(2k + 3) - 2(k - 8) > 3(2k + 4) + k - 2$

54. $2(3z - 5) + 4(z + 6) \geq 2(3z + 2) + 3z - 15$

Write a three-part inequality involving the variable x that describes each set of numbers graphed. See Example 1(b).

55. 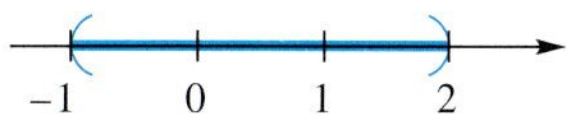

56. 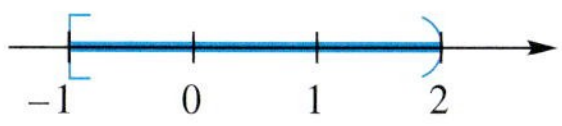

57. 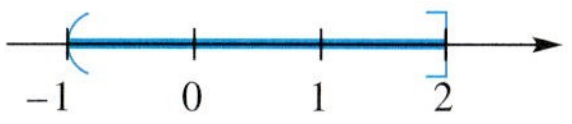

58. 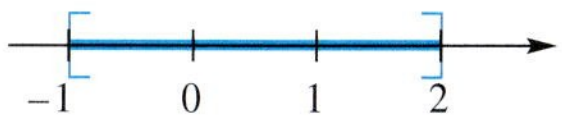

Solve each inequality. Write the solution set in interval notation and graph it. See Example 7.

59. $-5 \leq 2x - 3 \leq 9$ **60.** $-7 \leq 3x - 4 \leq 8$

61. $5 < 1 - 6m < 12$ **62.** $-1 \leq 1 - 5q \leq 16$

63. $10 < 7p + 3 < 24$ **64.** $-8 \leq 3r - 1 \leq -1$

65. $-12 \leq \frac{1}{2}z + 1 \leq 4$ **66.** $-6 \leq 3 + \frac{1}{3}a \leq 5$

67. $1 \leq 3 + \frac{2}{3}p \leq 7$ **68.** $2 < 6 + \frac{3}{4}y < 12$

69. $-7 \leq \frac{5}{4}r - 1 \leq -1$ **70.** $-12 \leq \frac{3}{7}a + 2 \leq -4$

RELATING CONCEPTS (EXERCISES 71–76)

The methods for solving linear equations and linear inequalities are quite similar. In Exercises 71–76, we show how the solutions of an inequality are closely connected to the solution of the corresponding equation.

Work these exercises in order.

71. Solve the equation $3x + 2 = 14$ and graph the solution set as a single point on the number line.

72. Solve the inequality $3x + 2 > 14$ and graph the solution set as an interval on the number line. How does this result compare to the one in Exercise 71?

(continued)

RELATING CONCEPTS (EXERCISES 71–76) (CONTINUED)

73. Solve the inequality $3x + 2 < 14$ and graph the solution set as an interval on the number line. How does this result compare to the one in Exercise 72?

74. If you were to graph all the solution sets from Exercises 71–73 on the same number line, what would the graph be? (This is called the *union* of all the solution sets.)

75. Based on your results from Exercises 71–74, if you were to graph the union of the solution sets of

$$-4x + 3 = -1, \qquad -4x + 3 > -1, \qquad \text{and} \qquad -4x + 3 < -1,$$

what do you think the graph would be?

76. Comment on the following statement: *Equality* is the boundary between *less than* and *greater than.*

Did you make the connection that the value that satisfies an equation separates the values that satisfy the corresponding less than and greater than inequalities?

Solve each problem by writing and solving an inequality. See Example 6.

77. Inkie Landry has grades of 76 and 81 on her first two algebra tests. If she wants an average of at least 80 after her third test, what possible scores can she make on her third test?

78. Mabimi Pampo has grades of 96 and 86 on his first two geometry tests. What possible scores can he make on his third test so that his average is at least 90?

79. The formula for converting Fahrenheit temperature to Celsius is

$$C = \frac{5}{9}(F - 32).$$

If the Celsius temperature on a certain summer day in Toledo is never more than 30°, how would you describe the corresponding Fahrenheit temperatures?

80. The formula for converting Celsius temperature to Fahrenheit is

$$F = \frac{9}{5}C + 32.$$

The Fahrenheit temperature of Key West, Florida, has never exceeded 95°. How would you describe this using Celsius temperature?

81. A product will break even or produce a profit if the revenue R from selling the product is at least equal to the cost C of producing it. Suppose that the cost C (in dollars) to produce x units of bicycle helmets is $C = 50x + 5000$, while the revenue R (in dollars) collected from the sale of x units is $R = 60x$. For what values of x does the product break even or produce a profit?

82. (See Exercise 81.) If the cost to produce x units of basketball cards is $C = 100x + 6000$ (in dollars), and the revenue collected from selling x units is $R = 500x$ (in dollars), for what values of x does the product break even or produce a profit?

83. For what values of x would the rectangle have perimeter of at least 400?

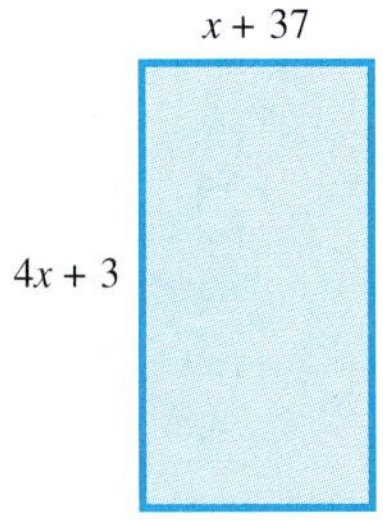

84. For what values of x would the triangle have perimeter of at least 72?

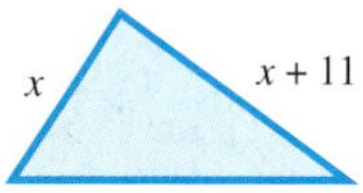

85. A long-distance phone call costs \$2.00 for the first three minutes plus \$.30 per minute for each minute or fractional part of a minute after the first three minutes. If x represents the number of minutes of the length of the call after the first three minutes, then $2 + .30x$ represents the cost of the call. If Jorge has \$5.60 to spend on a call, what is the maximum total time he can use the phone?

86. If the call described in Exercise 85 costs between \$5.60 and \$6.50, what are the possible total time lengths for the call?

RELATING CONCEPTS (EXERCISES 87–98)

The words *and* and *or* are very important in the context of inequalities. We use them to combine simple inequalities. For instance, the statement $-1 \le x < 3$ means $-1 \le x$ *and* $x < 3$, so that only values of x that satisfy *both* conditions are in the solution set, the interval $[-1, 3)$. On the other hand, the statement $x < -1$ *or* $x > 3$, that uses the connective *or,* will be true if $x < -1$ or if $x > 3$. Only one of the conditions must be true for the statement to be true. The two intervals that make this *or* statement true are combined, using the union symbol $\cup$, as $(-\infty, -1) \cup (3, \infty)$.

Work Exercises 87–98 in order.

87. Is the statement "Today is Tuesday *or* today is Wednesday" true if today is actually Wednesday? Is it true if it is actually Tuesday?

88. Is the statement "Today is Tuesday *and* today is Wednesday" true if today is actually Wednesday? Is it true if it is actually Tuesday?

89. Is the statement $x > 3$ and $x < 8$ true for $x = 5$? For $x = 10$? For $x = 3$?

90. Graph $x > 3$.

91. Graph $x < 8$.

92. Compare the graphs from Exercises 90 and 91. Give the values of x in interval notation that make *both* inequalities true.

93. Graph the solution set of $x > 3$ and $x < 8$. How does the graph compare to your answer to Exercise 92?

94. Give the solution set of $x < -4$ and $x < 2$ in interval notation and graph it.

95. What is the solution set of $3 - x > 6$ and $2x > -4$?

96. Graph the solution set of $x > 3$ or $x < 8$. Compare the graph with your answers to Exercises 90 and 91.

97. Give the solution set of $x < -4$ or $x < 2$ in interval notation and graph it. Compare with the answer to Exercise 94.

98. Give the solution set in interval notation and graph the solution set of $3 - x > 6$ or $2x > -4$. Compare with the answer to Exercise 95.

Did you make the connection that when *and* is used to connect two inequalities, both statements must be true for the combined statement to be true, but when *or* is used as the connective, the combined statement is true if either one or both statements are true?

CHAPTER 2 GROUP ACTIVITY

Are You a Race-Walker?

Objective: Use proportions to calculate walking speeds.

Materials: Students will need a large area for walking. Each group will need a stopwatch.

Race-walking at speeds exceeding 8 miles per hour is a high fitness, long-distance competitive sport. The table below contains the gold medal winners of the 1996 Olympic race-walking competition. Complete the table by applying the proportion given below to find the race-walker's steps per minute.

Use 10 km ≈ 6.21 miles. (Round all answers except those for steps per minute to the nearest thousandths. Round steps per minute to the nearest whole number.)

$$\frac{70 \text{ steps per minute}}{2 \text{ miles per hour}} = \frac{x \text{ steps per minute}}{y \text{ miles per hour}}$$

Event	*Gold Medal Winner*	*Country*	*Time in Hours: Minutes: Seconds*	*Time in Minutes*	*Time in Hours*	*y Miles per Hour*	*x Steps per Minute*
10 km Walk, Women	Yelena Nikolayeva	Russia	0:41:49				
20 km Walk, Men	Jefferson Perez	Ecuador	1:20:07				
50 km Walk, Men	Robert Korzeniowski	Poland	3:43:30				

Source: 1999 World Almanac.

A. Using a stopwatch, take turns counting how many steps each member of the group takes in one minute while walking at a normal pace. Record the results in the chart below. Then do it again at a fast pace. Record these results.

	Normal Pace		*Fast Pace*	
Name	*x* Steps per Minute	*y* Miles per Hour	*x* Steps per Minute	*y* Miles per Hour

B. Use the proportion above to convert the numbers from part A to miles per hour and complete the chart.

1. Find the average speed for the group at a normal pace and at a fast pace.
2. What is the minimum number of steps per minute you would have to take to be a race-walker?
3. At a fast pace did anyone in the group walk fast enough to be a race-walker? Explain how you decided.

CHAPTER 2 SUMMARY

KEY TERMS

2.1 linear equation in one variable
solution set
equivalent equations

2.2 empty (null) set

2.3 degree
complementary angles
supplementary angles
consecutive integers

2.4 perimeter
circumference
area
vertical angles
straight angle

2.5 ratio
proportion
cross products
vary directly as (is proportional to)

2.7 interval on the number line
interval notation
linear inequality
three-part inequality

NEW SYMBOLS

Symbol	Meaning
$\emptyset$	empty set
1°	one degree
a to b, $a:b$, or $\frac{a}{b}$	the ratio of a to b
(a, b)	interval notation for $a < x < b$
$[a, b]$	interval notation for $a \le x \le b$
∞	infinity
$-\infty$	negative infinity
$(-\infty, \infty)$	set of all real numbers

TEST YOUR WORD POWER

See how well you have learned the vocabulary in this chapter. Answers, with examples, are given at the bottom of the page.

1. A **solution set** is the set of numbers that
(a) make an expression undefined
(b) make an equation false
(c) make an equation true
(d) make an expression equal to 0.

2. The **empty set** is a set
(a) with 0 as its only element
(b) with an infinite number of elements
(c) with no elements
(d) of ideas.

3. Complementary angles are angles
(a) formed by two parallel lines
(b) whose sum is 90°
(c) whose sum is 180°
(d) formed by perpendicular lines.

4. Supplementary angles are angles
(a) formed by two parallel lines
(b) whose sum is 90°
(c) whose sum is 180°
(d) formed by perpendicular lines.

5. A **ratio**
(a) compares two quantities using a quotient
(b) says that two quotients are equal
(c) is a product of two quantities
(d) is a difference between two quantities.

6. A **proportion**
(a) compares two quantities using a quotient
(b) says that two quotients are equal
(c) is a product of two quantities
(d) is a difference between two quantities.

7. An **inequality** is
(a) a statement that two algebraic expressions are equal
(b) a point on a number line
(c) an equation with no solutions
(d) a statement with algebraic expressions related by $<$, $\le$, $>$, or $\ge$.

8. Interval notation is
(a) a portion of a number line
(b) a special notation for describing a point on a number line
(c) a way to use symbols to describe an interval on a number line
(d) a notation to describe unequal quantities.

Answers to Test Your Word Power

1. (c) *Example:* $\{8\}$ is the solution set of $2x + 5 = 21$. **2.** (c) *Example:* The empty set $\emptyset$ is the solution set of $5x + 3 = 5x + 4$. **3.** (b) *Example:* Angles with measures 35° and 55° are complementary angles. **4.** (c) *Example:* Angles with measures 112° and 68° are supplementary angles. **5.** (a) *Example:* $\frac{7 \text{ inches}}{12 \text{ inches}} = \frac{7}{12}$ **6.** (b) *Example:* $\frac{2}{3} = \frac{8}{12}$ **7.** (d) *Examples:* $x < 5$, $7 + 2y \ge 11$, $-5 < 2z - 1 \le 3$ **8.** (c) *Examples:* $(-\infty, 5]$, $(1, \infty)$, $[-3, 3)$

QUICK REVIEW

CONCEPTS	EXAMPLES
2.1 THE ADDITION AND MULTIPLICATION PROPERTIES OF EQUALITY	
The same expression may be added to (or subtracted from) each side of an equation without changing the solution.	Solve $x - 6 = 12$. $x - 6 + 6 = 12 + 6$ Add 6. $x = 18$ Combine terms. Solution set: $\{18\}$
Each side of an equation may be multiplied (or divided) by the same nonzero expression without changing the solution.	Solve $\frac{3}{4}x = -9$. $\frac{4}{3} \cdot \frac{3}{4}x = \frac{4}{3}(-9)$ Multiply by $\frac{4}{3}$. $x = -12$ Solution set: $\{-12\}$
2.2 MORE ON SOLVING LINEAR EQUATIONS	
Solving a Linear Equation	Solve the equation $2x + 3(x + 1) = 38$.
1. Clear parentheses and combine like terms to simplify each side.	$2x + 3x + 3 = 38$ Clear parentheses. $5x + 3 = 38$ Combine like terms.
2. Get the variable term on one side, a number on the other.	$5x + 3 - 3 = 38 - 3$ Subtract 3. $5x = 35$ Combine terms.
3. Get the equation into the form $x =$ a number.	$\frac{5x}{5} = \frac{35}{5}$ Divide by 5. $x = 7$
4. Check by substituting the result into the original equation.	$2x + 3(x + 1) = 38$ Check. $2(7) + 3(7 + 1) = 38$? Let $x = 7$. $14 + 24 = 38$? Multiply. $38 = 38$ True Solution set: $\{7\}$
2.3 AN INTRODUCTION TO APPLICATIONS OF LINEAR EQUATIONS	
Solving an Applied Problem Using the Six-Step Method	One number is 5 more than another. Their sum is 21. Find both numbers.
Step 1 Choose a variable to represent the unknown.	Let x be the smaller number.
Step 2 Determine expressions for any other unknown quantities, using the variable. Draw figures or diagrams and use charts if they apply.	Let $x + 5$ be the larger number.
Step 3 Write an equation.	$x + (x + 5) = 21$
Step 4 Solve the equation.	$2x + 5 = 21$ Combine terms. $2x + 5 - 5 = 21 - 5$ Subtract 5. $2x = 16$ Combine terms. $x = 8$ Divide by 2.

CONCEPTS	EXAMPLES
Step 5 Answer the question(s) asked in the problem.	The numbers are 8 and 13.
Step 6 Check your solution by using the original words of the problem. Be sure that the answer is appropriate and makes sense.	13 is 5 more than 8, and $8 + 13 = 21$. It checks.

2.4 FORMULAS AND APPLICATIONS FROM GEOMETRY

To find the values of one of the variables in a formula given values for the others, substitute the known values into the formula.

Find L if $A = LW$, given that $A = 24$ and $W = 3$.

$$24 = L \cdot 3 \qquad A = 24,\ W = 3$$

$$\frac{24}{3} = \frac{L \cdot 3}{3} \qquad \text{Divide by 3.}$$

$$8 = L$$

To solve a formula for one of the variables, isolate that variable by treating the other variables as numbers and using the steps for solving equations.

Solve $A = \frac{1}{2}bh$ for b.

$$2A = 2\left(\frac{1}{2}bh\right) \qquad \text{Multiply by 2.}$$

$$2A = bh$$

$$\frac{2A}{h} = b \qquad \text{Divide by } h.$$

2.5 RATIOS AND PROPORTIONS

To write a ratio, express quantities in the same units.

Express as a ratio: 4 feet to 8 inches.

4 feet to 8 inches = 48 inches to 8 inches

$$= \frac{48}{8} = \frac{6}{1} \quad \text{or} \quad 6 \text{ to } 1 \quad \text{or} \quad 6:1$$

To solve a proportion, use cross products.

Solve $\frac{x}{12} = \frac{35}{60}$.

$$60x = 12 \cdot 35 \qquad \text{Cross products}$$

$$60x = 420 \qquad \text{Multiply.}$$

$$\frac{60x}{60} = \frac{420}{60} \qquad \text{Divide by 60.}$$

$$x = 7$$

Solution set: $\{7\}$

Solving Direct Variation Problems

1. Write the variation equation using $y = kx$.
2. Find k by substituting the given values of x and y into the equation.
3. Write the equation with the value of k from Step 2 and the given value of x or y. Solve for the remaining variable.

At a constant speed (or rate) distance varies directly as time. Find the time to travel 85 miles at a constant speed if it takes 1.5 hours to go 67.5 miles at the same constant speed.

Let t represent the unknown time. The direct variation equation is $d = kt$. Substitute $d = 67.5$ and $t = 1.5$ to find k.

$$67.5 = k(1.5)$$

$$45 = k \qquad \text{Divide by 1.5.}$$

Now substitute $k = 45$ and $d = 85$ in the equation $d = kt$.

$$85 = 45t$$

$$1.89 = t \qquad \text{Divide by 45; round up.}$$

It will take approximately 1.89 hours to travel 85 miles.

CONCEPTS	EXAMPLES

2.6 MORE ABOUT PROBLEM SOLVING

Problems involving applications of percent can be solved using charts.

A sum of money is invested at simple interest in two ways. Part is invested at 12%, and $20,000 less than that amount is invested at 10%. If the total interest for one year is $9000, find the amount invested at each rate.

Let x = amount invested at 12%;

$x - 20{,}000$ = amount invested at 10%.

Dollars Invested	Rate of Interest	Interest for One Year
x	.12	$.12x$
$x - 20{,}000$	.10	$.10(x - 20{,}000)$

$$.12x + .10(x - 20{,}000) = 9000$$

$$.12x + .10x + .10(-20{,}000) = 9000 \quad \text{Distributive property}$$

$$12x + 10x + 10(-20{,}000) = 900{,}000 \quad \text{Multiply by 100.}$$

$$12x + 10x - 200{,}000 = 900{,}000$$

$$22x - 200{,}000 = 900{,}000 \quad \text{Combine terms.}$$

$$22x = 1{,}100{,}000 \quad \text{Add 200,000.}$$

$$x = 50{,}000 \quad \text{Divide by 22.}$$

$50,000 is invested at 12% and $30,000 is invested at 10%.

The three forms of the formula relating distance, rate, and time, are $d = rt$, $r = \frac{d}{t}$, and $t = \frac{d}{r}$.

Two cars leave from the same point, traveling in opposite directions. One travels at 45 miles per hour and the other at 60 miles per hour. How long will it take them to be 210 miles apart?

Let t = time it takes for them to be 210 miles apart.

To solve a problem about distance, set up a sketch showing what is happening in the problem.

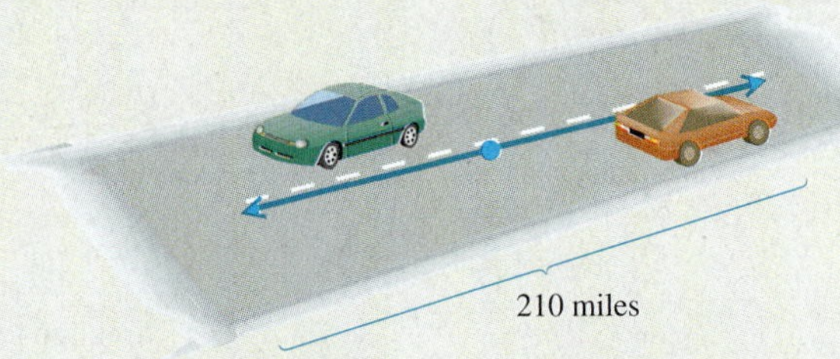

Make a chart using the information given in the problem, along with the unknown quantities.

The chart gives the information from the problem, with expressions for distance obtained by using $d = rt$.

	Rate	Time	Distance
One car	45	t	$45t$
Other car	60	t	$60t$

The sum of the distances, $45t$ and $60t$, must be 210 miles.

$$45t + 60t = 210$$

$$105t = 210 \quad \text{Combine like terms.}$$

$$t = 2 \quad \text{Divide by 2.}$$

It will take them 2 hours to be 210 miles apart.

CONCEPTS	EXAMPLES

2.7 THE ADDITION AND MULTIPLICATION PROPERTIES OF INEQUALITY

To solve an inequality:

1. Clear parentheses and combine like terms.
2. Add or subtract the same expression on each side to get the variable term on one side and a number on the other side.
3. Multiply or divide by the same expression on each side to get the form $x > a$ or $x < a$. (When multiplying or dividing by a negative expression, reverse the direction of the inequality symbol.)

Solve $3(1 - x) + 5 - 2x > 9 - 6$.

$3 - 3x + 5 - 2x > 9 - 6$ Clear parentheses.

$8 - 5x > 3$ Combine terms.

$8 - 5x - 8 > 3 - 8$ Subtract 8.

$-5x > -5$ Combine terms.

$\frac{-5x}{-5} < \frac{-5}{-5}$ Divide by -5; change $>$ to $<$.

$x < 1$ Lowest terms

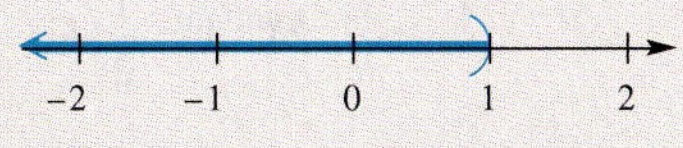

Solution set: $(-\infty, 1)$

To solve an inequality such as

$$4 < 2x + 6 < 8$$

work with all three expressions at the same time.

Solve $4 < 2x + 6 < 8$.

$4 - 6 < 2x + 6 - 6 < 8 - 6$ Subtract 6.

$-2 < 2x < 2$ Combine terms.

$\frac{-2}{2} < \frac{2x}{2} < \frac{2}{2}$ Divide by 2.

$-1 < x < 1$

Solution set: $(-1, 1)$

CHAPTER 2 REVIEW EXERCISES

[2.1–2.2] *Solve each equation.*

1. $m - 5 = 1$
2. $y + 8 = -4$
3. $3k + 1 = 2k + 8$
4. $5k = 4k + \frac{2}{3}$
5. $(4r - 2) - (3r + 1) = 8$
6. $3(2y - 5) = 2 + 5y$
7. $7k = 35$
8. $12r = -48$
9. $2p - 7p + 8p = 15$
10. $\frac{m}{12} = -1$
11. $\frac{5}{8}k = 8$
12. $12m + 11 = 59$
13. $3(2x + 6) - 5(x + 8) = x - 22$
14. $5x + 9 - (2x - 3) = 2x - 7$
15. $\frac{1}{2}r - \frac{r}{3} = \frac{r}{6}$
16. $.10(x + 80) + .20x = 14$
17. $3x - (-2x + 6) = 4(x - 4) + x$
18. $2(y - 3) - 4(y + 12) = -2(y + 27)$

[2.3] *Use the six-step method to solve each problem.*

19. Soccer is the world's most popular sport, with over 20 million participants. The number of participating nations in the 1990 World Cup was 6 less than in the 1986 World Cup. In 1994, 28 more nations participated than in 1986. How many nations participated in 1986 if there was a total of 358 participating nations in those three years? (*Source:* FIFA.)

53. Sue Fredine drove from Louisville to Dallas, a distance of 819 miles, averaging 63 miles per hour. What was her driving time?

54. Two planes leave St. Louis at the same time. One flies north at 350 miles per hour and the other flies south at 420 miles per hour. In how many hours will they be 1925 miles apart?

55. Jim leaves his house on his bicycle and averages 5 miles per hour. His wife, Annie, leaves $\frac{1}{2}$ hour later, following the same path and averaging 8 miles per hour. How long will it take for Annie to catch up to Jim?

56. The circumference of a circle varies directly as its radius. A circle with a circumference of 5 inches has a radius of approximately .796 inches. Find the radius of a circle with a circumference of 17.5 inches.

57. Typically, the supply of an item varies directly as the price. When a computer game was priced at \$50, a merchant had a supply of 40 games. After the game went on sale for \$35, what was the supply?

58. The pressure exerted by a liquid at a given point is proportional to the depth of the point beneath the surface of the liquid. The pressure at 10 meters is $\frac{80}{3}$ newtons per square centimeter. What pressure is exerted at 30 meters?

59. Write your own applied problem, and use the six-step method to solve it.

[2.7] *Write each inequality in interval notation and graph it on a number line.*

60. $p \geq -4$

61. $x < 7$

62. $-5 \leq y < 6$

Solve each inequality and graph the solution set.

63. $y + 6 \geq 3$

64. $5t < 4t + 2$

65. $-6x \leq -18$

66. $8(k - 5) - (2 + 7k) \geq 4$

67. $4x - 3x > 10 - 4x + 7x$

68. $3(2w + 5) + 4(8 + 3w) < 5(3w + 2) + 2w$

69. $-3 \leq 2m + 1 \leq 4$

70. $9 < 3m + 5 \leq 20$

Solve each problem by writing an inequality.

71. In 1995 more than 6900 Americans were active water skiers. Of this group, the number of males was 1.4 times the number of females. Find the possible numbers of female water skiers. (*Source:* National Sporting Goods Association, Mount Prospect, Illinois.)

72. In 1995, the average National Football League quarterback's salary was \$1.307 million. In 1996, it rose to \$1.336 million. What salary in 1997 would make the average salary for these three years at least \$1.4 million? (*Source:* NFL Players Association.)

MIXED REVIEW EXERCISES*

Solve.

73. $\frac{y}{7} = \frac{y - 5}{2}$

74. $I = prt$ for r

75. $-2x > -4$

76. $2k - 5 = 4k + 13$

77. $.05x + .02x = 4.9$

78. $2 - 3(y - 5) = 4 + y$

*The order of the exercises in this final group does not correspond to the order in which topics occur in the chapter. This random ordering should help you in your preparation for the chapter test.

79. $9x - (7x + 2) = 3x + (2 - x)$

80. $\frac{1}{3}s + \frac{1}{2}s + 7 = \frac{5}{6}s + 5 + 2$

81. A student solved $3 - (8 + 4x) = 2x + 7$ and gave the answer as $\{6\}$. Verify that this answer is incorrect by checking it in the equation. Then explain the error and give the correct solution set. (*Hint:* The error involves the subtraction sign.)

82. A recipe for biscuit tortoni calls for $\frac{2}{3}$ cup of macaroon cookie crumbs. The recipe is for 8 servings. How many cups of macaroon cookie crumbs would be needed for 30 servings?

83. The Golden Gate Bridge in San Francisco is 2605 feet longer than the Brooklyn Bridge. Together, their spans total 5795 feet. How long is each bridge? (*Source: World Almanac and Book of Facts.*)

84. A student solves $\frac{3}{x} = \frac{5}{x}$ by cross multiplying and gets 0 as an answer. What is wrong with this answer?

85. Which is the best buy for apple juice?
32-ounce size: \$1.19
48-ounce size: \$1.79
64-ounce size: \$1.99

86. If 1 quart of oil must be mixed with 24 quarts of gasoline, how much oil would be needed for 192 quarts of gasoline?

87. Two trains are 390 miles apart. They start at the same time and travel toward one another, meeting 3 hours later. If the speed of one train is 30 miles per hour more than the speed of the other train, find the speed of each train.

88. One side of a triangle is 3 centimeters longer than the shortest side. The third side is twice as long as the shortest side. If the perimeter of the triangle cannot exceed 39 centimeters, find all possible lengths for the shortest side.

89. The shorter base of a trapezoid is 42 centimeters long, and the longer base is 48 centimeters long. The area of the trapezoid is 360 square centimeters. Find the height of the trapezoid.

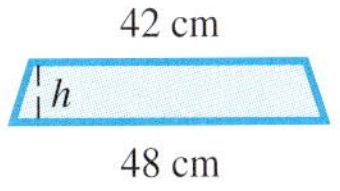

$A = 360$ square centimeters

90. The perimeter of a square cannot be greater than 200 meters. Find the possible values for the length of a side.

91. Can $4 + \frac{5}{x} = \frac{2}{3}$ be solved directly by cross multiplication? Explain your answer.

CHAPTER 2 TEST

Solve each equation.

1. $5x + 9 = 7x + 21$

2. $-\frac{4}{7}x = -12$

3. $7 - (m - 4) = -3m + 2(m + 1)$

4. $.06(x + 20) + .08(x - 10) = 4.6$

5. $-8(2x + 4) = -4(4x + 8)$

Solve each problem.

22. For a woven hanging, Miguel Hidalgo needs three pieces of yarn, which he will cut from a 40-centimeter piece. The longest piece is to be 3 times as long as the middle-sized piece, and the shortest piece is to be 5 centimeters shorter than the middle-sized piece. What lengths should he cut?

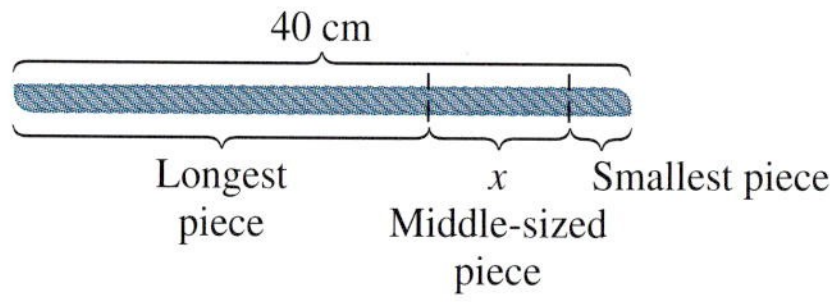

23. A radio telescope at the Max Planck Institute in Germany has a 328-foot diameter. What is the circumference of this telescope? (Use $\pi = 3.14$.) (*Source: The Guinness Book of World Records.*)

24. A cook wants to increase a recipe that serves 6 to make enough for 20 people. The recipe calls for $1\frac{1}{4}$ cups of grated cheese. How much cheese will be needed to serve 20?

25. Two cars are 400 miles apart. Both start at the same time and travel toward one another. They meet 4 hours later. If the speed of one car is 20 miles per hour faster than the other, what is the speed of each car?

Linear Equations in Two Variables

3

Business

Calvin Coolidge, 30th president of the United States, once said "The chief business of America is business." Small businesses account for 99 percent of the 19 million nonfarm businesses in the United States today. Small businesses employ 55 percent of the private workforce, make 44 percent of all sales in America, and produce 38 percent of the nation's gross national product.*

The number of new consumer packaged-goods products introduced in the United States from 1991–1996 is illustrated in the line graph. The graph shows that, with the exception of 1994, this number has increased from year to year at a steady pace. What might account for the one-year decrease? We will return to this graph in the exercises for Section 3.1.

Source: Marketing Intelligence Service, Ltd.

Mathematics plays an important role in the business world. Preparing a business plan, pricing, borrowing money, determining market share, and tracking costs, revenue, and profit all require a good understanding of mathematics. Throughout this chapter many of the examples and exercises will further illustrate the use of mathematics in business management.

*Data from *The Universal Almanac,* 1997, John W. Wright, General Editor.

INTERNET Visit our Web site at www.LialAlgebra.com

3.1 Linear Equations in Two Variables

OBJECTIVES

1. Interpret graphs.
2. Write a solution as an ordered pair.
3. Decide whether a given ordered pair is a solution of a given equation.
4. Complete ordered pairs for a given equation.
5. Plot ordered pairs.

FOR EXTRA HELP

SSG Sec. 3.1
SSM Sec. 3.1
Pass the Test Software
InterAct Math Tutorial Software
Video 5

Graphs are prevalent in our society. Pie charts (circle graphs) and bar graphs were introduced in Chapter 1, and we have seen many examples of them in the first two chapters of this book. It is important to be able to interpret graphs correctly.

OBJECTIVE 1 **Interpret graphs.** We begin with a bar graph where we must estimate the heights of the bars.

EXAMPLE 1 **Interpreting Bar Graphs**

Venture capital is money invested in new, often speculative, business enterprises. The amount of venture capital invested in companies has risen during the decade of the 1990s, as shown in the graph in Figure 1. Use the graph to determine the following.

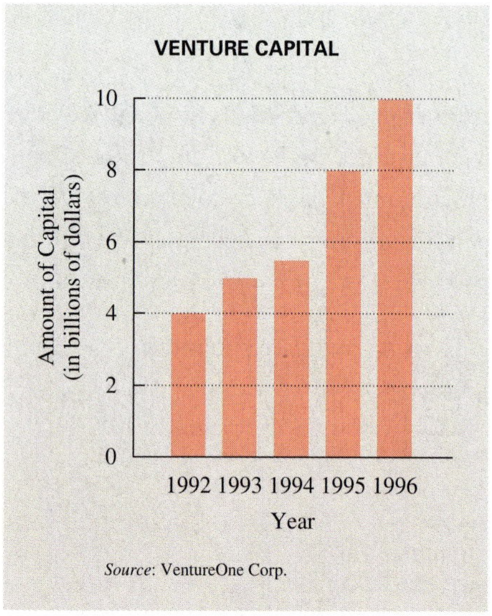

Figure 1

(a) What amount of venture capital was provided in 1992?

Move horizontally from the top of the bar for 1992 to the scale on the left to see that about \$4 billion was provided.

(b) In what year was that amount doubled?

Follow the line for \$8 billion across to the right. The bar for 1995 just touches that line, so the amount provided in 1992 was doubled in 1995.

EXAMPLE 2 Interpreting Line Graphs

The line graph in Figure 2 shows the total number of deals that involved venture capital in the 1990s. Use the graph to estimate the number of deals in 1993 and in 1995.

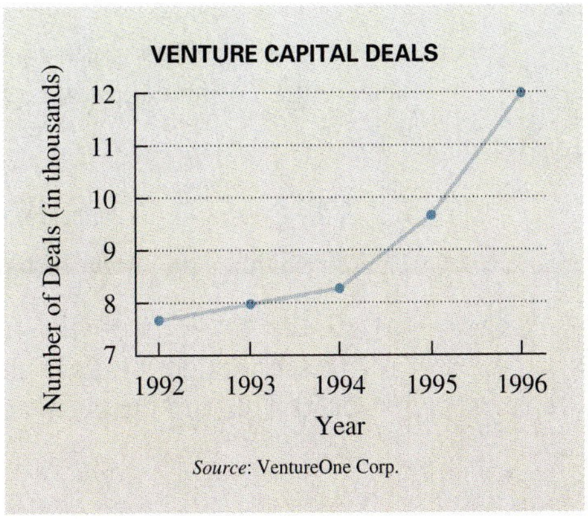

Figure 2

In 1993 the point lies on the line that corresponds to 8000, so 8000 venture capital deals were closed in 1993. The point for 1995 is about $\frac{2}{3}$, or .7, of the way between the lines for 9 and 10. Thus, we estimate 9700 deals were made in 1995.

We solved linear equations in one variable and explored their applications in Chapter 2. Now we want to extend those ideas to *linear equations in two variables.*

Linear Equation

A **linear equation in two variables** is an equation that can be put in the form

$$Ax + By = C,$$

where A, B, and C are real numbers and A and B are not both 0.

OBJECTIVE 2 Write a solution as an ordered pair. A solution of a linear equation in *two* variables requires *two* numbers, one for each variable. For example, the equation $y = 4x + 5$ is satisfied if x is replaced with 2 and y is replaced with 13, since

$$13 = 4(2) + 5. \qquad \text{Let } x = 2;\ y = 13.$$

The pair of numbers $x = 2$ and $y = 13$ gives a solution of the equation $y = 4x + 5$. The phrase "$x = 2$ and $y = 13$" is abbreviated

x-value → ← y-value

(2, 13)

Ordered pair

with the x-value, 2, and the y-value, 13, given as a pair of numbers written inside parentheses. *The x-value is always given first.* A pair of numbers such as (2, 13) is called an

ordered pair. As the name indicates, the order in which the numbers are written is important. The ordered pairs (2, 13) and (13, 2) are not the same. The second pair indicates that $x = 13$ and $y = 2$. (Of course, letters other than x and y may be used in the equation with the numbers.)

OBJECTIVE 3 **Decide whether a given ordered pair is a solution of a given equation.** An ordered pair that is a solution of an equation is said to *satisfy* the equation.

EXAMPLE 3 Deciding Whether an Ordered Pair Satisfies an Equation

Decide whether the given ordered pair is a solution of the given equation.

(a) (3, 2); $2x + 3y = 12$

To see whether (3, 2) is a solution of the equation $2x + 3y = 12$, we substitute 3 for x and 2 for y in the given equation.

$$2x + 3y = 12$$

$$2(3) + 3(2) = 12 \quad ? \qquad \text{Let } x = 3; \text{ let } y = 2.$$

$$6 + 6 = 12 \quad ?$$

$$12 = 12 \qquad \text{True}$$

This result is true, so (3, 2) satisfies $2x + 3y = 12$.

(b) $(-2, -7)$; $m + 5n = 33$

$$(-2) + 5(-7) = 33 \quad ? \qquad \text{Let } m = -2; \text{ let } n = -7.$$

$$-2 + (-35) = 33 \quad ?$$

$$-37 = 33 \qquad \text{False}$$

This result is false, so $(-2, -7)$ is *not* a solution of $m + 5n = 33$.

OBJECTIVE 4 **Complete ordered pairs for a given equation.** Choosing a number for one variable in a linear equation makes it possible to find the value of the other variable, as shown in the next example.

EXAMPLE 4 Completing an Ordered Pair

Complete the ordered pair (7,) for the equation $y = 4x + 5$.

In this ordered pair, $x = 7$. (Remember that x always comes first.) To find the corresponding value of y, replace x with 7 in the equation $y = 4x + 5$.

$$y = 4(7) + 5 = 28 + 5 = 33$$

The ordered pair is (7, 33).

Ordered pairs often are displayed in a **table of values** as in the next example. The table may be written either vertically or horizontally.

EXAMPLE 5 **Completing a Table of Values**

Complete the given table of values for each equation. Then write the results as ordered pairs.

(a) $x - 2y = 8$

x	y
2	
10	
	0
	-2

To complete the first two ordered pairs, let $x = 2$ and $x = 10$, respectively.

If	$x = 2,$	If	$x = 10,$
then	$x - 2y = 8$	then	$x - 2y = 8$
becomes	$2 - 2y = 8$	becomes	$10 - 2y = 8$
	$-2y = 6$		$-2y = -2$
	$y = -3.$		$y = 1.$

Now complete the last two ordered pairs by letting $y = 0$ and $y = -2$, respectively.

If	$y = 0,$	If	$y = -2,$
then	$x - 2y = 8$	then	$x - 2y = 8$
becomes	$x - 2(0) = 8$	becomes	$x - 2(-2) = 8$
	$x - 0 = 8$		$x + 4 = 8$
	$x = 8.$		$x = 4.$

The completed table of values is as follows.

x	y
2	-3
10	1
8	0
4	-2

The corresponding ordered pairs are $(2, -3)$, $(10, 1)$, $(8, 0)$, and $(4, -2)$.

(b) $x = 5$

x	y
	-2
	6
	3

The given equation is $x = 5$. No matter which value of y might be chosen, the value of x is always the same, 5.

x	y
5	-2
5	6
5	3

The ordered pairs are $(5, -2)$, $(5, 6)$, and $(5, 3)$.

NOTE We can think of $x = 5$ in Example 5(b) as an equation in two variables by rewriting $x = 5$ as $x + 0y = 5$. This form of the equation shows that for any value of y, the value of x is 5. Similarly, $y = -2$ is the same as $0x + y = -2$.

Earlier in this book, we saw that linear equations in *one* variable had either one, zero, or an infinite number of real number solutions. Every linear equation in *two* variables has an infinite number of ordered pairs as solutions. Each choice of a number for one variable leads to a particular real number for the other variable.

To graph these solutions, represented as the ordered pairs (x, y), we need *two* number lines, one for each variable. These two number lines are drawn as shown in Figure 3. The horizontal number line is called the ***x*-axis.** The vertical line is called the ***y*-axis.** Together, the x-axis and y-axis form a **rectangular coordinate system.** It is also called the **Cartesian coordinate system,** in honor of René Descartes.

The coordinate system is divided into four regions, called **quadrants.** These quadrants are numbered counterclockwise, as shown in Figure 3. Points on the axes themselves are not in any quadrant. The point at which the x-axis and y-axis meet is called the **origin.** The origin, labeled 0 in Figure 3, is the point corresponding to (0, 0).

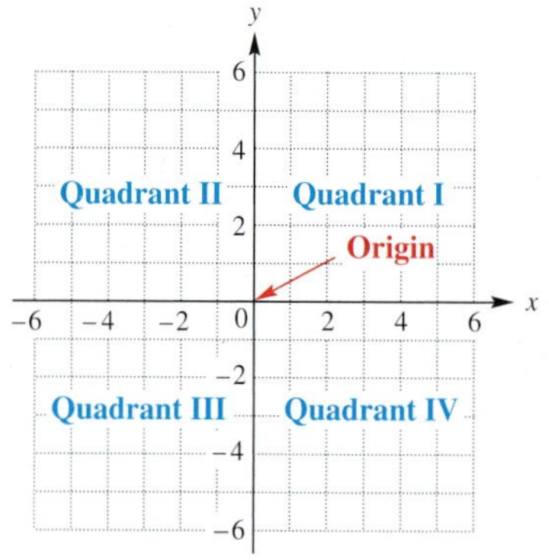

Figure 3

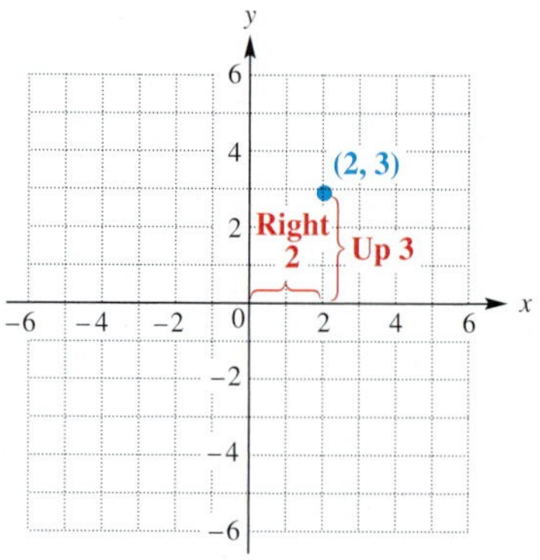

Figure 4

CONNECTIONS

The coordinate system that we use to plot points is credited to René Descartes (1596–1650), a French mathematician of the seventeenth century. It has been said that he developed the coordinate system while lying in bed, watching an insect move across the ceiling. He realized that he could locate the position of the insect at any given time by finding its distance from each of two perpendicular walls.

OBJECTIVE 5 Plot ordered pairs. By referring to the two axes, every point on the plane can be associated with an ordered pair. The numbers in the ordered pair are called the **coordinates** of the point. For example, locate the point associated with the ordered pair (2, 3) by starting at the origin. Since the x-coordinate is 2, go 2 units to the right along the x-axis. Then, since the y-coordinate is 3, turn and go up 3 units on a line parallel to the y-axis. This is called **plotting** the point (2, 3). (See Figure 4.) From now on we refer to the point with x-coordinate 2 and y-coordinate 3 as the point (2, 3).

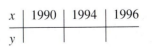

NOTE: When we graphed on a number line, one number corresponded to each point. On a plane, however, both numbers in the ordered pair are needed to locate a point. The ordered pair is a name for the point.

EXAMPLE 6 Plotting Ordered Pairs

Plot the given points on a coordinate system.

(a) $(1, 5)$ **(b)** $(-2, 3)$ **(c)** $(-1, -4)$

(d) $(7, -2)$ **(e)** $\left(\frac{3}{2}, 2\right)$ **(f)** $(5, 0)$

Locate the point $(-1, -4)$, for example, by first going 1 unit to the left along the x-axis. Then turn and go 4 units down, parallel to the y-axis. Plot the point $\left(\frac{3}{2}, 2\right)$, by going $\frac{3}{2}$ $\left(\text{or } 1\frac{1}{2}\right)$ units to the right along the x-axis. Then turn and go 2 units up, parallel to the y-axis. Figure 5 shows the graphs of the points in this example.

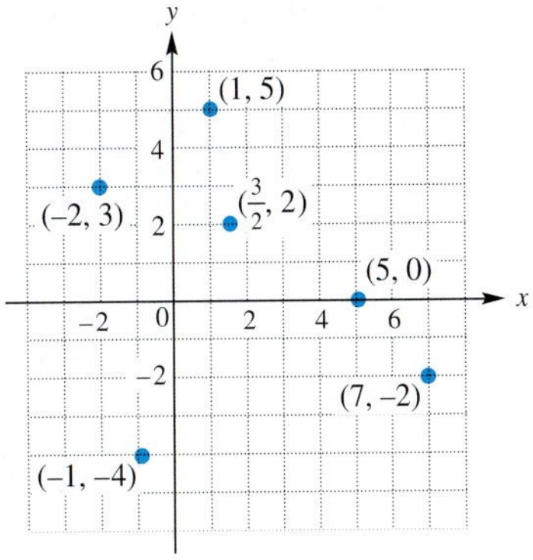

Figure 5

EXAMPLE 7 Completing Ordered Pairs to Estimate the Number of Business Incorporations

New business incorporations have increased steadily from 1990–1996. Their number can be closely approximated by the equation

$$y = 25.59x - 50{,}280,$$

where x is the year, and y is the number of incorporations in thousands.* This approximation is only valid for the years under examination, 1990–1996, since it was based on data for those years.

(a) Complete the table of ordered pairs for this linear equation. Round the y-values to the nearest thousand.

x	1990	1994	1996
y			

To find y when $x = 1990$, substitute into the equation.

$$y = 25.59(\mathbf{1990}) - 50{,}280 \qquad \text{Let } x = 1990.$$
$$= 644 \text{ (rounded)}$$

*Based on data from Dun & Bradstreet.

This means that, in 1990, there were 644,000 new incorporations. Find the y-values for 1994 and 1996 similarly. The completed table follows.

x	1990	1994	1996
y	644	746	798

(b) Interpret the ordered pair with $x = 1996$; with $x = 1994.5$.

The ordered pair (1996, 798) means that in 1996, 798,000 businesses were incorporated. We can interpret 1994.5 to mean halfway through 1994.

$$y = 25.59(1994.5) - 50{,}280 \qquad \text{Let } x = 1994.5.$$
$$= 759$$

There were approximately 759,000 incorporations.

(c) Graph the ordered pairs found in part (a).

The ordered pairs are graphed in Figure 6. Notice how the axes are labeled. In this application, x represents the year, and y represents the number of corporations in thousands. Different scales are used on the two axes because the two sets of numbers differ so much in size. Here, each square represents one unit in the horizontal direction and 50 units in the vertical direction. Because the numbers in the first ordered pair are quite large, we show a break in both axes near the origin.

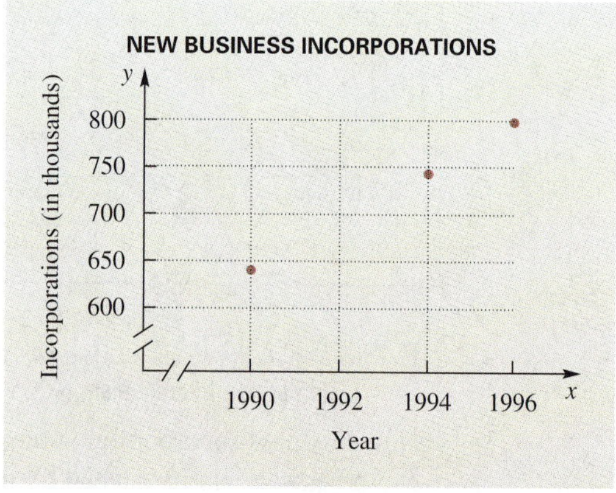

Figure 6

We can also think of ordered pairs as representing an input value x and an output value y. If we input x into the equation, the output is y. For instance, in Example 7, if we input the year into the equation

$$y = 25.59x - 50{,}280,$$

we get the number of incorporations (in thousands) in that year as the output. We encounter many examples of this type of relationship every day.

- The cost to fill the tank with gasoline depends on how many gallons are needed; the number of gallons is the input, and the cost is the output.
- The distance traveled depends on the traveling time; input a time, and the output is a distance.
- The growth of a plant depends on the amount of sun it gets; the input is the amount of sun, and the output is the growth.

This idea is illustrated in Figure 7 with an input-output "machine."

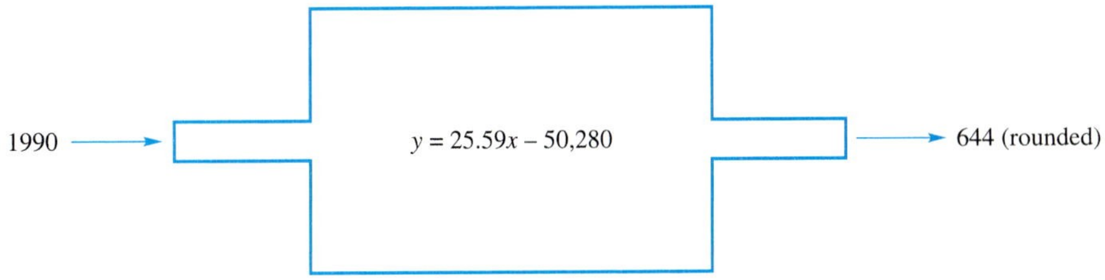

An input-output machine

Figure 7

3.1 EXERCISES

Fill in each blank with the correct response.

1. The symbol (x, y) ____________ (does/does not) represent an ordered pair, while the symbols $[x, y]$ and $\{x, y\}$ ____________ (do/do not) represent ordered pairs.

2. The ordered pair (3, 2) is a solution of the equation $2x - 5y =$ ______.

3. The point whose graph has coordinates $(-4, 2)$ is in quadrant ______.

4. The point whose graph has coordinates (0, 5) lies along the ______-axis.

5. The ordered pair (4, ______) is a solution of the equation $y = 3$.

6. The ordered pair (______, -2) is a solution of the equation $x = 6$.

Use the bar graph to respond to each statement or question. See Example 1.

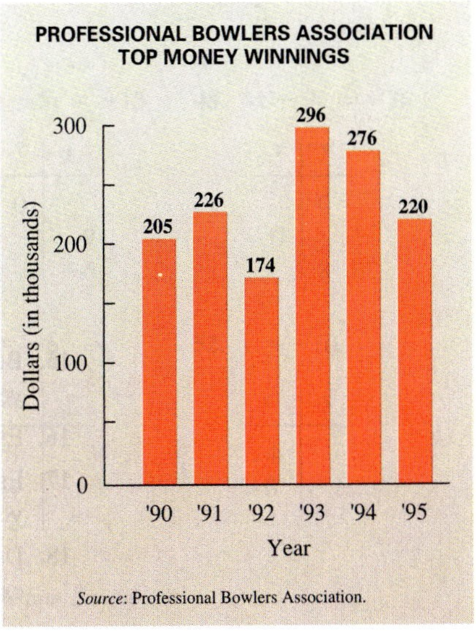

7. Between which pairs of consecutive years did the winnings increase? How much was the increase in each case?

8. How much did winnings decline between 1991 and 1992?

9. Which year had the greatest winnings?

10. Which year had the least winnings?

Fill in each blank with the word positive *or the word* negative.

The point with coordinates (x, y) is in

61. quadrant III if x is __________ and y is __________.

62. quadrant II if x is __________ and y is __________.

63. quadrant IV if x is __________ and y is __________.

64. quadrant I if x is __________ and y is __________.

Complete each table of values and then plot the ordered pairs. See Examples 5 and 6.

65. $x - 2y = 6$

x	y
0	
	0
2	
	−1

66. $2x - y = 4$

x	y
0	
	0
1	
	−6

67. $3x - 4y = 12$

x	y
0	
	0
−4	
	−4

68. $2x - 5y = 10$

x	y
0	
	0
−5	
	−3

69. $y + 4 = 0$

x	y
0	
5	
−2	
−3	

70. $x - 5 = 0$

x	y
	1
	0
	6
	−4

Solve each problem. See Example 7.

71. Suppose that it costs \$5000 to start up a business selling snow cones. Furthermore, it costs \$.50 per cone in labor, ice, syrup, and overhead. Then the cost to make x snow cones is given by y dollars, where $y = .50x + 5000$. Express as an ordered pair each of the following.

(a) When 100 snow cones are made, the cost is \$5050. (*Hint:* What does x represent? What does y represent?)

(b) When the cost is \$6000, the number of snow cones made is 2000.

72. It costs a flat fee of \$20 plus \$5 per day to rent a pressure washer. Therefore, the cost to rent the pressure washer for x days is given by $y = 5x + 20$, where y is in dollars. Express as an ordered pair each of the following.

(a) When the washer is rented for 5 days, the cost is \$45.

(b) I paid \$50 when I returned the washer, so I must have rented it for 6 days.

73. In statistics, ordered pairs are used to decide whether two quantities are related in such a way that one can be predicted from the other. These ordered pairs are plotted on a graph, called a *scatter diagram.*

Some major league baseball fans are concerned by the increase in time to complete a game. Make a scatter diagram by plotting the following ordered pairs of years since 1980 and minutes beyond two hours to complete a game: (5, 40), (7, 48), (9, 46), (11, 49), (13, 48), (15, 57). As shown in the figure, the horizontal axis is used

LENGTHS OF BASEBALL GAMES

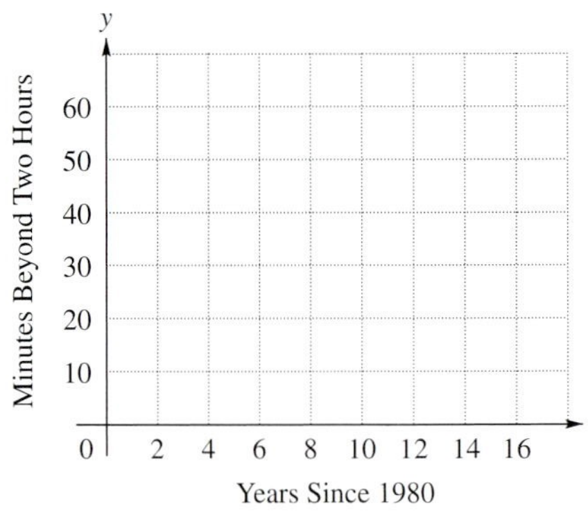

to represent the years since 1980, and the vertical axis represents the minutes beyond two hours.

(a) Graph these points on a similar grid. Do the points lie in an approximately linear pattern?

(b) Could the number of years since 1980 be used to predict the length of a game?

(c) What is the input? What is the output?

(*Source: The New York Times,* May 30, 1995, p. B9.)

74. The maximum benefit for the heart from exercising occurs if the heart rate is in the target heart rate zone. The line graph shows the upper limit of the zone (in beats per minute) for various ages.

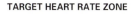

(a) Use the graph to estimate the ordered pairs in the form (x, y) that describe the upper limit of the target heart rate zone for ages 20, 40, 60, and 80.

(b) What is the input here? What is the output?

(*Source:* Robert V. Hockey, *Physical Fitness: The Pathway to Healthy Living,* Times Mirror/Mosby College Publishing, 1989, pp. 85–87.)

TARGET HEART RATE ZONE

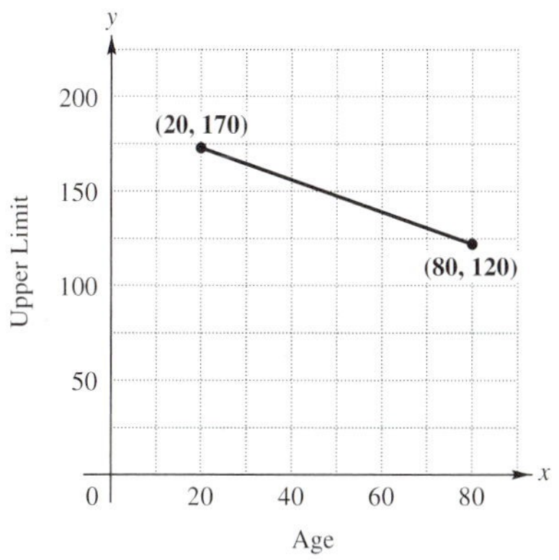

75. The lower limit of the target heart rate zone (see Exercise 74) is given in the accompanying graph of the equation $y = .7(220 - x)$, where the horizontal axis represents age and the vertical axis represents heart rate.

(a) Write ordered pairs (x, y) for ages 30, 40, 60, and 70 by estimating the y-values from the graph.

(b) What is the input here? What is the output?

TARGET HEART RATE ZONE

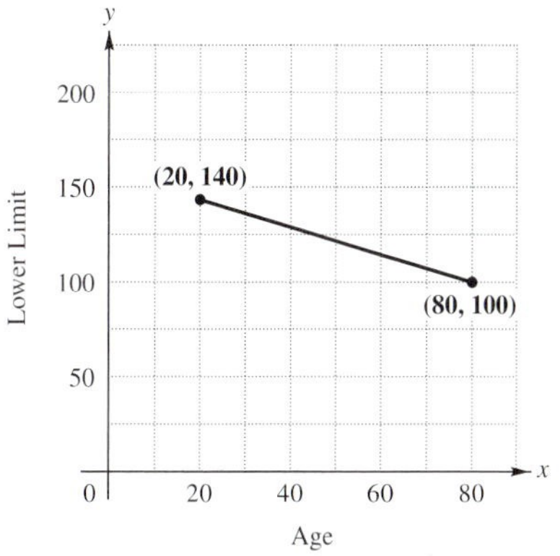

76. What is the heart rate zone for age 20? For age 40? (Refer to Exercises 74 and 75.)

77. Should the graph in Exercise 75 be used to estimate the lower limit of the target heart rate zone for ages below 20 or above 80? Why or why not?

OBJECTIVE 2 **Find intercepts.** In Figure 10 the graph intersects (crosses) the y-axis at (0, 3) and the x-axis at (2, 0). For this reason (0, 3) is called the **y-intercept** and (2, 0) is called the **x-intercept** of the graph. The intercepts are particularly useful for graphing linear equations, as in Example 1.

Finding Intercepts

We find the x-intercept by letting $y = 0$ in the given equation and solving for x.

We find the y-intercept by letting $x = 0$ in the given equation and solving for y.

EXAMPLE 2 **Finding Intercepts**

Find the intercepts for the graph of $2x + y = 4$. Draw the graph.

Find the y-intercept by letting $x = 0$; find the x-intercept by letting $y = 0$.

$$2x + y = 4 \qquad\qquad 2x + y = 4$$
$$2(0) + y = 4 \qquad\qquad 2x + 0 = 4$$
$$0 + y = 4 \qquad\qquad 2x = 4$$
$$y = 4 \qquad\qquad x = 2$$

The y-intercept is (0, 4). The x-intercept is (2, 0). The graph with the two intercepts shown in color is given in Figure 11. We get a third point as a check. For example, choosing $x = 1$ gives $y = 2$. These three ordered pairs are shown in the table with Figure 11. Plot (0, 4), (2, 0), and (1, 2) and draw a line through them. This line, shown in Figure 11, is the graph of $2x + y = 4$.

x	y
0	4
2	0
1	2

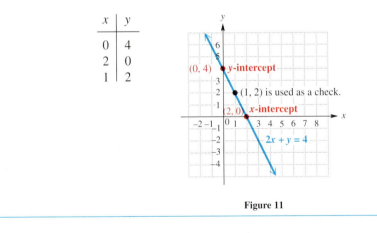

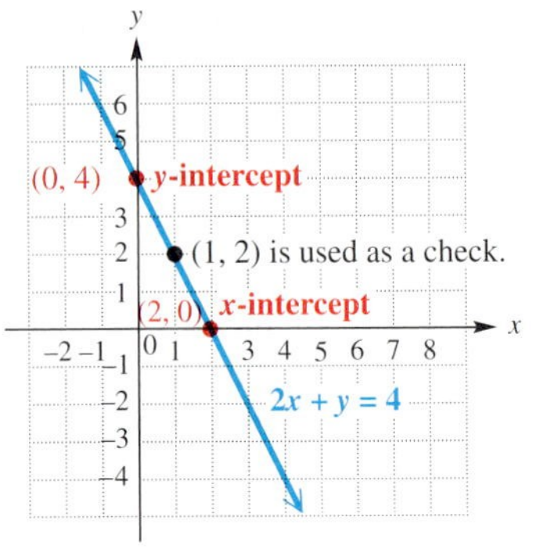

Figure 11

CAUTION When choosing x- or y-values to find ordered pairs to plot, be careful to choose so that the resulting points are not too close together. For example, using $(-1, -1)$, (0, 0), and (1, 1) may result in an inaccurate line. It is better to choose points where the x-values differ by at least 2.

OBJECTIVE 3 **Graph linear equations of the form $Ax + By = 0$.** In earlier examples, the x- and y-intercepts were used to help draw the graphs. This is not always possible, as the following examples show. Example 3 shows what to do when the x- and y-intercepts are the same point.

EXAMPLE 3 Graphing an Equation of the Form $Ax + By = 0$

Graph the linear equation $x - 3y = 0$.

If we let $x = 0$, then $y = 0$, giving the ordered pair (0, 0). Letting $y = 0$ also gives (0, 0). This is the same ordered pair, so choose two other values for x or y. Choosing 2 for y gives $x - 3 \cdot 2 = 0$, giving the ordered pair (6, 2). For a check point, we choose -6 for x getting -2 for y. This ordered pair, $(-6, -2)$, along with (0, 0) and (6, 2), was used to get the graph shown in Figure 12.

x	y
0	0
6	2
-6	-2

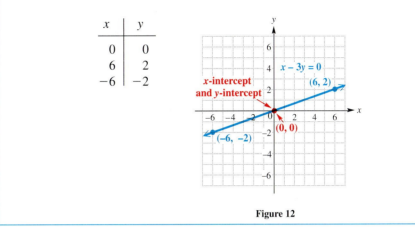

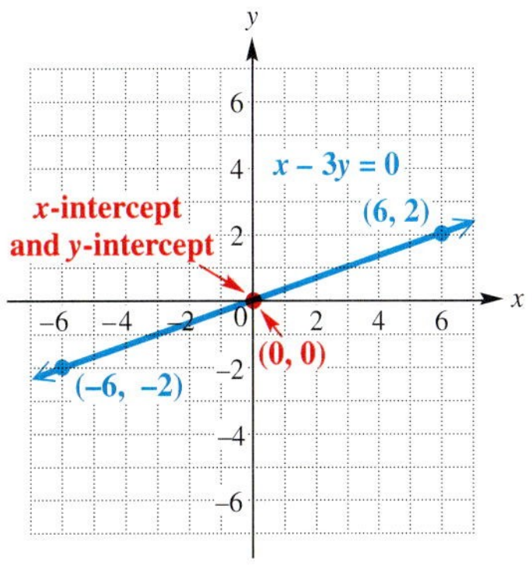

Figure 12

Example 3 can be generalized as follows.

Line through the Origin

If A and B are real numbers, the graph of a linear equation of the form

$$Ax + By = 0$$

goes through the origin (0, 0).

OBJECTIVE 4 Graph linear equations of the form $y = k$ or $x = k$. The equation $y = -4$ is a linear equation in which the coefficient of x is 0. (Write $y = -4$ as $0x + y = -4$ to see this.) Also, $x = 3$ is a linear equation in which the coefficient of y is 0. These equations lead to horizontal or vertical straight lines, as the next examples show.

EXAMPLE 4 Graphing an Equation of the Form $y = k$

Graph the linear equation $y = -4$.

As the equation states, for any value of x, y is always equal to -4. To get ordered pairs that are solutions of this equation, we choose any numbers for x, always using -4 for y. Three ordered pairs that satisfy the equation are shown in the table of values with Figure 13. Drawing a line through these points gives the horizontal line shown in Figure 13.

CONNECTIONS (CONTINUED)

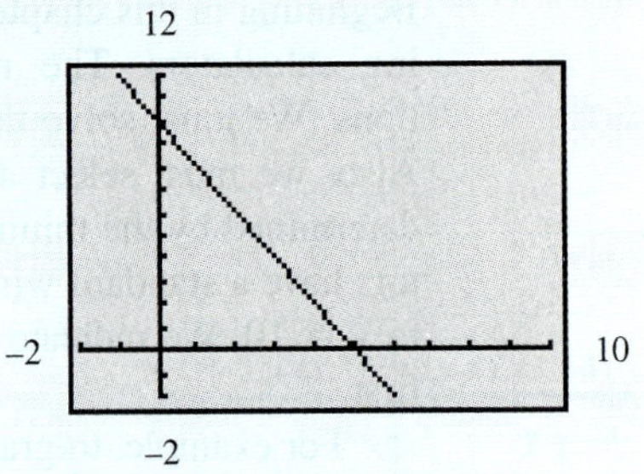

Third, use the graph to find the x-value of the point on the line where $y = 0$, the x-intercept. As shown in the figure, this occurs at $x = 5$. Verify this solution by substitution in the original equation.

FOR DISCUSSION OR WRITING

How would you rewrite these equations, with one side equal to 0, to input them into a graphing calculator for solution? (It is not necessary to clear parentheses or combine terms.)

1. $3x + 4 - 2x - 7 = 4x + 3$

2. $5x - 15 = 3(x - 2)$

The different forms of straight-line equations and the methods of graphing them are given in the following summary.

Graphing Straight Lines

Equation	*To Graph*	*Example*
$y = k$	Draw a horizontal line, through $(0, k)$.	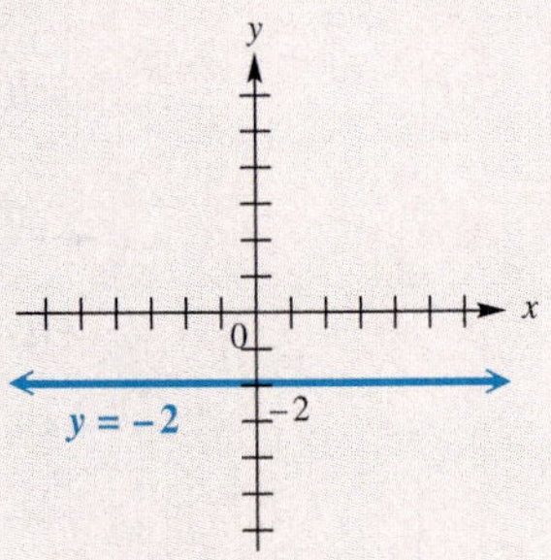
$x = k$	Draw a vertical line, through $(k, 0)$.	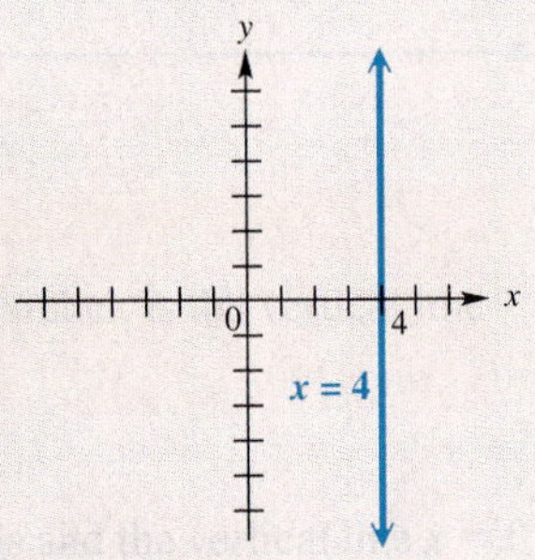

Graphing Straight Lines (continued)

Equation	To Graph	Example
$Ax + By = 0$	Graph goes through (0, 0). Get additional points that lie on the graph by choosing any value of x or y, except 0.	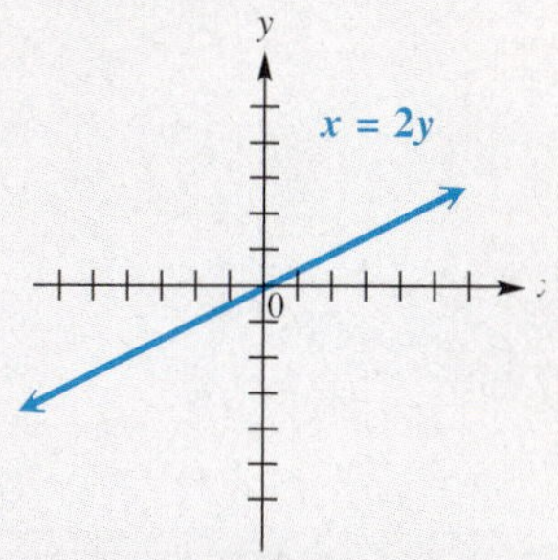
$Ax + By = C$ but not of the types above	Find any two points the line goes through. A good choice is to find the intercepts: let $x = 0$, and find the corresponding value of y; then let $y = 0$, and find x. As a check, get a third point by choosing a value of x or y that has not yet been used.	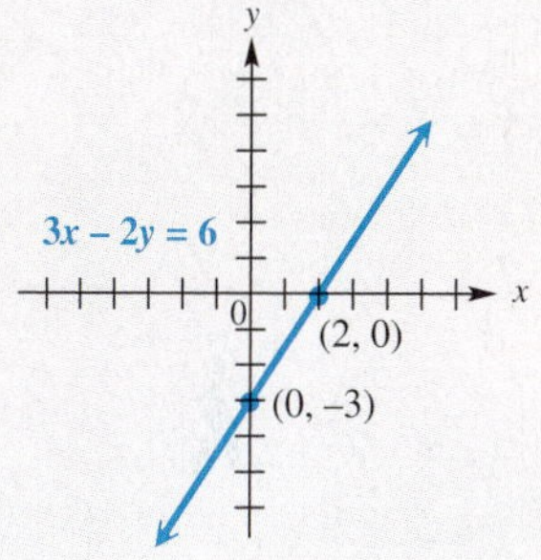

OBJECTIVE 5 **Define a function.** In the previous section, we saw several examples of relationships between two variables where each input value x produced an output value y. If each input x produces just *one* output y, we call the relationship a **function.** When a set of input values and a set of output values are related by a linear equation, $Ax + By = C$, each x-value corresponds to just one y-value, so linear equations define functions, unless $B = 0$. If $B = 0$, the equation becomes $Ax = C$ or $x = \frac{C}{A}$, where $\frac{C}{A}$ represents a real number. In this case, as we saw in Example 5, one input value x corresponds to an infinite number of output values y and the equation does not define a function.

EXAMPLE 6 **Deciding Whether a Set of Ordered Pairs Defines a Function**

Which of the following sets defines a function?

(a) $\{(1, -2), (2, 3), (0, 4), (4, 7)\}$

The input 1 is paired with the output -2, the input 2 is paired with the output 3, and so on. Every input value corresponds to exactly one output value, so the set defines a function.

(b) $\{(2, -4), (1, -1), (0, 0), (1, 1), (2, 4)\}$

Here, the input value 2 corresponds to the output values -4 and 4, so this set does not define a function. (What two output values correspond to the input value 1?)

(c) The ordered pairs that satisfy the equation $2x + 3y = 12$

Because this is a linear equation with $B \neq 0$, it defines a function.

(d) The ordered pairs that satisfy the equation $5x = 10$

In the context of two variables, when the coefficient of y is zero, as here, any value can be used for y. The only restriction imposed by the equation is that x must equal

2. Thus, (2, 0), (2, 5), (2, −1), and any ordered pair with $x = 2$ satisfies this relationship. This is not a function, however, because $x = 2$ is paired with more than one real number.

(e) The ordered pairs that satisfy the equation $4y = 16$

Every ordered pair in this set has a y-value of 4. Since there is no restriction here on x, every real number is paired with $y = 4$. For example, (−1, 4), (2.5, 4), (100, 4), and so on belong to this set, which defines a function, because every x corresponds to exactly one y, namely 4.

(f) The relationship between number of gallons needed to fill the gas tank and cost for a fill-up

This defines a function with ordered pairs of the form (number of gallons, cost). For a specific number of gallons, there is exactly one cost.

(g) The relationship between time spent traveling at a constant speed and number of miles traveled

Here the ordered pairs have the form (time, number of miles). A specific amount of time will correspond to exactly one number of miles, so this, too, defines a function.

Notice that in those sets that define functions, *no x-value appears more than once* (with a different y-value). The sets that were not functions had the same x as the input in at least two ordered pairs.

3.2 EXERCISES

In Exercises 1–6, match the information about the graphs with the linear equations in (a)–(e).

(a) $x = 5$ **(b)** $y = -3$ **(c)** $2x - 5y = 8$ **(d)** $x + 4y = 0$ **(e)** $3x + y = -4$

1. The graph of the equation has x-intercept (4, 0).

2. The graph of the equation has y-intercept (0, −4).

3. The graph of the equation goes through the origin.

4. The graph of the equation is a vertical line.

5. The graph of the equation is a horizontal line.

6. The graph of the equation goes through (9, 2).

Complete the given ordered pairs using the given equation. Then graph each equation by plotting the points and drawing a line through them. See Examples 1 and 2.

7. $y = -x + 5$
(0,), (, 0), (2,)

8. $y = x - 2$
(0,), (, 0), (5,)

9. $y = \frac{2}{3}x + 1$
(0,), (3,), (−3,)

10. $y = -\frac{3}{4}x + 2$
(0,), (4,), (−4,)

11. $3x = -y - 6$
(0,), (, 0), $\left(-\frac{1}{3}, \quad\right)$

12. $x = 2y + 3$
(, 0), (0,), $\left(\quad , \frac{1}{2}\right)$

Find the x-intercept and the y-intercept for the graph of each equation. See Example 2.

13. $2x - 3y = 24$ **14.** $-3x + 8y = 48$ **15.** $x + 6y = 0$ **16.** $3x - y = 0$

17. What is the equation of the x-axis?

18. What is the equation of the y-axis?

19. A student attempted to graph $4x + 5y = 0$ by finding intercepts. She first let $x = 0$ and found y; then she let $y = 0$ and found x. In both cases, the resulting point was $(0, 0)$. She knew that she needed at least two different points to graph the line, but was unsure what to do next since finding intercepts gave her only one point. How would you explain to her what to do next?

20. Write a paragraph summarizing how to graph a linear equation in two variables.

Graph each linear equation. See Examples 1–5.

21. $x = y + 2$	**22.** $x = -y + 6$	**23.** $x - y = 4$	**24.** $x - y = 5$
25. $2x + y = 6$	**26.** $-3x + y = -6$	**27.** $3x + 7y = 14$	**28.** $6x - 5y = 18$
29. $y - 2x = 0$	**30.** $y + 3x = 0$	**31.** $y = -6x$	**32.** $y = 4x$
33. $y + 1 = 0$	**34.** $y - 3 = 0$	**35.** $x = -2$	**36.** $x = 4$

TECHNOLOGY INSIGHTS (EXERCISES 37–42)

In each exercise below, a calculator-generated graph of a linear equation in one variable with one side equal to 0 is shown. Accompanying the graph is the equation itself, where y is expressed in terms of x on the left side. Solve the equation using the methods of Section 2.2, and show that the solution you get is the same as the x-intercept (labeled "zero") on the calculator screen.

37. $8 - 2(3x - 4) - 2x = 0$

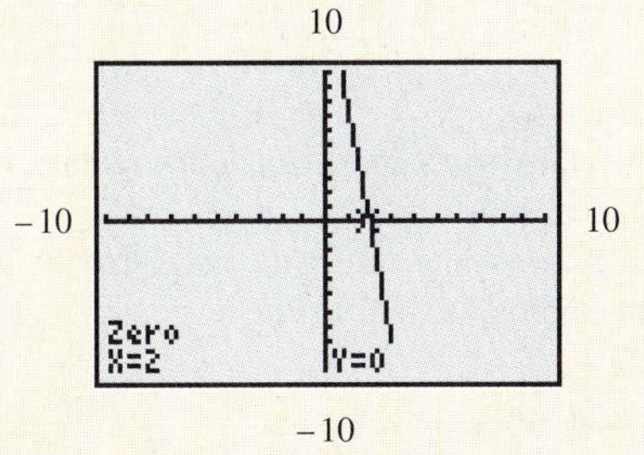

38. $5(2x - 1) - 4(2x + 1) - 7 = 0$

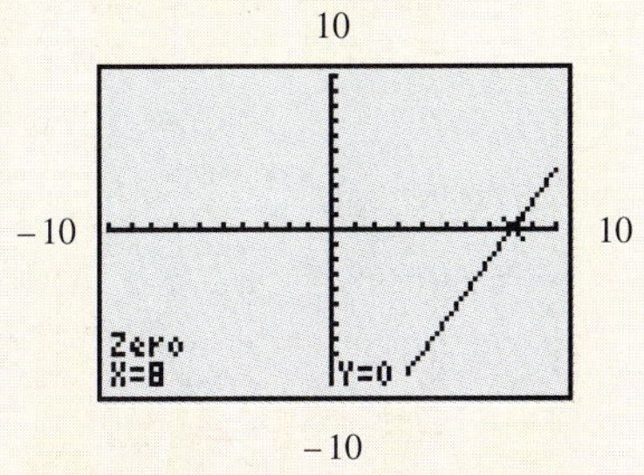

39. $.6x - .1x - x + 2.5 = 0$

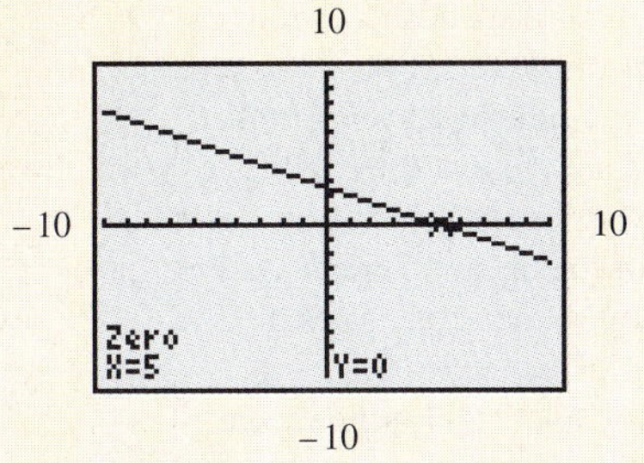

40. $-\frac{2}{7}x + 2x - \frac{1}{2}x - \frac{17}{2} = 0$

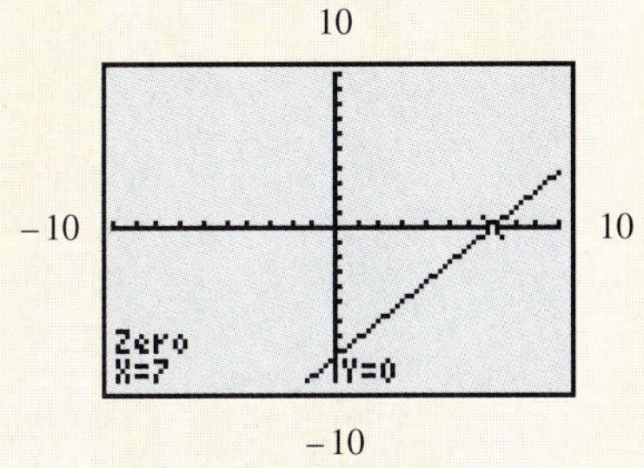

41. Use the results of Exercises 37–40 to explain how the x-intercept of the graph of an equation in two variables corresponds to the solution of an equation in one variable.

42. A horizontal line has no x-intercept. If you try to solve $5x - (3x + 2x) + 4 = 0$, you get no solution. What would the graph of $y = 5x - (3x + 2x) + 4$ look like on a graphing calculator?

Solve each problem.

43. The number of master's degrees earned in business management increased during the years 1971–1994 as shown in the figure. If $x = 0$ represents 1970, $x = 5$ represents 1975, $x = 10$ represents 1980, and so on, the number of master's degrees can be approximated by

$$y = 2.81x + 24.1,$$

where y is in thousands. (This is a *linear model* for the data.)

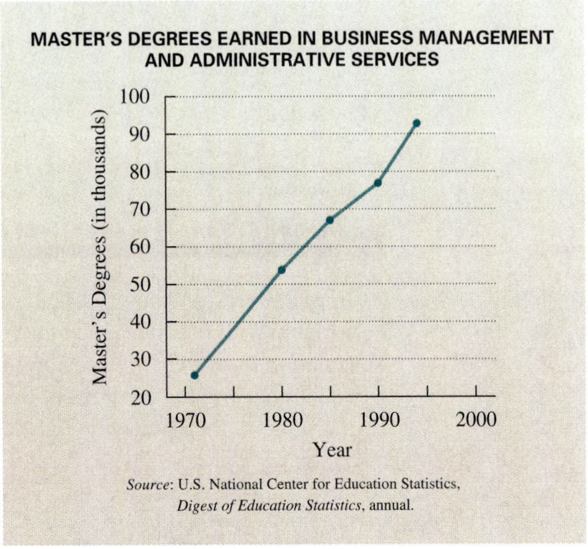

(a) Use the equation to approximate the number of such degrees in the years 1980, 1985, 1990, and 1994.

(b) Estimate the y-values from the graph for the same years.

(c) Are the approximations using the equation close to the values you read from the graph?

44. Sporting goods sales (in billions of dollars) from 1988–1996 are approximated by the equation

$$y = 1.625x + 40.75,$$

where $x = 0$ corresponds to 1986, $x = 2$ corresponds to 1988, and so on. Sales for even-numbered years in that time period are plotted in the accompanying figure, which also shows the graph of the linear equation. As the graph indicates, the actual sales in 1992 (when the U.S. economy was depressed) of about \$47 billion were less than the sales approximated by the equation.

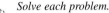

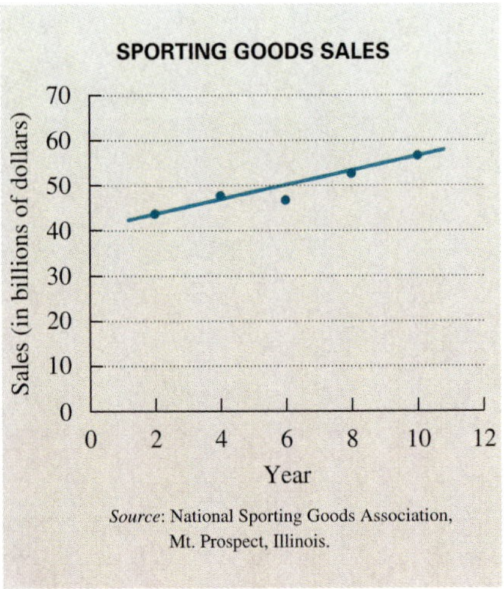

(a) Use the *equation* to approximate the sales in each of the even-numbered years.

(b) Does this equation define a function? Why or why not?

45. The height y of a woman (in centimeters) is a function of the length of her radius bone x (from the wrist to the elbow) and is defined by $y = 73.5 + 3.9x$. Estimate the heights of women with radius bones of the following lengths.

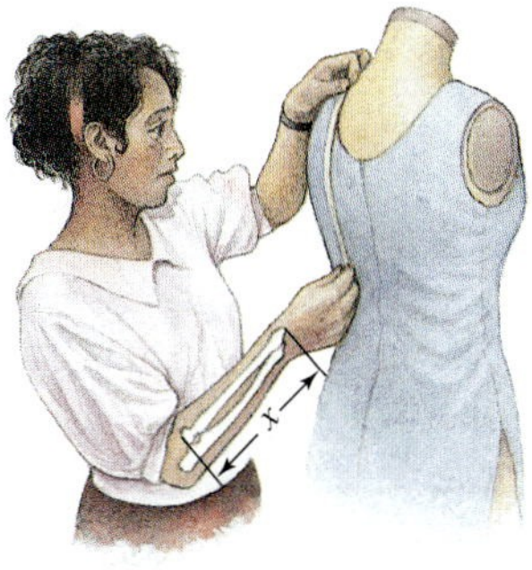

(a) 23 centimeters
(b) 25 centimeters
(c) 20 centimeters
(d) Graph $y = 73.5 + 3.9x$.

46. As a rough estimate, the weight of a man taller than about 60 inches is a linear function approximated by $y = 5.5x - 220$, where x is the height of the person in inches, and y is the weight in pounds. Estimate the weights of men whose heights are as follows.
(a) 62 inches **(b)** 64 inches
(c) 68 inches **(d)** 72 inches
(e) Graph $y = 5.5x - 220$.

47. The graph shows that the value of a certain automobile over its first five years is a function of the year. Use the graph to estimate the depreciation (loss in value) during the following years.

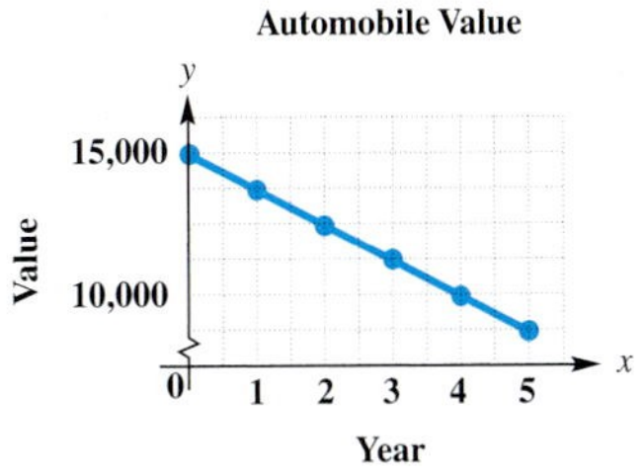

(a) First **(b)** Second **(c)** Fifth
(d) What is the total depreciation over the 5-year period?

48. The demand for an item is a function of its price. As price goes up, demand goes down. On the other hand, when price goes down, demand goes up. Suppose the demand for a certain Beanie Baby is 1000 when its price is \$30 and 8000 when it costs \$15.
(a) Let x be the price and y be the demand for the Beanie Baby. Graph the two given pairs of prices and demands.
(b) Assume the relationship is linear. Draw a line through the two points from part (a). From your graph estimate the demand if the price drops to \$10.
(c) Use the graph to estimate the price if the demand is 4000.

Decide whether each set of ordered pairs defines a function. See Example 6.

49. $\{(0, 5), (2, 3), (4, 1), (6, -1), (8, -3)\}$

50. $\{(1, 3), (2, 3), (3, 3), (4, 3)\}$

51. The ordered pairs that satisfy $-x + 2y = 9$

52. The ordered pairs that satisfy $y = 4$

53. The ordered pairs that satisfy $x = 8$

54. The ordered pairs that satisfy $5x + y = 7$

55. The relationship in Exercise 43 between year and number of master's degrees

56. The relationship in Exercise 44 between year and sporting goods sales

3.3 The Slope of a Line

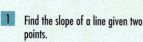

OBJECTIVES

1. Find the slope of a line given two points.
2. Find the slope from the equation of a line.
3. Use the slope to determine whether two lines are parallel, perpendicular, or neither.

FOR EXTRA HELP

SSG Sec. 3.3
SSM Sec. 3.3
Pass the Test Software
InterAct Math Tutorial Software
Video 5

When two variables are related in such a way that the value of one depends on the value of the other, we can form ordered pairs of corresponding numbers. For example, if $x + y = 7$, then when $x = 2$, $y = 5$, and when $x = -3$, $y = 10$. We write these pairs as (2, 5) and (−3, 10), respectively, with the understanding that the first number in the ordered pair represents x and the second number represents y. To indicate two nonspecific ordered pairs that satisfy a particular equation relating x and y, we use *subscript notation.* We write the pairs as (x_1, y_1) and (x_2, y_2). (Read x_1 as "x-sub-one" and x_2 as "x-sub-two.")

We can graph a straight line if at least two different points on the line are known. A line also can be graphed by using just one point on the line if the "steepness" of the line is known.

OBJECTIVE 1 Find the slope of a line given two points. One way to measure the steepness of a line is to compare the vertical change in the line (the rise) to the horizontal change (the run) while moving along the line from one fixed point to another. This measure of steepness is called the *slope* of the line.

Figure 15 shows a line with the points (x_1, y_1) and (x_2, y_2). As we move along the line from the point (x_1, y_1) to the point (x_2, y_2), y changes by $y_2 - y_1$ units. This is the vertical change. Similarly, x changes by $x_2 - x_1$ units, the horizontal change. The ratio of the change in y to the change in x gives the slope of the line. We usually denote slope with the letter m. The slope of a line is defined as follows.

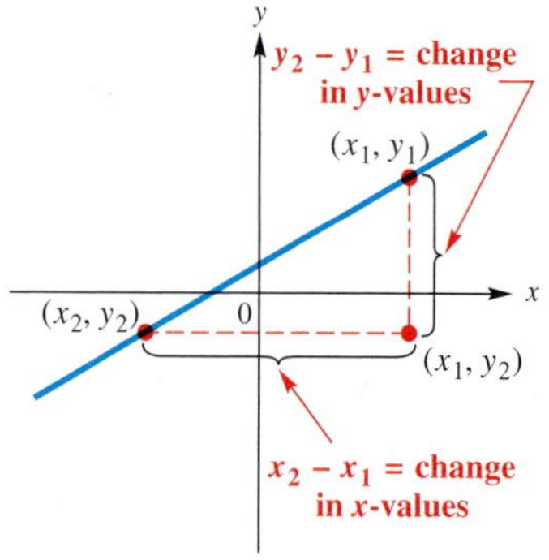

Figure 15

Slope Formula

The **slope** of the line through the points (x_1, y_1) and (x_2, y_2) is

$$m = \frac{\text{change in } y}{\text{change in } x} = \frac{y_2 - y_1}{x_2 - x_1} \quad \text{if } x_1 \neq x_2.$$

The slope of a line tells how fast y changes for each unit of change in x; that is, the slope gives the ratio of the change in y to the change in x. The change in y is called the **rise,** and the change in x is called the **run.**

CONNECTIONS

The idea of slope is used in many everyday situations. For example, because $10\% = \frac{1}{10}$, a highway with a 10% grade (or slope) rises one meter for every 10 horizontal meters. The highway sign shown below is used to warn of a downgrade ahead that may be long or steep. Architects specify the pitch of a roof using slope; a $\frac{5}{12}$ roof means that the roof rises 5 feet for every 12 feet in the horizontal direction. The slope of a stairwell also indicates the ratio of the vertical rise to the horizontal run. The slope of the stairs in the figure is $\frac{8}{14}$.

Hill

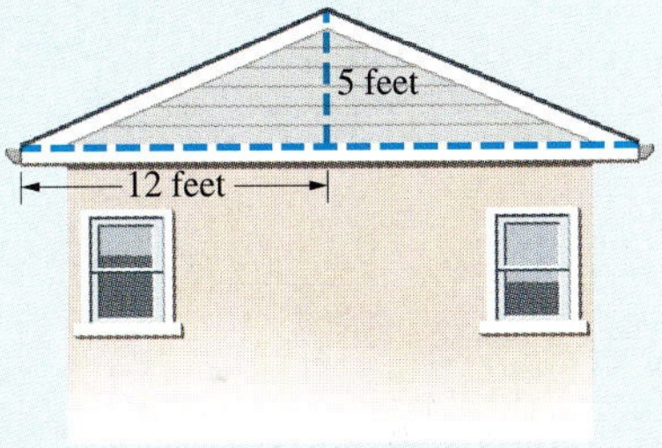

$\frac{5}{12}$ roof pitch

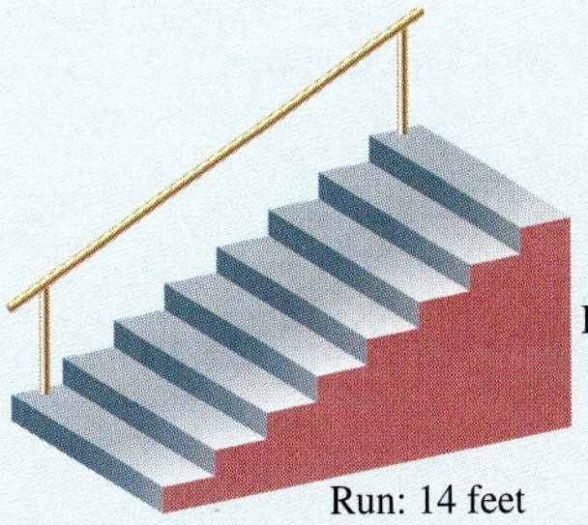

Slope of a stairwell

FOR DISCUSSION OR WRITING

Describe some other everyday examples of slope.

EXAMPLE 1 Finding the Slope of a Line

Find the slope of each of the following lines.

(a) The line through $(-4, 7)$ and $(1, -2)$
Use the definition of slope. Let $(-4, 7) = (x_2, y_2)$ and $(1, -2) = (x_1, y_1)$. Then

$$\text{slope} = \frac{\text{change in } y}{\text{change in } x}$$

$$m = \frac{y_2 - y_1}{x_2 - x_1}$$

$$= \frac{7 - (-2)}{-4 - 1} = \frac{9}{-5} = -\frac{9}{5}.$$

As Figure 16 shows, this line has a slope of $-\frac{9}{5}$ (which can also be written as $\frac{-9}{5}$ or $\frac{9}{-5}$). One way of interpreting this is that the line drops vertically 9 units for a horizontal change of 5 units to the right.

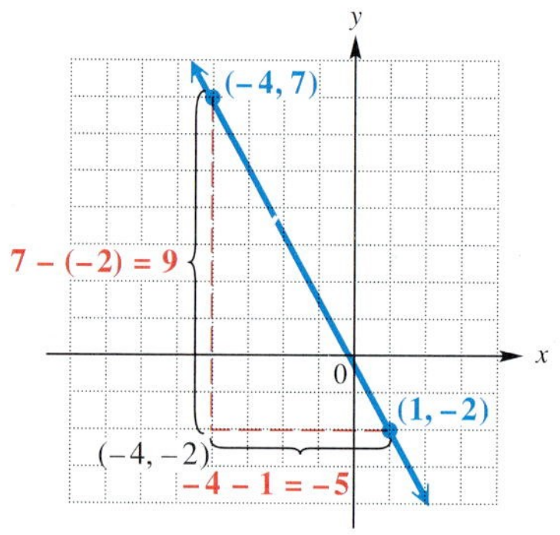

Figure 16

(b) The line through $(12, -5)$ and $(-9, -2)$

$$m = \frac{-5 - (-2)}{12 - (-9)} = \frac{-3}{21} = -\frac{1}{7}$$

The same slope is found by subtracting in reverse order.

$$\frac{-2 - (-5)}{-9 - 12} = \frac{3}{-21} = -\frac{1}{7}$$

CAUTION It makes no difference which point is (x_1, y_1) or (x_2, y_2); however, it is important to be consistent. Start with the x- and y-values of one point (either one) and subtract the corresponding values of the other point.

In Example 1(a) the slope is negative and the corresponding line in Figure 16 falls from left to right. As Figure 17(a) shows, this is generally true of lines with negative slopes. Lines with positive slopes go up (rise) from left to right, as shown in Figure 17(b).

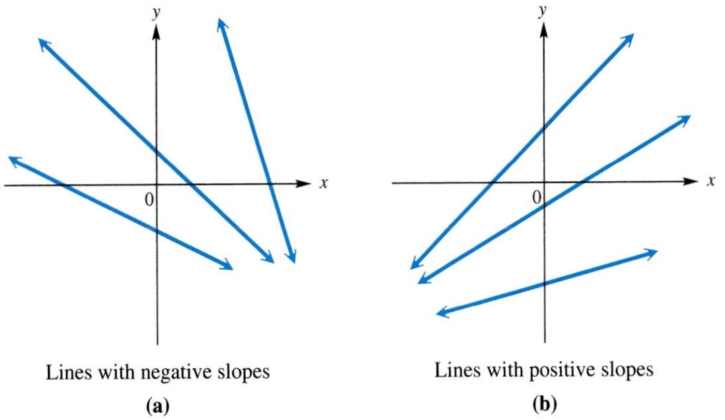

Figure 17

Positive and Negative Slopes

A line with positive slope rises from left to right.

A line with negative slope falls from left to right.

EXAMPLE 2 **Finding the Slope of a Horizontal Line**

Find the slope of the line through $(-8, 4)$ and $(2, 4)$.

Use the definition of slope.

$$m = \frac{4 - 4}{-8 - 2} = \frac{0}{-10} = 0$$

As shown in Figure 18, the line through these two points is horizontal, with equation $y = 4$. *All horizontal lines have a slope of 0,* since the difference in y-values is always 0.

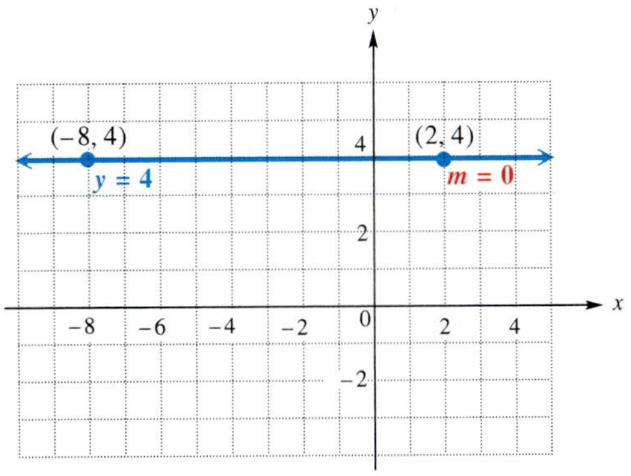

Figure 18

EXAMPLE 3 Finding the Slope of a Vertical Line

Find the slope of the line through (6, 2) and (6, −9).

$$m = \frac{2 - (-9)}{6 - 6} = \frac{11}{0} \quad \text{Undefined}$$

Since division by 0 is undefined, the slope is undefined. The graph in Figure 19 shows that the line through these two points is vertical, with equation $x = 6$. All points on a vertical line have the same x-value, so *the slope of any vertical line is undefined.*

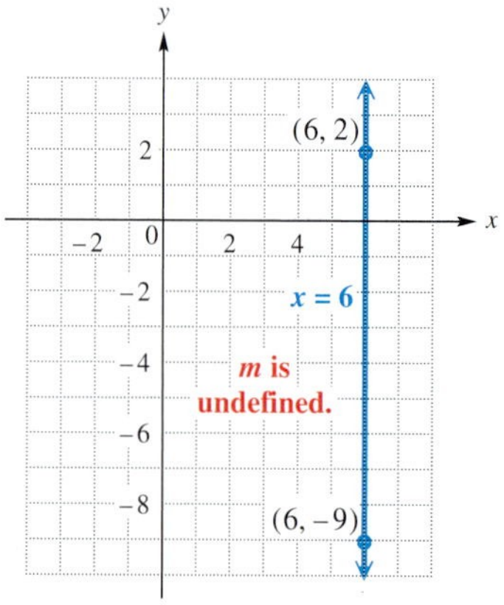

Figure 19

Slopes of Horizontal and Vertical Lines

Horizontal lines, with equations of the form $y = k$, **have slope 0.**

Vertical lines, with equations of the form $x = k$, **have undefined slope.**

OBJECTIVE 2 **Find the slope from the equation of a line.** The slope of a line also can be found directly from its equation. For example, the slope of the line

$$y = -3x + 5$$

can be found using any two points on the line. Get two points by choosing two different values of x, say −2 and 4, and finding the corresponding y-values.

If $x = -2$:	If $x = 4$:
$y = -3(-2) + 5$	$y = -3(4) + 5$
$y = 6 + 5$	$y = -12 + 5$
$y = 11.$	$y = -7.$

The ordered pairs are (−2, 11) and (4, −7). Now use the slope formula to find the slope.

$$m = \frac{11 - (-7)}{-2 - 4} = \frac{18}{-6} = -3$$

The slope, −3, is the same number as the coefficient of x in the equation $y = -3x + 5$. It can be shown that this always happens, *as long as the equation is solved for y.* This fact is used to find the slope of a line from its equation.

Finding the Slope of a Line from its Equation

Step 1 Solve the equation for y.

Step 2 The slope is given by the coefficient of x.

EXAMPLE 4 Finding Slope from an Equation

Find the slope of each of the following lines.

(a) $2x - 5y = 4$

Solve the equation for y.

$$2x - 5y = 4$$

$$-5y = -2x + 4 \quad \text{Subtract } 2x \text{ from each side.}$$

$$y = \frac{2}{5}x - \frac{4}{5} \quad \text{Divide each side by } -5.$$

The slope is given by the coefficient of x, so the slope is $m = \frac{2}{5}$.

(b) $8x + 4y = 1$

Solve the equation for y.

$$8x + 4y = 1$$

$$4y = -8x + 1 \quad \text{Subtract } 8x \text{ from each side.}$$

$$y = -2x + \frac{1}{4} \quad \text{Divide each side by } 4.$$

The slope of this line is given by the coefficient of x, -2.

OBJECTIVE 3 Use the slope to determine whether two lines are parallel, perpendicular, or neither. Two lines in a plane that never intersect are **parallel.** We use slopes to tell whether two lines are parallel. For example, Figure 20 shows the graph of $x + 2y = 4$ and the graph of $x + 2y = -6$. These lines appear to be parallel. Solve for y to find that both $x + 2y = 4$ and $x + 2y = -6$ have a slope of $-\frac{1}{2}$. Nonvertical parallel lines always have equal slopes.

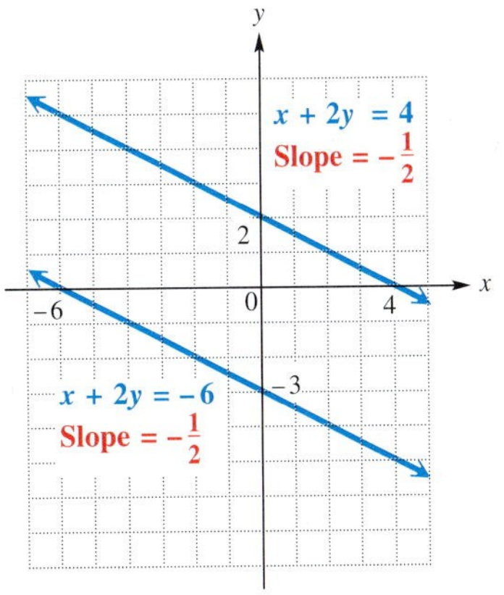

Figure 20

Figure 21 shows the graph of $x + 2y = 4$ and the graph of $2x - y = 6$. These lines appear to be **perpendicular** (meet at a 90° angle). Solving for y shows that the slope of $x + 2y = 4$ is $-\frac{1}{2}$, while the slope of $2x - y = 6$ is 2. The product of $-\frac{1}{2}$ and 2 is

$$-\frac{1}{2}(2) = -1.$$

This is true in general; the product of the slopes of two perpendicular lines, neither of which is vertical, is always -1. This means that the slopes of perpendicular lines are negative reciprocals: if one slope is the nonzero number a, the other is $-\frac{1}{a}$.

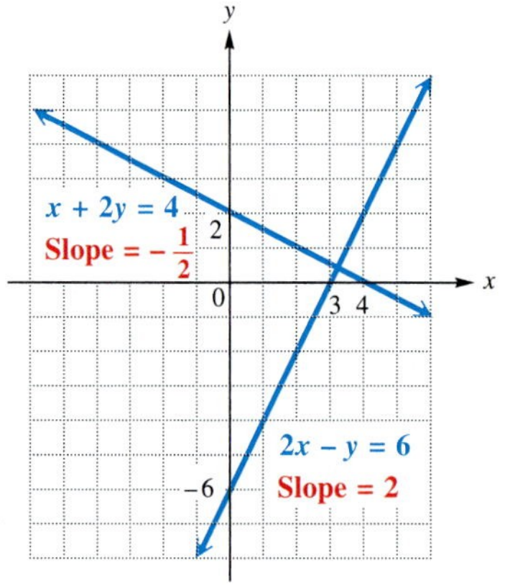

Figure 21

Parallel and Perpendicular Lines

Two nonvertical lines with the same slope are parallel; two perpendicular lines, neither of which is vertical, have slopes that are negative reciprocals of each other.

EXAMPLE 5 Deciding Whether Two Lines Are Parallel or Perpendicular

Decide whether the lines are *parallel, perpendicular,* or *neither.*

(a) $x + 2y = 7$
$-2x + y = 3$

Find the slope of each line by first solving each equation for y.

$$\begin{aligned} x + 2y &= 7 \\ 2y &= -x + 7 \\ y &= -\frac{1}{2}x + \frac{7}{2} \end{aligned} \qquad\qquad \begin{aligned} -2x + y &= 3 \\ y &= 2x + 3 \end{aligned}$$

Slope: $-\frac{1}{2}$ Slope: 2

Since the slopes are not equal, the lines are not parallel. Check the product of the slopes: $-\frac{1}{2}(2) = -1$. The two lines are perpendicular because the product of their slopes is -1, indicating that the slopes are negative reciprocals.

(b) $3x - \ \ y = 4$
$6x - 2y = 9$
Find the slopes. Both lines have a slope of 3, so the lines are parallel.

(c) $4x + 3y = 6$
$2x - \ \ y = 5$
Here the slopes are $-\frac{4}{3}$ and 2. These two straight lines are neither parallel nor perpendicular.

CONNECTIONS

Because the viewing window of a graphing calculator is a rectangle, the graphs of perpendicular lines will not appear perpendicular unless appropriate intervals are used for x and y. Graphing calculators usually have a key to select a "square" window automatically. In a square window, the x-interval is about 1.5 times the y-interval. The equations used in Figure 21 are graphed with the standard (nonsquare) window and then with a square window in the screens below.

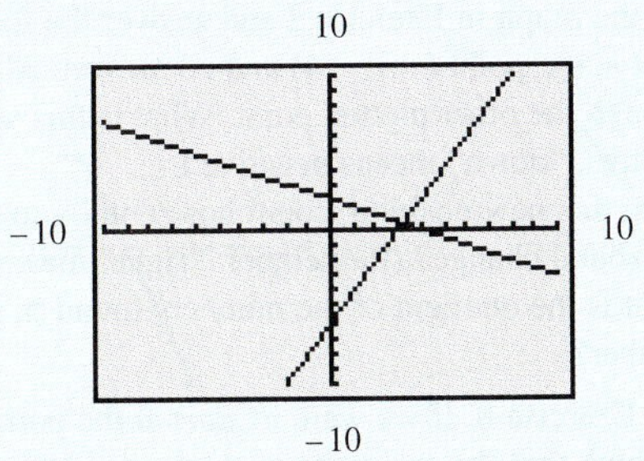

A standard (nonsquare) window
[−10, 10] by [−10, 10]
Lines do not appear perpendicular.

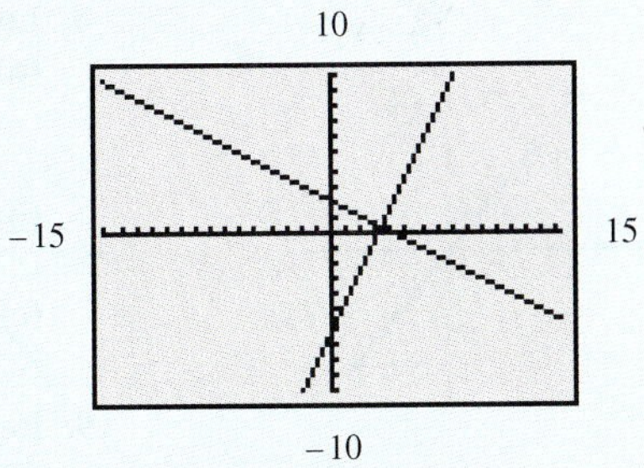

A square window
[−15, 15] by [−10, 10]
Lines appear perpendicular.

3.3 EXERCISES

1. What is meant by "rise"? What is meant by "run"?

Use the coordinates of the indicated points to find the slope of each line. See Example 1.

2.

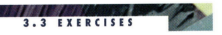

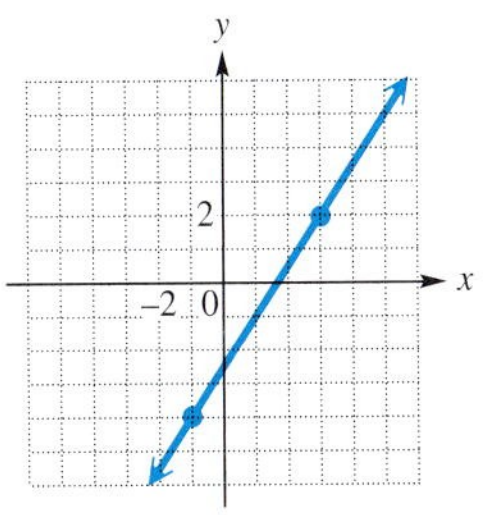

3.

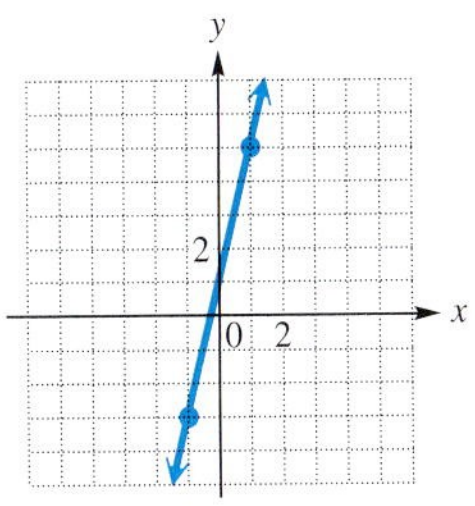

For Exercises 49 and 50, you may wish to refer to the Connections box preceding Example 1.

49. What is the slope (or pitch) of this roof?

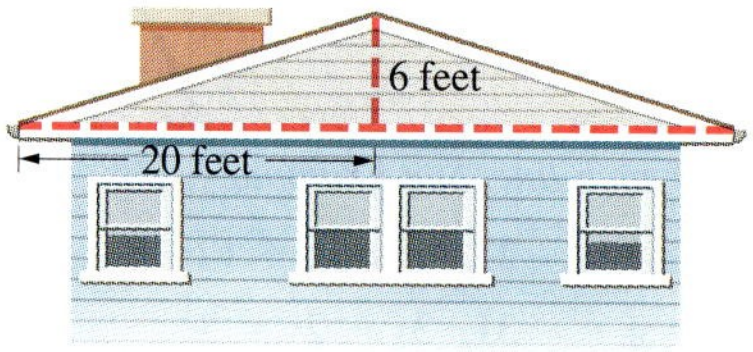

50. What is the slope (or grade) of this hill?

RELATING CONCEPTS (EXERCISES 51–56)

Figure A depicts public school enrollment (in thousands) in grades 9–12 in the United States. Figure B gives the (average) number of public school students per computer.

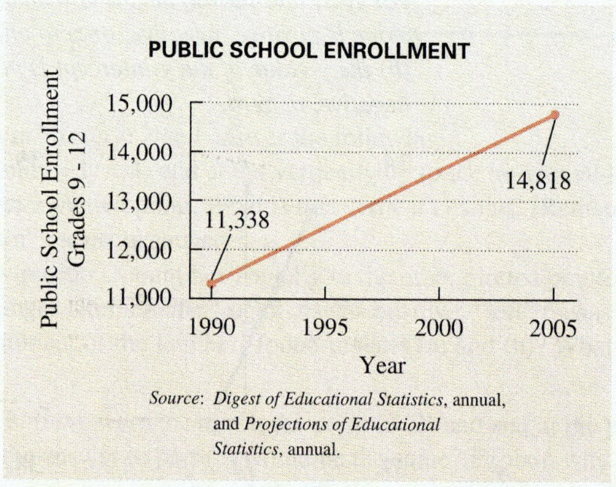

Figure A

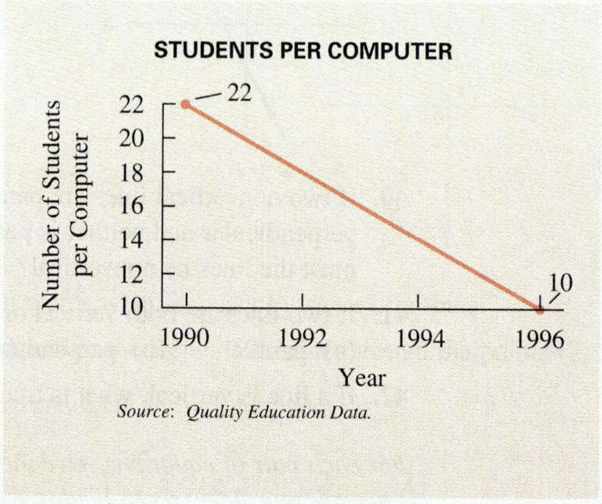

Figure B

Work Exercises 51–56 in order.

51. Use the ordered pairs (1990, 11,338) and (2005, 14,818) to find the slope of the line in Figure A.

RELATING CONCEPTS (EXERCISES 51–56) (CONTINUED)

52. The slope of the line in Figure A is ______________ (positive/negative). This means that during the period represented, enrollment ______________ (increased/decreased).

53. The slope of a line represents its *rate of change.* Based on Figure A, what was the increase in students *per year* during the period shown?

54. Use the given ordered pairs to find the slope of the line in Figure B.

55. The slope of the line in Figure B is ______________ (positive/negative). This means that during the period represented, the number of students per computer ______________ (increased/decreased).

56. Based on Figure B, what was the decrease in students per computer *per year* during the period shown?

Did you make the connection between the sign of the slope of the line and the increase or decrease in the quantity represented by y?

Work each business-related problem.

57. The table gives the number of shopping centers in the years 1991–1996. The equation

$$y = .797x - 1549,$$

where x represents the year and y is the number of shopping centers in thousands, approximates these data quite well.

Number of Shopping Centers (in thousands)

Year	1991	1992	1993	1994	1995	1996
Number of Centers	38.0	39.0	39.6	40.4	41.2	42.1

Source: International Council of Shopping Centers.

(a) Give three ordered pairs for this *equation* and indicate the input and output numbers for each.

(b) Compare the y-values you find with the values given for the number of centers in the table. Are the approximations close?

58. The line graph shows the points for the ordered pairs from the table in Exercise 57. A straight line graph of the equation in Exercise 57 is also shown. Except for one point, the straight line is almost the same as the line graph. Which point is farthest from the straight line?

59. The growth in retail square footage, in billions, is shown in the line graph. This graph looks like a straight line. If the change in square footage each year is the same, then it is a straight line. Find the change in square footage for the years shown in the graph. (*Hint:* To find the change in square footage from 1991 to 1992, subtract the y-value for 1991 from the y-value for 1992.) Is the graph a straight line?

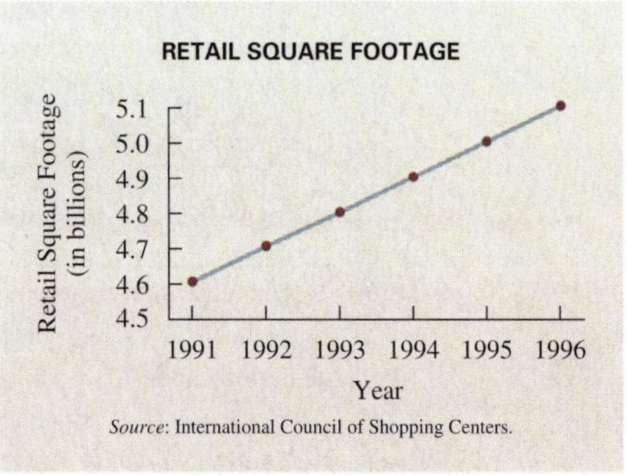

60. Find the slope of the line in Exercise 59 by using any two of the points shown on the line. How does the slope compare with the yearly change in square footage?

TECHNOLOGY INSIGHTS (EXERCISES 61–66)

61. Two views of the same line are shown in the accompanying calculator screen, along with coordinates of two points displayed at the bottoms. What is the slope of this line?

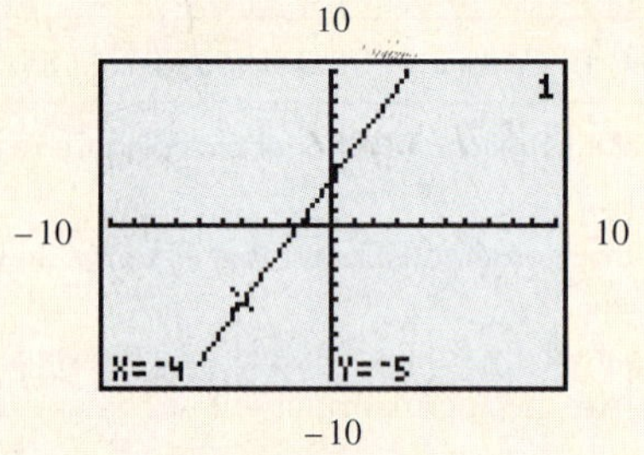

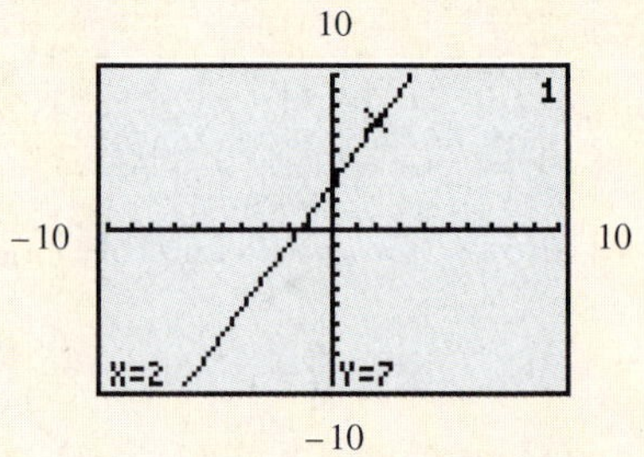

62. Repeat Exercise 61 for the line shown here.

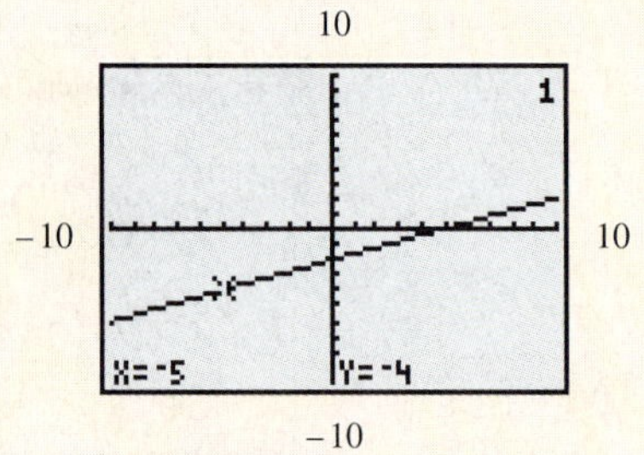

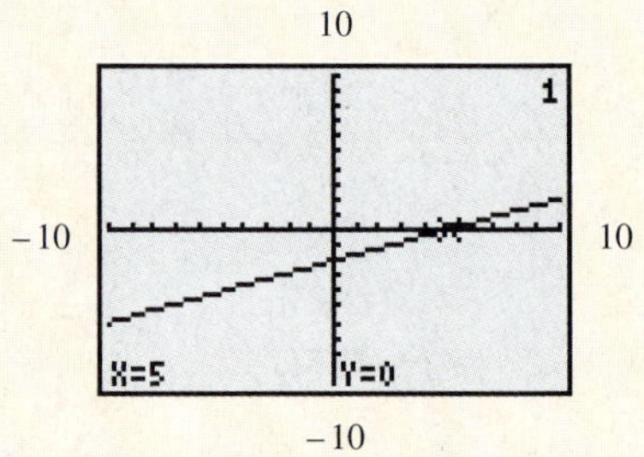

TECHNOLOGY INSIGHTS (EXERCISES 61–66) (CONTINUED)

Some graphing calculators have the capability of displaying a table of points for a graph. The table shown here gives several points that lie on a line designated Y_1.

X	Y1
-12	-.8
-10	0
-8	.8
-6	1.6
-4	2.4
-2	3.2
0	4

X= -12

63. Use any pair of points displayed to find the slope of the line.
64. What is the x-intercept of the line?
65. What is the y-intercept of the line?
66. Which one of the two lines shown is the graph of Y_1?

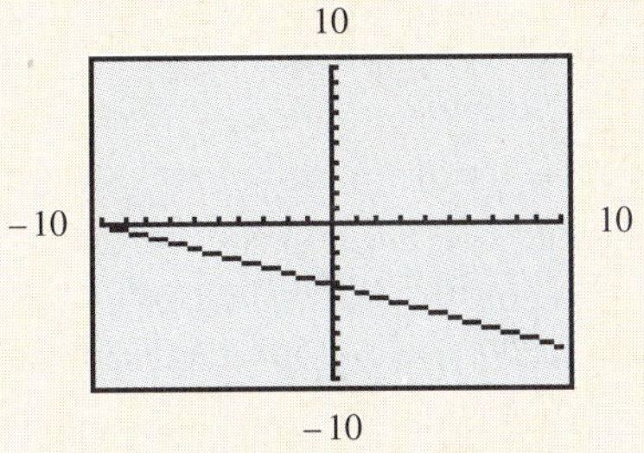

(a)

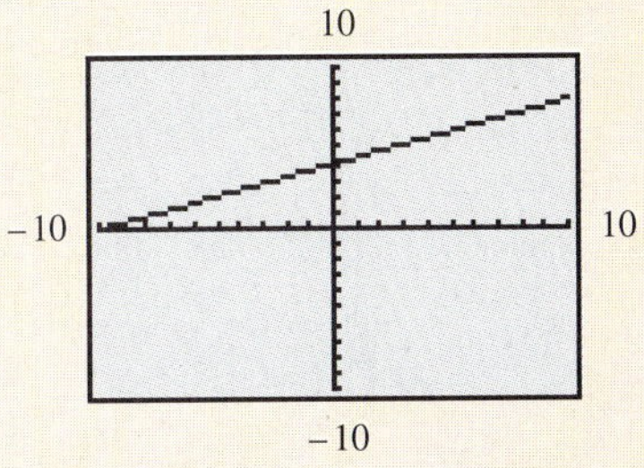

(b)

CHAPTER 3 GROUP ACTIVITY

Determining Business Profit

Objective: Use equations, tables, and graphs to make business decisions.

Graphs are used by businesses to analyze and make decisions. This activity will explore such a process.

The Parent Teacher Organization of a school decides to sell bags of popcorn as a fund-raiser. The parents want to estimate their profit. Costs include \$14 for popcorn (enough for 680 bags) and \$7 for bags to hold the popcorn.

A. Write a profit formula.

1. From the information above, determine the total costs. (Assume the organization is buying enough supplies for 680 bags of popcorn.)
2. If the bags sell for \$.25 each, write an expression for the total sales of n bags of popcorn.
3. Since Profit = Sales − Cost, write an equation that represents the profit P for n bags of popcorn sold.
4. Work in pairs. One person should use the profit equation from above to complete the table on the next page. The other person should decide on an appropriate scale and graph the profit equation.

(continued)

n	P
0	
	0
100	

B. Choose a different price for a bag of popcorn (between \$.20 and \$.75).

1. Write the profit equation for this cost.
2. Switch roles, that is, if you drew the graph in part A, now make a table of values for this equation. Have your partner graph the profit equation on the same coordinate system you used in part A.

C. Compare your findings and answer the following questions.

1. What is the break-even point (that is, when profits are 0 or sales equal costs) for each equation?
2. What does it mean if you end up with a negative value for P?
3. If you sell all 680 bags, what will your profits be for the two different prices? Explain how you would estimate this from your graph.
4. What are the advantages and/or disadvantages of charging a higher price?
5. Together, decide on the price you would charge for a bag of popcorn. Explain why you chose this price.

CHAPTER 3 SUMMARY

KEY TERMS

3.1 **linear equation**, **ordered pair**, **table of values**, **x-axis**, **y-axis**, **rectangular (Cartesian) coordinate system**, **quadrants**, **origin**, **coordinates**, **plot**

3.2 **graph, graphing**, **y-intercept**, **x-intercept**, **function**

3.3 **subscript notation**, **slope**, **rise**, **run**, **parallel lines**, **perpendicular lines**

NEW SYMBOLS

(a, b) an ordered pair

(x_1, y_1) x-sub-one, y-sub-one

m slope

TEST YOUR WORD POWER

See how well you have learned the vocabulary in this chapter. Answers, with examples, are given at the bottom of the next page.

1. An **ordered pair** is a pair of numbers written
   (a) in numerical order between brackets
   (b) between parentheses or brackets
   (c) between parentheses in which order is important
   (d) between parentheses in which order does not matter.

TEST YOUR WORD POWER (CONTINUED)

2. The **coordinates** of a point are
(a) the numbers in the corresponding ordered pair
(b) the solution of an equation
(c) the values of the *x*- and *y*-intercepts
(d) the graph of the point.

3. An **intercept** is
(a) the point where the *x*-axis and *y*-axis intersect
(b) a pair of numbers written in parentheses in which order matters
(c) one of the four regions determined by a rectangular coordinate system
(d) the point where a graph intersects the *x*-axis or the *y*-axis.

4. A **function** is
(a) the numbers in an ordered pair
(b) a set of ordered pairs in which each *x*-value corresponds to exactly one *y*-value
(c) a pair of numbers written between parentheses in which order matters
(d) the set of all ordered pairs that satisfy an equation.

5. The **slope** of a line is
(a) the measure of the run over the rise of the line
(b) the distance between two points on the line
(c) the ratio of the change in *y* to the change in *x* along the line
(d) the horizontal change compared to the vertical change of two points on the line.

6. Two lines in a plane are **parallel** if
(a) they represent the same line
(b) they never intersect
(c) they intersect at a 90° angle
(d) one has a positive slope and one has a negative slope.

7. Two lines in a plane are **perpendicular** if
(a) they represent the same line
(b) they never intersect
(c) they intersect at a 90° angle
(d) one has a positive slope and one has a negative slope.

QUICK REVIEW

CONCEPTS	EXAMPLES
3.1 LINEAR EQUATIONS IN TWO VARIABLES	
An ordered pair is a solution of an equation if it satisfies the equation.	Is $(2, -5)$ or $(0, -6)$ a solution of $4x - 3y = 18$? $4(2) - 3(-5) = 23 \neq 18$ — $(2, -5)$ is not a solution. $4(0) - 3(-6) = 18$ — $(0, -6)$ is a solution.
If a value of either variable in an equation is given, the other variable can be found by substitution.	Complete the ordered pair $(0, \quad)$ for $3x = y + 4$. $3(0) = y + 4$ $0 = y + 4$ $-4 = y$ The ordered pair is $(0, -4)$.
Plot the ordered pair $(-2, 4)$ by starting at the origin, going 2 units to the left, then going 4 units up.	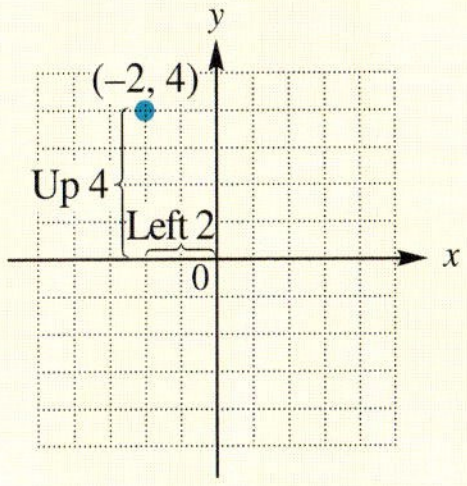

Answers to Test Your Word Power

1. (c) *Examples:* (0, 3), (3, 8), (4, 0) **2.** (a) *Example:* The point associated with the ordered pair (1, 2) has *x*-coordinate 1 and *y*-coordinate 2. **3.** (d) *Example:* The graph of the equation $4x - 3y = 12$ has *x*-intercept at (3, 0) and *y*-intercept at (0, −4). **4.** (b) *Example:* The set {(2, 0), (4, 3), (6, 6), (8, 9)} defines a function. **5.** (c) *Example:* The line through (3, 6) and (5, 4) has slope $\frac{4-6}{5-3} = \frac{-2}{2} = -1$. **6.** (b) *Example:* See Figure 20 in Section 3.3. **7.** (c) *Example:* See Figure 21 in Section 3.3.

27.

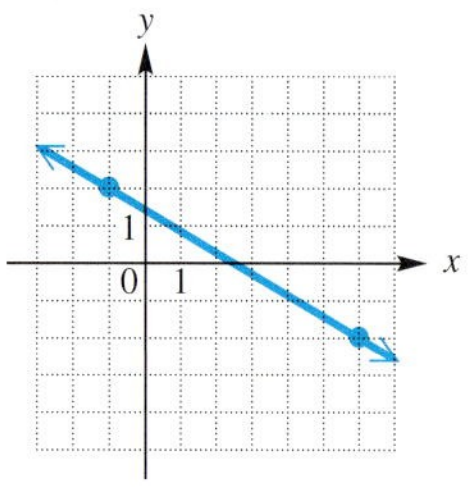

28. $y = 5$

29. A line perpendicular to the graph of $y = -3x + 3$

30. Explain why the signs of the slopes of perpendicular lines (neither of which is vertical) cannot be the same.

Decide whether the lines in each pair are parallel, perpendicular, or neither.

31. $3x + 2y = 6$
$6x + 4y = 8$

32. $x - 3y = 1$
$3x + y = 4$

33. $x - 2y = 8$
$x + 2y = 8$

MIXED REVIEW EXERCISES

34. Use the bar graph to estimate the growth rates for motor vehicles and parts sales for the years 1992–1994. (The rate for 1992, for example, represents the percent increase in sales from 1991 to 1992.) Sales in 1993 were approximately \$456 billion. Use the growth rate in 1994 to estimate sales in that year.

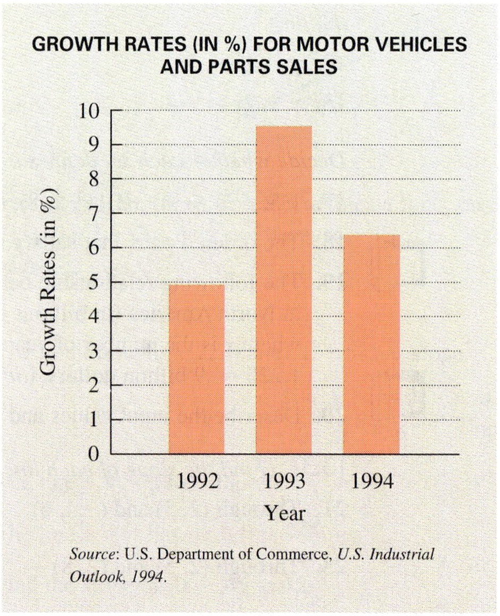

Source: U.S. Department of Commerce, *U.S. Industrial Outlook, 1994.*

35. Complete these ordered pairs for $x = 3y$: (0,), (8,), (, −3).

In Exercises 36–38, find the x- and y-intercepts and slope.

36. $x = -2$

37. $y = 2x + 3$

38. $11x - 3y = 4$

39. Find the slope of the line through $(4, -1)$ and $(-2, -3)$.

40. What is the slope of a line perpendicular to a line with undefined slope?

41. Is $\left(0, -\frac{16}{3}\right)$ a solution of $5x - 3y = 16$?

Graph each equation given in Exercises 42–44.

42. $x + 3y = 9$ **43.** $x - 5 = 0$ **44.** $2x - y = 3$

45. Two points determine a line. Explain why it is a good idea to plot three points before drawing the line.

46. The set of ordered pairs gives the five U.S. cities with the largest Hispanic populations in 1990. Data is in millions. (*Source:* U.S. Census Bureau.)

{(New York, 1.78), (Los Angeles, 1.39), (Chicago, .55), (San Antonio, .52), (Houston, .45)}

Does this set define a function? What are the inputs and the outputs?

CHAPTER 3 TEST

The graph shows the cost of 30 seconds of advertising time on the annual Super Bowl telecast. Use the graph to respond to Exercises 1 and 2.

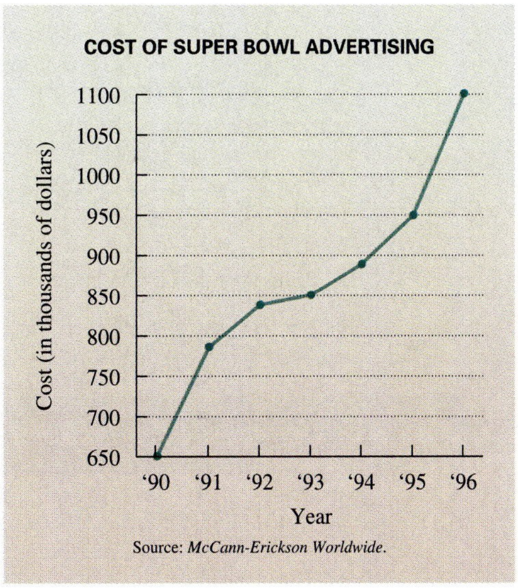

1. Which one of the following is the best estimate for the advertising cost in 1993?
(a) \$820,000 **(b)** \$848,000 **(c)** \$910,000

2. To the nearest percent, what was the percent increase in the advertising cost from 1995 to 1996?

Complete the ordered pairs for the given equation.

3. $3x + 5y = -30$; $(0, \quad)$, $(\quad, 0)$, $(\quad, 3)$

4. $y + 12 = 0$; $(0, \quad)$, $(-4, \quad)$, $\left(\frac{5}{2}, \quad\right)$

5. Is $(4, -1)$ a solution of $4x - 7y = 9$?

6. How do you find the x-intercept and the y-intercept for a linear equation in two variables?

Graph each linear equation. Give the x- and y-intercepts.

7. $3x + y = 6$ **8.** $y + 3 = 0$

9. Does the set of ordered pairs {(90, 650), (91, 780), (92, 830), (93, 850), (94, 880), (95, 950), (96, 1100)}, shown as points on the graph for Exercises 1 and 2, define a function? Explain why or why not.

10. Describe the inputs and outputs for the set of ordered pairs in Exercise 9.

Find the slope of each line in Exercises 11–16.

11. Through $(-4, 6)$ and $(-1, -2)$

12. $2x + y = 10$

13.

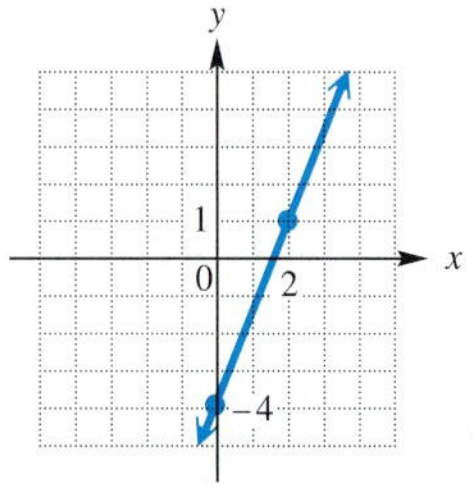

14. (These are two different views of the same line.)

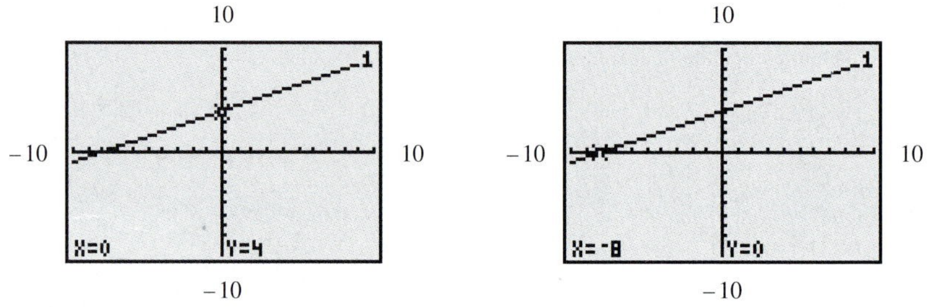

15. A line parallel to the graph of $y = -4x + 6$

16. A line perpendicular to the graph of $y = -4x + 6$

CUMULATIVE REVIEW EXERCISES CHAPTERS 1–3

Perform the indicated operations.

1. $16\frac{7}{8} - 3\frac{1}{10}$

2. $\frac{3}{4} \div \frac{5}{8}$

3. $-11 + 20 + (-2)$

4. $\dfrac{(-3)^2 - (-4)(2^4)}{5 \cdot 2 - (-2)^3}$

5. Rafael and Elda Muñoz are painting a bedroom for their new baby. They painted $\frac{1}{4}$ of the room on Saturday and $\frac{1}{3}$ of the room on Sunday. How much of the room is still unpainted?

6. True or false? $\dfrac{4(3 - 9)}{2 - 6} \geq 6$

7. Find the value of $xz^3 - 5y^2$ when $x = -2$, $y = -4$, and $z = 3$.

8. What property does $3(-2 + x) = -6 + 3x$ illustrate?

9. Simplify $-4p - 6 + 3p + 8$ by combining terms.

Solve.

10. $V = \frac{1}{3}\pi r^2 h$ for h

11. $6 - 3(1 + a) = 2(a + 5) - 2$

12. $-(m - 1) = 3 - 2m$

13. $\frac{y - 2}{3} = \frac{2y + 1}{5}$

14. $-5z \geq 4z - 18$

15. $2 < -6(z + 1) < 10$

Solve each problem.

16. Kimshana Lavoris earned \$200 working part time during July, \$375 during August, and \$325 during September. If her average income for the four months from July through October must be at least \$300, what possible amounts could she earn in October?

17. Mount Mayon in the Philippines is the most perfectly shaped conical volcano in the world. Its base is a perfect circle with a 39-mile circumference and it has a height of 8000 feet. (One mile is 5280 feet.) Find the radius of the circular base to the nearest mile. (*Hint:* This problem has some unneeded information.)

18. How much of a 20% chemical solution must be mixed with 30 liters of a 60% solution to get a 50% mixture?

19. The winning times in seconds for the women's 1000-meter speed skating event in the Winter Olympics for the years from 1960 to 1998 can be closely approximated by the equation

$$y = -.4685x + 95.07,$$

where x is the number of years since 1960. That is, $x = 5$ represents 1965, $x = 10$ represents 1970, and so on. Complete the table of ordered pairs for this linear equation. Round the y-values to the nearest hundredth of a second. (*Source: The Universal Almanac,* 1997, John W. Wright, General Editor.)

x	y
12	
28	
36	

20. Baby boomers are expected to inherit \$10.4 trillion from their parents over the next 45 years, an average of \$50,000 each. The pie chart shows how they plan to spend their inheritance. How much of the \$50,000 is expected to go toward the purchase of a home? How much to retirement?

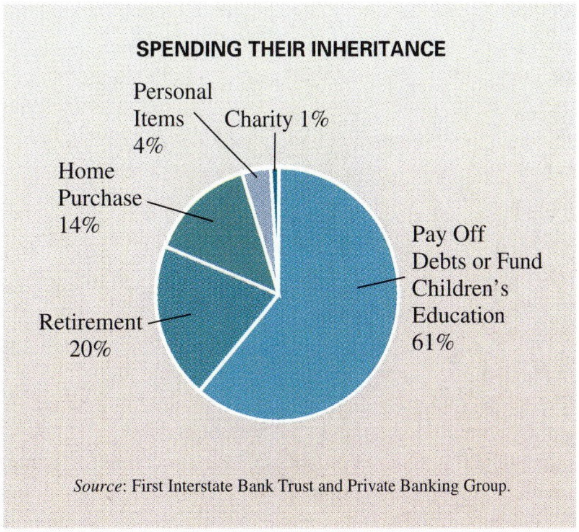

Consider the linear equation $3x + 2y = 12$. *Find the following.*

21. The x- and y-intercepts **22.** The graph **23.** The slope

24. Are the lines with equations $x + 5y = -6$ and $y = 5x - 8$ parallel, perpendicular, or neither?

25. California's exports to Pacific Rim nations, with the exception of China, fell during the first half of 1998, as shown in the bar graph. Is the set whose ordered pairs represent the data from the graph as (importer, % change) a function? Give the ordered pairs with the smallest change and the largest change. How is absolute value used in your answer?

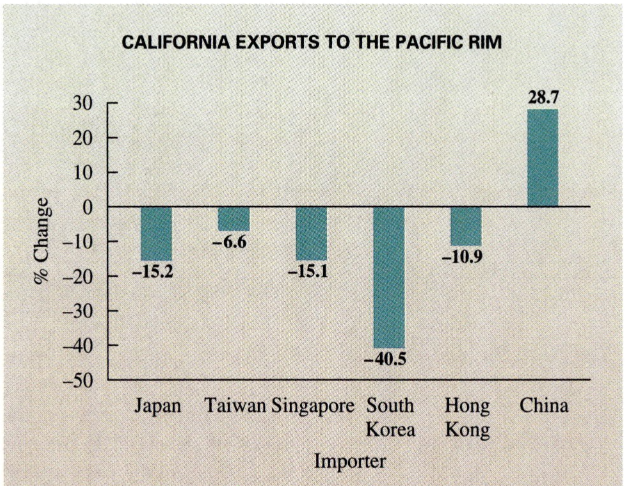

Source: California Trade and Commerce Agency.

Polynomials and Exponents 4

The number of passengers traveling by air has increased rapidly in the last decade. (See Exercise 46 in Section 4.1.) From 1985 to 1995, scheduled air carriers enjoyed an increase in net profits of $1,514,000,000. Surprisingly, in spite of more passengers traveling in planes that are consistently full, consumer complaints against U.S. airlines have generally decreased, as shown in the figure. There is one year, however, when all three graphs indicate an increase in each category from one year to the next, rather than a decrease. In which year did this occur?

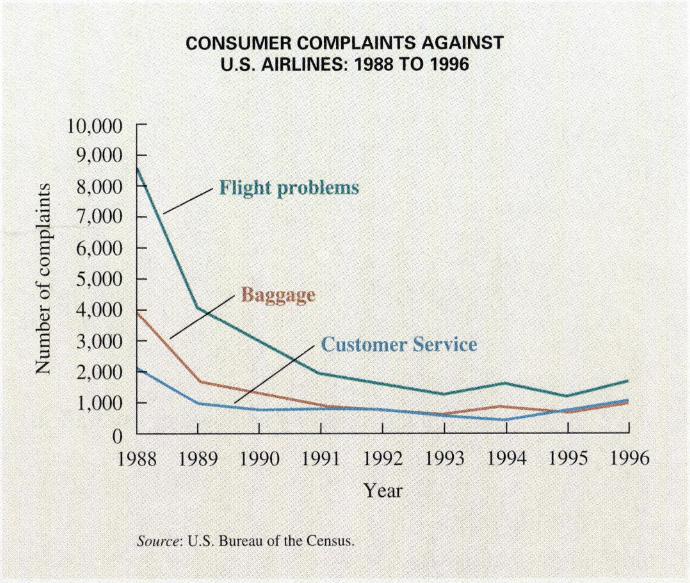

The graphs in the figure can each be approximated by a *polynomial function.* Polynomials are one of the topics studied in this chapter. Throughout the chapter we will see other examples of polynomials that describe information about the aeronautics industry.

Aeronautics

4.1 Addition and Subtraction of Polynomials; Graphing Simple Polynomials

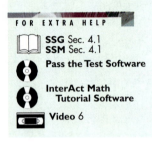

OBJECTIVES

1. Identify terms and coefficients.
2. Add like terms.
3. Know the vocabulary for polynomials.
4. Evaluate polynomials.
5. Add and subtract polynomials.
6. Graph equations defined by polynomials with degree 2.

FOR EXTRA HELP

SSG Sec. 4.1
SSM Sec. 4.1
Pass the Test Software
InterAct Math Tutorial Software
Video 6

OBJECTIVE 1 Identify terms and coefficients. In Chapter 1 we saw that in an expression such as

$$4x^3 + 6x^2 + 5x + 8,$$

the quantities $4x^3$, $6x^2$, $5x$, and 8 are called *terms.* As mentioned earlier, in the term $4x^3$, the number 4 is called the *numerical coefficient,* or simply the *coefficient,* of x^3. In the same way, 6 is the coefficient of x^2 in the term $6x^2$, 5 is the coefficient of x in the term $5x$, and 8 is the coefficient in the term 8. A constant term, like 8 in the expression above, can be thought of as $8x^0$, where

$$x^0 \text{ is defined to equal } 1.$$

We explain the reason for this definition later in this chapter.

EXAMPLE 1 Identifying Coefficients

Name the (numerical) coefficient of each term in these expressions.

(a) $4x^3$
The coefficient is 4.

(b) $x - 6x^4$
The coefficient of x is 1 because $x = 1 \cdot x$. The coefficient of x^4 is -6 since $x - 6x^4$ can be written as the sum $x + (-6x^4)$.

(c) $5 - v^3$
The coefficient of the term 5 is 5 because $5 = 5v^0$. By writing $5 - v^3$ as a sum, $5 + (-v^3)$, or $5 + (-1v^3)$, the coefficient of v^3 can be identified as -1.

OBJECTIVE 2 Add like terms. Recall from Section 1.8 that *like terms* have exactly the same combination of variables with the same exponents on the variables. Only the coefficients may differ. Examples of like terms are

$$19m^5 \quad \text{and} \quad 14m^5,$$
$$6y^9, \quad -37y^9, \quad \text{and} \quad y^9,$$
$$3pq \quad \text{and} \quad -2pq,$$
$$2xy^2 \quad \text{and} \quad -xy^2.$$

Using the distributive property, we add like terms by adding their coefficients.

EXAMPLE 2 Adding Like Terms

Simplify each expression by adding like terms.

(a) $-4x^3 + 6x^3 = (-4 + 6)x^3 = 2x^3$ Distributive property

(b) $9x^6 - 14x^6 + x^6 = (9 - 14 + 1)x^6 = -4x^6$

(c) $12m^2 + 5m + 4m^2 = (12 + 4)m^2 + 5m = 16m^2 + 5m$

(d) $3x^2y + 4x^2y - x^2y = (3 + 4 - 1)x^2y = 6x^2y$

In Example 2(c), we cannot combine $16m^2$ and $5m$. These two terms are unlike because the exponents on the variables are different. *Unlike terms* have different variables or different exponents on the same variables.

OBJECTIVE 3 Know the vocabulary for polynomials. A **polynomial in x** is a term or the sum of a finite number of terms of the form ax^n, for any real number a and any whole number n. For example,

$$16x^8 - 7x^6 + 5x^4 - 3x^2 + 4$$

is a polynomial in x (the 4 can be written as $4x^0$). This polynomial is written in **descending powers** of the variable, since the exponents on x decrease from left to right. On the other hand,

$$2x^3 - x^2 + \frac{4}{x}$$

is not a polynomial in x, since a variable appears in a denominator. Of course, we could define *polynomial* using any variable and not just x, as in Example 2(c). In fact, polynomials may have terms with *more* than one variable, as in Example 2(d).

The **degree of a term** is the sum of the exponents on the variables. For example, $3x^4$ has degree 4, while $6x^{17}$ has degree 17. The term $5x$ has degree 1, -7 has degree 0 (since -7 can be written as $-7x^0$), and $2x^2y$ has degree $2 + 1 = 3$ (y has an exponent of 1.) The **degree of a polynomial** is the highest degree of any nonzero term of the polynomial. For example, $3x^4 - 5x^2 + 6$ is of degree 4, the polynomial $5x + 7$ is of degree 1, 3 (or $3x^0$) is of degree 0, and $x^2y + xy - 5xy^2$ is of degree 3.

Three types of polynomials are very common and are given special names. A polynomial with exactly three terms is called a **trinomial.** (*Tri-* means "three," as in *tri*angle.) Examples are

$$9m^3 - 4m^2 + 6, \qquad 19y^2 + 8y + 5, \qquad \text{and} \qquad -3m^5n^2 + 2n^3 - m^4.$$

A polynomial with exactly two terms is called a **binomial.** (*Bi-* means "two," as in *bi*cycle.) Examples are

$$-9x^4 + 9x^3, \qquad 8m^2 + 6m, \qquad \text{and} \qquad 3m^5n^2 - 9m^2n^4.$$

A polynomial with only one term is called a **monomial.** (*Mon(o)-* means "one," as in *mono*rail.) Examples are

$$9m, \qquad -6y^5, \qquad a^2b^2, \qquad \text{and} \qquad 6.$$

EXAMPLE 3 Classifying Polynomials

For each polynomial, first simplify if possible by combining like terms. Then give the degree and tell whether it is a monomial, a binomial, a trinomial, or none of these.

(a) $2x^3 + 5$

The polynomial cannot be simplified. The degree is 3. The polynomial is a binomial.

(b) $4xy - 5xy + 2xy$

Add like terms to simplify: $4xy - 5xy + 2xy = xy$, which is a monomial of degree 2.

OBJECTIVE 4 Evaluate polynomials. A polynomial usually represents different numbers for different values of the variable, as shown in the next example.

EXAMPLE 4 Evaluating a Polynomial

Find the value of $3x^4 + 5x^3 - 4x - 4$ when $x = -2$ and when $x = 3$.

First, substitute -2 for x.

$$\begin{aligned} 3x^4 + 5x^3 - 4x - 4 &= 3(-2)^4 + 5(-2)^3 - 4(-2) - 4 \\ &= 3 \cdot 16 + 5 \cdot (-8) + 8 - 4 \\ &= 48 - 40 + 8 - 4 \\ &= 12 \end{aligned}$$

Next, replace x with 3.

$$\begin{aligned} 3x^4 + 5x^3 - 4x - 4 &= 3(3)^4 + 5(3)^3 - 4(3) - 4 \\ &= 3 \cdot 81 + 5 \cdot 27 - 12 - 4 \\ &= 362 \end{aligned}$$

CAUTION

Notice the use of parentheses around the numbers that are substituted for the variable in Example 4. This is particularly important when substituting a negative number for a variable that is raised to a power, so that the sign of the product is correct.

CONNECTIONS

In Section 3.2 we introduced the idea of a function: for every input x, there is one output y. Polynomials often provide a way of defining functions that approximate data collected over a period of time. For example, according to the U.S. National Aeronautics and Space Administration (NASA), the budget in millions of dollars for space station research for 1996–2001 can be approximated by the polynomial equation

$$y = -10.25x^2 - 126.04x + 5730.21,$$

where $x = 0$ represents 1996, $x = 1$ represents 1997, and so on, up to $x = 5$ representing 2001. The actual budget for 1998 was 5327 million dollars; an input of $x = 2$ (for 1998) gives approximately $y = 5437$. Considering the magnitude of the numbers, this is a very good approximation.

FOR DISCUSSION OR WRITING

Use the given polynomial equation to approximate the budget in other years between 1996 and 2001. Compare to the actual figures given here.

Year	Budget (in millions of dollars)
1996	5710
1997	5675
1998	5327
1999	5306
2000	5077
2001	4832

OBJECTIVE 5 Add and subtract polynomials. Polynomials may be added, subtracted, multiplied, and divided.

Adding Polynomials

To add two polynomials, add like terms.

EXAMPLE 5 Adding Polynomials Vertically

Add $6x^3 - 4x^2 + 3$ and $-2x^3 + 7x^2 - 5$.

Write like terms in columns.

$$\begin{array}{r} 6x^3 - 4x^2 + 3 \\ -2x^3 + 7x^2 - 5 \\ \hline \end{array}$$

Now add, column by column.

$$\begin{array}{r} 6x^3 \\ -2x^3 \\ \hline \mathbf{4x^3} \end{array} \qquad \begin{array}{r} -4x^2 \\ 7x^2 \\ \hline \mathbf{3x^2} \end{array} \qquad \begin{array}{r} 3 \\ -5 \\ \hline \mathbf{-2} \end{array}$$

Add the three sums together.

$$4x^3 + 3x^2 + (-2) = 4x^3 + 3x^2 - 2$$

The polynomials in Example 5 also can be added horizontally, as shown in the next example.

EXAMPLE 6 Adding Polynomials Horizontally

Add $6x^3 - 4x^2 + 3$ and $-2x^3 + 7x^2 - 5$.

Write the sum as

$$(6x^3 - 4x^2 + 3) + (-2x^3 + 7x^2 - 5).$$

Use the associative and commutative properties to rewrite this sum with the parentheses removed and with the subtractions changed to additions of inverses.

$$6x^3 + (-4x^2) + 3 + (-2x^3) + 7x^2 + (-5)$$

Place like terms together.

$$6x^3 + (-2x^3) + (-4x^2) + 7x^2 + 3 + (-5)$$

Combine like terms to get

$$4x^3 + 3x^2 + (-2), \qquad \text{or simply} \qquad 4x^3 + 3x^2 - 2,$$

the same answer found in Example 5.

Earlier, we defined the difference $x - y$ as $x + (-y)$. (We find the difference $x - y$ by adding x and the opposite of y.) For example,

$$7 - 2 = 7 + (-2) = 5 \qquad \text{and} \qquad -8 - (-2) = -8 + 2 = -6.$$

A similar method is used to subtract polynomials.

Subtracting Polynomials

To subtract two polynomials, change all the signs on the second polynomial and add the result to the first polynomial.

EXAMPLE 7 Subtracting Polynomials

(a) Perform the subtraction $(5x - 2) - (3x - 8)$.
By the definition of subtraction,

$$(5x - 2) - (3x - 8) = (5x - 2) + [-(3x - 8)].$$

As shown in Chapter 1, the distributive property gives

$$-(3x - 8) = \mathbf{-1}(3x - 8) = -3x + 8,$$

so

$$(5x - 2) - (3x - 8) = (5x - 2) + (-3x + 8) = 2x + 6.$$

(b) Subtract $6x^3 - 4x^2 + 2$ from $11x^3 + 2x^2 - 8$.
Write the problem.

$$(11x^3 + 2x^2 - 8) - (6x^3 - 4x^2 + 2)$$

Change all the signs in the second polynomial and add the two polynomials.

$$(11x^3 + 2x^2 - 8) + (-6x^3 + 4x^2 - 2) = 5x^3 + 6x^2 - 10$$

To check a subtraction problem, use the fact that if $a - b = c$, then $a = b + c$. For example, $6 - 2 = 4$, so we check by writing $6 = 2 + 4$, which is correct. Check the polynomial subtraction above by adding $6x^3 - 4x^2 + 2$ and $5x^3 + 6x^2 - 10$. Since the sum is $11x^3 + 2x^2 - 8$, the subtraction was performed correctly.

Subtraction also can be done in columns (vertically). We will use vertical subtraction in Section 4.6 when we study polynomial division.

EXAMPLE 8 Subtracting Polynomials Vertically

Use the method of subtracting by columns to find

$$(14y^3 - 6y^2 + 2y - 5) - (2y^3 - 7y^2 - 4y + 6).$$

Arrange like terms in columns.

$$\begin{array}{r} 14y^3 - 6y^2 + 2y - 5 \\ 2y^3 - 7y^2 - 4y + 6 \\ \hline \end{array}$$

Change all signs in the second row, and then add.

$$\begin{array}{rl} 14y^3 - 6y^2 + 2y - 5 & \\ -2y^3 + 7y^2 + 4y - 6 & \text{Change all signs.} \\ \hline 12y^3 + y^2 + 6y - 11 & \text{Add.} \end{array}$$

Either the horizontal or the vertical method may be used to add and subtract polynomials.

Polynomials in more than one variable are added and subtracted by combining like terms, just as with single variable polynomials.

EXAMPLE 9 Adding and Subtracting Polynomials with More Than One Variable

Add or subtract as indicated.

(a) $(4a + 2ab - b) + (3a - ab + b)$

$$(4a + 2ab - b) + (3a - ab + b) = 4a + 2ab - b + 3a - ab + b$$
$$= 7a + ab$$

(b) $(2x^2y + 3xy + y^2) - (3x^2y - xy - 2y^2)$

$$(2x^2y + 3xy + y^2) - (3x^2y - xy - 2y^2)$$
$$= 2x^2y + 3xy + y^2 - 3x^2y + xy + 2y^2$$
$$= -x^2y + 4xy + 3y^2$$

OBJECTIVE 6 Graph equations defined by polynomials with degree 2. In Chapter 3 we introduced graphs of straight lines. These graphs were defined by linear equations (which are actually polynomial equations of degree 1). By selective point-plotting, we can find the graphs of polynomial equations of degree 2.

EXAMPLE 10 Graphing Equations Defined by Polynomials with Degree 2

(a) Graph $y = x^2$.

Select several values for x; then find the corresponding y-values. For example, selecting $x = 2$ gives

$$y = 2^2 = 4,$$

and so the point (2, 4) is on the graph of $y = x^2$. (Recall that in an ordered pair such as (2, 4), the x-value comes first and the y-value second.) We show some ordered pairs that satisfy $y = x^2$ in a table next to Figure 1. If the ordered pairs from the table are plotted on a coordinate system and a smooth curve drawn through them, the graph is as shown in Figure 1.

x	y
3	9
2	4
1	1
0	0
−1	1
−2	4
−3	9

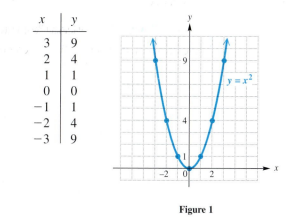

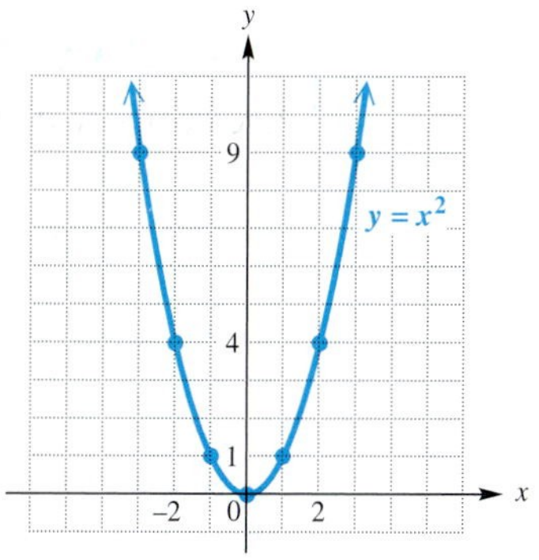

Figure 1

The graph of $y = x^2$ is the graph of a function, since each input x is related to just one output y. The curve in Figure 1 is called a **parabola.** The point (0, 0), the lowest point on this graph, is called the **vertex** of the parabola. The vertical line through the vertex (the y-axis here) is called the **axis** of the parabola. The axis of a parabola is a **line of symmetry** for the graph. If the graph is folded on this line, the two halves will match.

(b) Graph $y = -x^2 + 3$.
Once again plot points to obtain the graph. For example, if $x = -2$,

$$y = -(-2)^2 + 3 = -4 + 3 = -1.$$

This point and several others are shown in the table that accompanies the graph in Figure 2. The vertex of this parabola is (0, 3). This time the vertex is the *highest* point of the graph. The graph opens downward because x^2 has a negative coefficient.

x	y
−2	−1
−1	2
0	3
1	2
2	−1

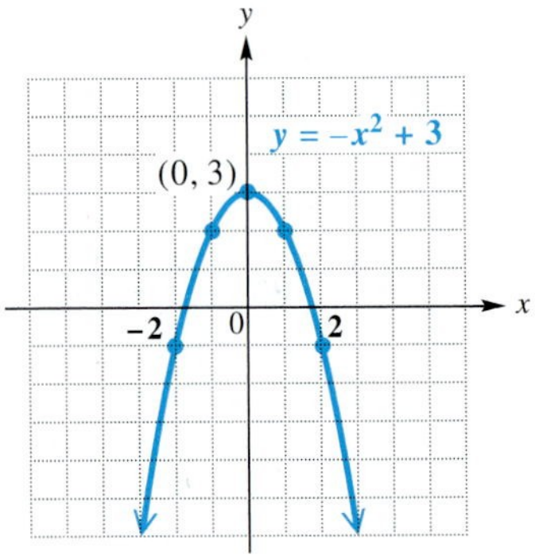

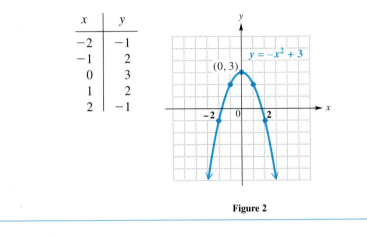

Figure 2

All polynomials of degree 2 have parabolas as their graphs. When graphing by plotting points, it is necessary to continue finding points until the vertex and points on either side of it are located. (In this section, all parabolas have their vertices on the x-axis or the y-axis.)

4.1 EXERCISES

Fill in each blank with the correct response.

1. In the term $7x^5$, the coefficient is _____ and the exponent is _____.
2. The expression $5x^3 - 4x^2$ has ___________ term(s). (how many?)
3. The degree of the term $-4x^8$ is _____.
4. The polynomial $4x^2 - y^2$ ___________ an example of a trinomial. (is/is not)
5. When $x^2 + 10$ is evaluated for $x = 4$, the result is _____.
6. $5x$ — $+ 3x^3 - 7x$ is a trinomial of degree 4.
7. $3xy + 2xy - 5xy =$ _____.
8. _____ is an example of a monomial with coefficient 5, in the variable x, having degree 9.

For each polynomial, determine the number of terms and name the coefficients of the terms. See Example 1.

9. $6x^4$
10. $-9y^5$
11. t^4
12. s^7
13. $-19r^2 - r$
14. $2y^3 - y$
15. $x + 8x^2 + 5x^3$
16. $v - 2v^3 - v^7$

In each polynomial add like terms whenever possible. Write the result in descending powers of the variable. See Example 2.

17. $-3m^5 + 5m^5$

18. $-4y^3 + 3y^3$

19. $2r^5 + (-3r^5)$

20. $-19y^2 + 9y^2$

21. $.2m^5 - .5m^2$

22. $-.9y + .9y^2$

23. $-3x^5 + 2x^5 - 4x^5$

24. $6x^3 - 8x^3 + 9x^3$

25. $-4p^7 + 8p^7 + 5p^9$

26. $-3a^8 + 4a^8 - 3a^2$

27. $-4y^2 + 3y^2 - 2y^2 + y^2$

28. $3r^5 - 8r^5 + r^5 + 2r^5$

For each polynomial first simplify, if possible, and write it in descending powers of the variable. Then give the degree of the resulting polynomial and tell whether it is a monomial, a binomial, a trinomial, or none of these. See Example 3.

29. $6x^4 - 9x$

30. $7t^3 - 3t$

31. $5m^4 - 3m^2 + 6m^4 - 7m^3$

32. $6p^5 + 4p^3 - 8p^5 + 10p^2$

33. $\frac{5}{3}x^4 - \frac{2}{3}x^4$

34. $\frac{4}{5}r^6 + \frac{1}{5}r^6$

35. $.8x^4 - .3x^4 - .5x^4 + 7$

36. $1.2t^3 - .9t^3 - .3t^3 + 9$

*Find the value of each polynomial when **(a)** $x = 2$ and when **(b)** $x = -1$. See Example 4.*

37. $2x^5 - 4x^4 + 5x^3 - x^2$

38. $2x^2 + 5x + 1$

39. $-3x^2 + 14x - 2$

40. $-2x^2 + 3$

TECHNOLOGY INSIGHTS (EXERCISES 41–42)

The graphing calculator screen shown here indicates that 1 has been stored into the memory location designated X, and then the polynomial $X^2 - 3X + 6$ has been evaluated. The result is 4. (This can be verified by hand using direct substitution, as shown in Example 4.)

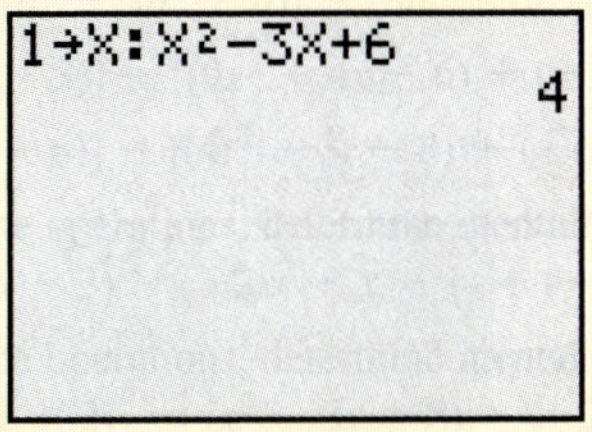

Predict the result the calculator will give for the following screens.

41.

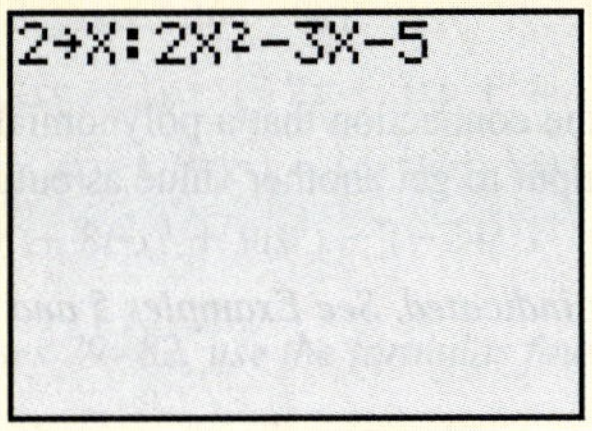

42.

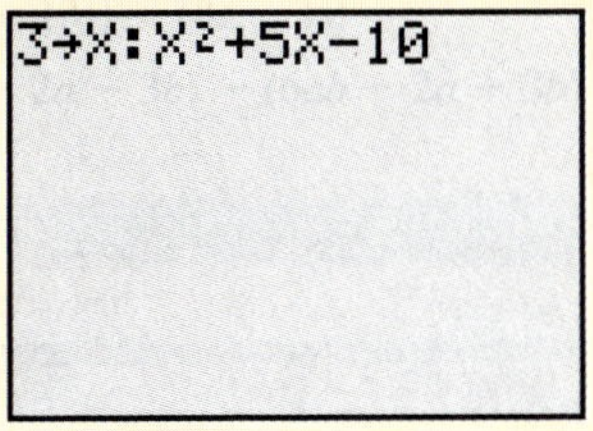

Step 3 Add like terms.

$$
\begin{array}{rrrrr}
 & x^3 + & 2x^2 + & 4x + & 1 \\
 & & & 3x + & 5 \\
\hline
 & 5x^3 + & 10x^2 + & 20x + & 5 \\
3x^4 + & 6x^3 + & 12x^2 + & 3x & \\
\hline
3x^4 + & 11x^3 + & 22x^2 + & 23x + & 5
\end{array}
$$

The product is $3x^4 + 11x^3 + 22x^2 + 23x + 5$.

OBJECTIVE 3 **Multiply binomials by the FOIL method.** In algebra, many of the polynomials to be multiplied are both binomials (with just two terms). For these products a shortcut that eliminates the need to write out all the steps is used. To develop this shortcut, multiply $x + 3$ and $x + 5$ using the distributive property.

$$
\begin{aligned}
(x + 3)(x + 5) &= x(x + 5) + 3(x + 5) \\
&= x(x) + x(5) + 3(x) + 3(5) \\
&= x^2 + 5x + 3x + 15 \\
&= x^2 + 8x + 15
\end{aligned}
$$

The first term in the second line, $x(x)$, is the product of the first terms of the two binomials.

$(x + 3)(x + 5)$ Multiply the first terms: $x(x)$.

The term $x(5)$ is the product of the first term of the first binomial and the last term of the second binomial. This is the **outer product.**

$(x + 3)(x + 5)$ Multiply the outer terms: $x(5)$.

The term $3(x)$ is the product of the last term of the first binomial and the first term of the second binomial. The product of these middle terms is the **inner product.**

$(x + 3)(x + 5)$ Multiply the inner terms: $3(x)$.

Finally, $3(5)$ is the product of the last terms of the two binomials.

$(x + 3)(x + 5)$ Multiply the last terms: $3(5)$.

The inner product and the outer product should be added mentally, so that the three terms of the answer can be written without extra steps as

$$(x + 3)(x + 5) = x^2 + 8x + 15.$$

A summary of these steps is given below. This procedure is sometimes called the **FOIL method,** which comes from the abbreviation for *First, Outer, Inner, Last.*

Multiplying Binomials by the FOIL Method

Step 1 **Multiply the first terms.** Multiply the two first terms of the binomials to get the first term of the answer.

Step 2 **Find the outer and inner products.** Find the outer product and the inner product and add them (mentally if possible) to get the middle term of the answer.

Step 3 **Multiply the last terms.** Multiply the two last terms of the binomials to get the last term of the answer.

EXAMPLE 4 Using the FOIL Method

Find the product $(x + 8)(x - 6)$ by the FOIL method.

Step 1 **F** Multiply the *first* terms: $x(x) = x^2$.

Step 2 **O** Find the product of the *outer* terms: $x(-6) = -6x$.

I Find the product of the *inner* terms: $8(x) = 8x$.

Add the outer and inner products mentally: $-6x + 8x = 2x$.

Step 3 **L** Multiply the *last* terms: $8(-6) = -48$.

The product $(x + 8)(x - 6)$ is the sum of the terms found in the three steps above, so

$$(x + 8)(x - 6) = x^2 - 6x + 8x - 48 = x^2 + 2x - 48.$$

As a shortcut, this product can be found in the following manner.

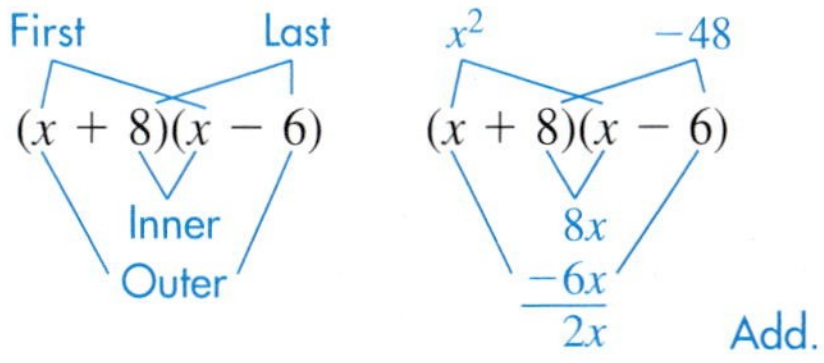

It is not possible to add the inner and outer products of the FOIL method if unlike terms result, as shown in the next example.

EXAMPLE 5 Using the FOIL Method

Multiply $(9x - 2)(3y + 1)$.

First	$(9x - 2)(3y + 1)$	$27xy$	
Outer	$(9x - 2)(3y + 1)$	$9x$	Unlike terms
Inner	$(9x - 2)(3y + 1)$	$-6y$	
Last	$(9x - 2)(3y + 1)$	-2	

F O I L

$$(9x - 2)(3y + 1) = 27xy + 9x - 6y - 2$$

EXAMPLE 6 Using the FOIL Method

Find the following products.

$$\begin{aligned}\textbf{(a)}\ (2k+5y)(k+3y) &= \overset{F}{(2k)(k)} + \overset{O}{(2k)(3y)} + \overset{I}{(5y)(k)} + \overset{L}{(5y)(3y)}\\ &= 2k^2 + 6ky + 5ky + 15y^2\\ &= 2k^2 + 11ky + 15y^2 \quad \text{Combine like terms.}\end{aligned}$$

(b) $(7p + 2q)(3p - q) = 21p^2 - pq - 2q^2$ FOIL

$$\begin{aligned}\textbf{(c)}\ 2x^2(x-3)(3x+4) &= 2x^2(3x^2 - 5x - 12) \quad \text{FOIL}\\ &= 6x^4 - 10x^3 - 24x^2 \quad \text{Distributive property}\end{aligned}$$

4.3 EXERCISES

1. Match each product in Column I with the correct monomial in Column II.

I	II
(a) $(5x^3)(6x^5)$	**A.** $125x^{15}$
(b) $(-5x^5)(6x^3)$	**B.** $30x^8$
(c) $(5x^5)^3$	**C.** $-216x^9$
(d) $(-6x^3)^3$	**D.** $-30x^8$

2. Match each product in Column I with the correct polynomial in Column II.

I	II
(a) $(x-5)(x+3)$	**A.** $x^2 + 8x + 15$
(b) $(x+5)(x+3)$	**B.** $x^2 - 8x + 15$
(c) $(x-5)(x-3)$	**C.** $x^2 - 2x - 15$
(d) $(x+5)(x-3)$	**D.** $x^2 + 2x - 15$

Find each product. See Example 1.

3. $(-5a^9)(-8a^5)$

4. $(-3m^6)(-5m^4)$

5. $-2m(3m + 2)$

6. $-5p(6 + 3p)$

7. $3p(8 - 6p + 12p^3)$

8. $4x(3 + 2x + 5x^3)$

9. $-8z(2z + 3z^2 + 3z^3)$

10. $-7y(3 + 5y^2 - 2y^3)$

11. $7x^2y(2x^3y^2 + 3xy - 4y)$

12. $9xy^3(-3x^2y^4 + 6xy - 2x)$

Find each product. See Examples 2 and 3.

13. $(6x + 1)(2x^2 + 4x + 1)$

14. $(9y - 2)(8y^2 - 6y + 1)$

15. $(4m + 3)(5m^3 - 4m^2 + m - 5)$

16. $(y + 4)(3y^3 - 2y^2 + y + 3)$

17. $(2x - 1)(3x^5 - 2x^3 + x^2 - 2x + 3)$

18. $(2a + 3)(a^4 - a^3 + a^2 - a + 1)$

19. $(5x^2 + 2x + 1)(x^2 - 3x + 5)$

20. $(2m^2 + m - 3)(m^2 - 4m + 5)$

Find each product. Use the FOIL method. See Examples 4–6.

21. $(n - 2)(n + 3)$

22. $(r - 6)(r + 8)$

23. $(x + 6)(x - 6)$

24. $(y + 9)(y - 9)$

25. $(4r + 1)(2r - 3)$

26. $(5x + 2)(2x - 7)$

27. $(3x + 2)(3x - 2)$

28. $(7x + 3)(7x - 3)$

29. $(3q + 1)(3q + 1)$

30. $(4w + 7)(4w + 7)$

31. $(3x + y)(x - 2y)$

32. $(5p + m)(2p - 3m)$

33. $(-3t + 4)(t + 6)$

34. $(-5x + 9)(x - 2)$

35. $3y^3(2y + 3)(y - 5)$

36. $5t^4(t + 3)(3t - 1)$

37. Find a polynomial that represents the area of this square.

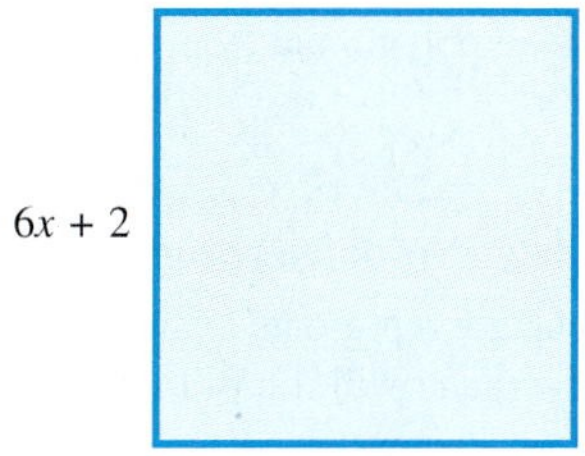

38. Find a polynomial that represents the area of this rectangle.

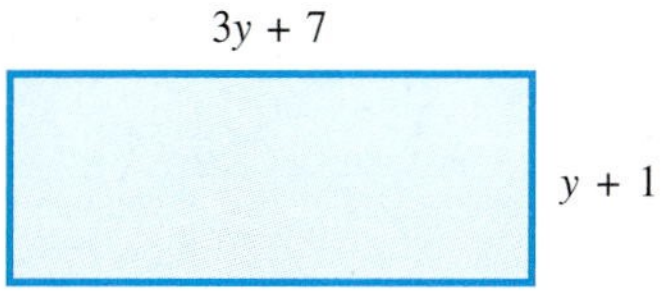

39. Perform the following multiplications:

$$(x+4)(x-4); \quad (y+2)(y-2); \quad (r+7)(r-7).$$

Observe your answers, and explain the pattern that can be found.

40. Repeat Exercise 39 for the following:

$$(x+4)(x+4); \quad (y-2)(y-2); \quad (r+7)(r+7).$$

Find each product.

41. $\left(3p+\frac{5}{4}q\right)\left(2p-\frac{5}{3}q\right)$

42. $\left(-x+\frac{2}{3}y\right)\left(3x-\frac{3}{4}y\right)$

43. $(m^3-4)(2m^3+3)$

44. $(4a^2+b^2)(a^2-2b^2)$

45. $(2k^3+h^2)(k^2-3h^2)$

46. $(4x^3-5y^4)(x^2+y)$

47. $3p^3(2p^2+5p)(p^3+2p+1)$

48. $5k^2(k^2-k+4)(k^3-3)$

49. $-2x^5(3x^2+2x-5)(4x+2)$

50. $-4x^3(3x^4+2x^2-x)(-2x+1)$

Find a polynomial that represents the area of each shaded region. In Exercises 53 and 54 leave π in your answer. Use the formulas found on the inside covers.

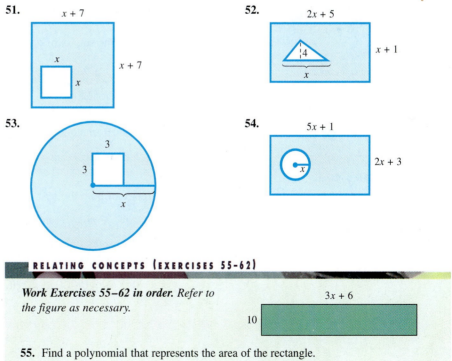

51.

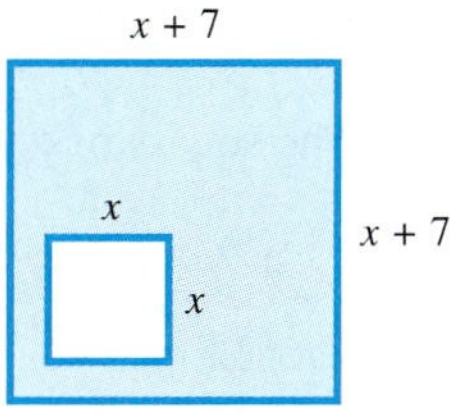

52.

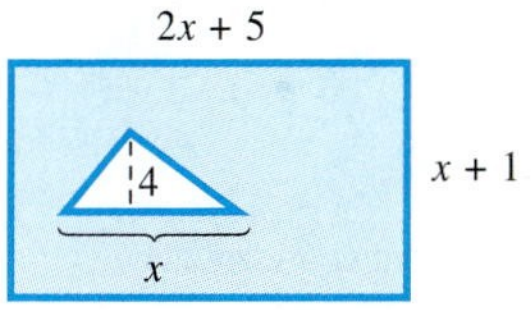

53.

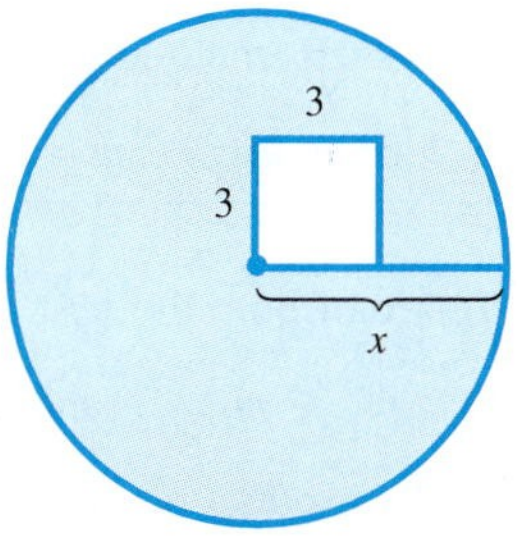

54.

RELATING CONCEPTS (EXERCISES 55–62)

Work Exercises 55–62 in order. *Refer to the figure as necessary.*

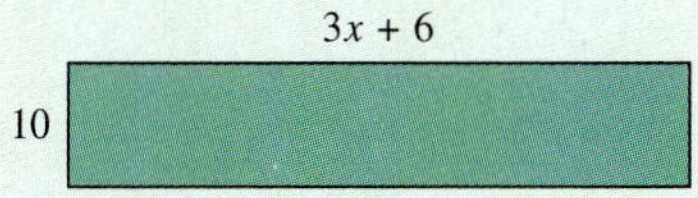

55. Find a polynomial that represents the area of the rectangle.

RELATING CONCEPTS (EXERCISES 55–62) (CONTINUED)

56. Suppose you know that the area of the rectangle is 600 square yards. Use this information and the polynomial from Exercise 55 to write an equation that allows you to solve for x.

57. Solve for x.

58. What are the dimensions of the rectangle (assume units are all in yards)?

59. Suppose the rectangle represents a lawn and it costs \$3.50 per square yard to lay sod on the lawn. How much will it cost to sod the entire lawn?

60. Use the result of Exercise 58 to find the perimeter of the lawn.

61. Again, suppose the rectangle represents a lawn and it costs \$9.00 per yard to fence the lawn. How much will it cost to fence the lawn?

62. **(a)** Suppose that it costs k dollars per square yard to sod the lawn. Determine a polynomial in the variables x and k that represents the cost to sod the entire lawn.

(b) Suppose that it costs r dollars per yard to fence the lawn. Determine a polynomial in the variables x and r that represents the cost to fence the lawn.

Did you make the connection that sodding requires knowing the area, while fencing requires knowing the perimeter?

63. Explain the FOIL method of multiplying two binomials. Give an example.

64. Why does the FOIL method not apply to the product of a binomial and a trinomial? Give an example.

4.4 Special Products

OBJECTIVES

1. Square binomials.
2. Find the product of the sum and difference of two terms.
3. Find higher powers of binomials.

FOR EXTRA HELP

SSG Sec. 4.4
SSM Sec. 4.4
Pass the Test Software
InterAct Math Tutorial Software
Video 6

In this section, we develop patterns for certain binomial products that occur frequently.

OBJECTIVE 1 Square binomials. The square of a binomial can be found quickly by using the method shown in Example 1.

EXAMPLE 1 Squaring a Binomial

Find $(m + 3)^2$.

Squaring $m + 3$ by the FOIL method gives

$$(m + 3)(m + 3) = m^2 + 3m + 3m + 9 = m^2 + 6m + 9.$$

The result has the square of both the first and the last terms of the binomial:

$$m^2 = m^2 \qquad \text{and} \qquad 3^2 = 9.$$

The middle term is twice the product of the two terms of the binomial, since both the outer and inner products are $(m)(3)$ and

$$(m)(3) + (m)(3) = 2(m)(3) = 6m.$$

This example suggests the following rule.

Square of a Binomial

The square of a binomial is a trinomial consisting of the square of the first term, plus twice the product of the two terms, plus the square of the last term of the binomial. For x and y,

$$(x + y)^2 = x^2 + 2xy + y^2$$

and

$$(x - y)^2 = x^2 - 2xy + y^2.$$

EXAMPLE 2 Squaring Binomials

Use the rule to find each product.

(a) $(5z - 1)^2 = (5z)^2 - 2(5z)(1) + 1^2 = 25z^2 - 10z + 1$
Recall that $(5z)^2 = 5^2z^2 = 25z^2$.

(b) $(3b + 5r)^2 = (3b)^2 + 2(3b)(5r) + (5r)^2 = 9b^2 + 30br + 25r^2$

(c) $(2a - 9x)^2 = 4a^2 - 36ax + 81x^2$

(d) $\left(4m + \frac{1}{2}\right)^2 = (4m)^2 + 2(4m)\left(\frac{1}{2}\right) + \left(\frac{1}{2}\right)^2 = 16m^2 + 4m + \frac{1}{4}$

(e) $t(a + 2b)^2 = t(a^2 + 4ab + 4b^2)$ Square the binomial.
$= a^2t + 4abt + 4b^2t$ Distributive property

CAUTION A common error when squaring a binomial is forgetting the middle term of the product. In general, $(x + y)^2 \neq x^2 + y^2$ and $(x - y)^2 \neq x^2 - y^2$.

OBJECTIVE 2 **Find the product of the sum and difference of two terms.** Binomial products of the form $(x + y)(x - y)$ also occur frequently. In these products, one binomial is the sum of two terms, and the other is the difference of the same two terms. As an example, the product of $a + 2$ and $a - 2$ is

$$(a + 2)(a - 2) = a^2 - 2a + 2a - 4 = a^2 - 4.$$

Using the FOIL method, the product of $x + y$ and $x - y$ is the difference of two squares.

Product of the Sum and Difference of Two Terms

The product of the sum and difference of the two terms x and y is

$$(x + y)(x - y) = x^2 - y^2.$$

CAUTION The product $(x + y)(x - y)$ cannot be written as $(x + y)^2$ or as $(x - y)^2$, since one factor involves addition and the other involves subtraction.

EXAMPLE 3 Finding the Product of the Sum and Difference of Two Terms

Find each product.

(a) $(x + 4)(x - 4)$

Use the pattern for the sum and difference of two terms.

$$(x + 4)(x - 4) = x^2 - 4^2 = x^2 - 16$$

(b) $(3 - w)(3 + w)$

By the commutative property, this product is the same as $(3 + w)(3 - w)$.

$$(3 - w)(3 + w) = (3 + w)(3 - w) = 3^2 - w^2 = 9 - w^2$$

(c) $(a - b)(a + b) = a^2 - b^2$

EXAMPLE 4 Finding the Product of the Sum and Difference of Two Terms

Find each product.

(a) $(5m + 3)(5m - 3)$

Use the rule for the product of the sum and difference of two terms.

$$(5m + 3)(5m - 3) = (5m)^2 - 3^2 = 25m^2 - 9$$

(b) $(4x + y)(4x - y) = (4x)^2 - y^2 = 16x^2 - y^2$

(c) $\left(z - \frac{1}{4}\right)\left(z + \frac{1}{4}\right) = z^2 - \frac{1}{16}$

(d) $r(x^2 + y)(x^2 - y) = r(x^4 - y^2)$
$= rx^4 - ry^2$

The product formulas of this section will be important later, particularly in Chapters 5 and 6. Therefore, it is important to memorize these formulas and practice using them.

OBJECTIVE 3 **Find higher powers of binomials.** The methods used in the previous section and this section can be combined to find higher powers of binomials.

EXAMPLE 5 Finding Higher Powers of Binomials

Find each product.

(a) $(x + 5)^3$

$$\begin{aligned}(x + 5)^3 &= (x + 5)^2(x + 5) && a^3 = a^2 \cdot a\\ &= (x^2 + 10x + 25)(x + 5) && \text{Square the binomial.}\\ &= x^3 + 10x^2 + 25x + 5x^2 + 50x + 125 && \text{Multiply polynomials.}\\ &= x^3 + 15x^2 + 75x + 125 && \text{Combine like terms.}\end{aligned}$$

(b) $(2y - 3)^4$

$$\begin{aligned}(2y - 3)^4 &= (2y - 3)^2(2y - 3)^2 && a^4 = a^2 \cdot a^2\\ &= (4y^2 - 12y + 9)(4y^2 - 12y + 9) && \text{Square each binomial.}\\ &= 16y^4 - 48y^3 + 36y^2 - 48y^3 + 144y^2 && \text{Multiply polynomials.}\\ &\quad - 108y + 36y^2 - 108y + 81 &&\\ &= 16y^4 - 96y^3 + 216y^2 - 216y + 81 && \text{Combine like terms.}\end{aligned}$$

4.4 EXERCISES

1. Consider the square $(2x + 3)^2$.
   (a) What is the square of the first term, $(2x)^2$?
   (b) What is twice the product of the two terms, $2(2x)(3)$?
   (c) What is the square of the last term, 3^2?
   (d) Write the final product, which is a trinomial, using your results in parts (a)–(c).

2. Explain in your own words how to square a binomial. Give an example.

Find each square. See Examples 1 and 2.

3. $(p + 2)^2$
4. $(r + 5)^2$
5. $(a - c)^2$
6. $(p - y)^2$
7. $(4x - 3)^2$
8. $(5y + 2)^2$
9. $(8t + 7s)^2$
10. $(7z - 3w)^2$
11. $\left(5x + \frac{2}{5}y\right)^2$
12. $\left(6m - \frac{4}{5}n\right)^2$
13. $x(2x + 5)^2$
14. $t(3t - 1)^2$
15. $-(4r - 2)^2$
16. $-(3y - 8)^2$

17. Consider the product $(7x + 3y)(7x - 3y)$.
    (a) What is the product of the first terms, $(7x)(7x)$?
    (b) Multiply the outer terms, $(7x)(-3y)$. Then multiply the inner terms, $(3y)(7x)$. Add the results. What is this sum?
    (c) What is the product of the last terms, $(3y)(-3y)$?
    (d) Write the complete product using your answers in parts (a) and (c). Why is the sum found in part (b) omitted here?

18. Explain in your own words how to find the product of the sum and the difference of two terms. Give an example.

Find each product. See Examples 3 and 4.

19. $(q + 2)(q - 2)$
20. $(x + 8)(x - 8)$
21. $(2w + 5)(2w - 5)$
22. $(3z + 8)(3z - 8)$
23. $(10x + 3y)(10x - 3y)$
24. $(13r + 2z)(13r - 2z)$
25. $(2x^2 - 5)(2x^2 + 5)$
26. $(9y^2 - 2)(9y^2 + 2)$
27. $\left(7x + \frac{3}{7}\right)\left(7x - \frac{3}{7}\right)$
28. $\left(9y + \frac{2}{3}\right)\left(9y - \frac{2}{3}\right)$
29. $p(3p + 7)(3p - 7)$
30. $q(5q - 1)(5q + 1)$

RELATING CONCEPTS (EXERCISES 31–40)

Special products can be illustrated by using areas of rectangles. Use the figure, and ***work Exercises 31–36 in order*** *to justify the special product $(a + b)^2 = a^2 + 2ab + b^2$.*

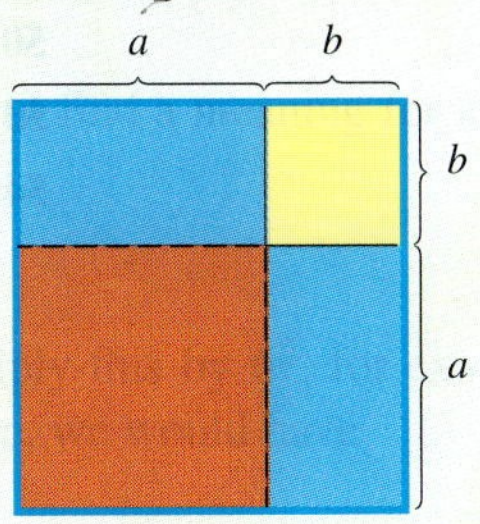

31. Express the area of the large square as the square of a binomial.

32. Give the monomial that represents the area of the red square.

Definitions and Rules for Exponents

If no denominators are zero, for any integers m and n:

			Examples
Product rule		$a^m \cdot a^n = a^{m+n}$	$7^4 \cdot 7^3 = 7^7$
Zero exponent		$a^0 = 1$	$(-3)^0 = 1$
Negative exponent		$a^{-n} = \dfrac{1}{a^n}$	$5^{-3} = \dfrac{1}{5^3}$
Quotient rule		$\dfrac{a^m}{a^n} = a^{m-n}$	$\dfrac{2^2}{2^5} = 2^{-3} = \dfrac{1}{2^3}$
Power rules	(a)	$(a^m)^n = a^{mn}$	$(4^2)^3 = 4^6$
	(b)	$(ab)^m = a^m b^m$	$(3k)^4 = 3^4k^4$
	(c)	$\left(\dfrac{a}{b}\right)^m = \dfrac{a^m}{b^m}$	$\left(\dfrac{2}{3}\right)^{10} = \dfrac{2^{10}}{3^{10}}$
	(d)	$\dfrac{a^{-m}}{b^{-n}} = \dfrac{b^n}{a^m}$	$\dfrac{5^{-3}}{3^{-5}} = \dfrac{3^5}{5^3}$
	(e)	$\left(\dfrac{a}{b}\right)^{-m} = \left(\dfrac{b}{a}\right)^m$	$\left(\dfrac{4}{7}\right)^{-2} = \left(\dfrac{7}{4}\right)^2$

OBJECTIVE 4 **Use combinations of rules.** As shown in the next example, sometimes we may need to use more than one rule to simplify an expression.

EXAMPLE 5 **Using a Combination of Rules**

Simplify each expression.

(a) $\dfrac{(4^2)^3}{4^5}$

Use power rule (a) and then the quotient rule.

$$\frac{(4^2)^3}{4^5} = \frac{4^6}{4^5} = 4^{6-5} = 4^1 = 4$$

(b) $(2x)^3(2x)^2$

Use the product rule first. Then use power rule (b).

$$(2x)^3(2x)^2 = (2x)^5 = 2^5x^5 \quad \text{or} \quad 32x^5$$

(c) $\left(\dfrac{2x^3}{5}\right)^{-4}$

Use power rules (a), (b), (c), and (e).

$$\left(\frac{2x^3}{5}\right)^{-4} = \left(\frac{5}{2x^3}\right)^4 \qquad \text{Power rule (e)}$$

$$= \frac{5^4}{2^4(x^3)^4} \qquad \text{Power rules (b) and (c)}$$

$$= \frac{5^4}{16x^{12}} \qquad \text{Power rule (a)}$$

(d) $\left(\dfrac{3x^{-2}}{4^{-1}y^{3}}\right)^{-3} = \dfrac{3^{-3}x^{6}}{4^{3}y^{-9}}$ Power rules

$= \dfrac{x^{6}y^{9}}{3^{3} \cdot 4^{3}}$ Power rule (d)

$= \dfrac{x^{6}y^{9}}{27(64)}$ Definition of exponent

$= \dfrac{x^{6}y^{9}}{1728}$

NOTE Since the steps can be done in different orders, there are many equally good ways to simplify a problem like Example 5(d).

CONNECTIONS

In Exercises 91–94 of Section 4.2, we gave an example of an exponential expression that modeled the growth of money left at compound interest. Exponential expressions can also model the way in which quantities grow or decay. (These are examples of functions.) For instance, U.S. commercial space revenues for 1990–1995 can be modeled by the exponential equation

$$y = 3501.09(1.17)^{x},$$

where y is in millions of dollars, and $x = 0$ corresponds to 1990, $x = 1$ corresponds to 1991, and so on, up to $x = 5$ for 1995. Using a calculator, we can find that in 1991 the revenues were approximately $3501.09(1.17)^{1} = 4096.28$ million dollars. (This is actually a bit lower than the true amount of 4370 million dollars.) (*Source:* U.S. Department of Commerce, International Trade Administration, *U.S. Industrial Outlook 1994*; and unpublished data.)

FOR DISCUSSION OR WRITING

Use the model and a calculator to find the approximate revenues for 1990, 1992, 1993, 1994, and 1995.

4.5 EXERCISES

Decide whether each statement is true or false.

1. $(-2)^{-4} < 0$ **2.** $-2^{-4} < 0$ **3.** $(-5)^{0} > 0$

4. $-5^{0} < 0$ **5.** $1 - 12^{0} = 1$ **6.** $-(-3)^{0} = -1$

The given expression is either equal to 0, 1, or −1. Decide which is correct. See Example 1.

7. 9^{0} **8.** 5^{0} **9.** $(-4)^{0}$ **10.** $(-10)^{0}$

11. -9^{0} **12.** -5^{0} **13.** $(-2)^{0} - 2^{0}$ **14.** $(-8)^{0} - 8^{0}$

15. $\dfrac{0^{10}}{10^{0}}$ **16.** $\dfrac{0^{5}}{5^{0}}$

63. One of your friends in class simplified $\dfrac{6x^2 - 12x}{6}$ as $x^2 - 12x$. Is this correct? If not, what error did your friend make and how would you explain the correct method of performing the division?

Perform each division.

64. $\dfrac{2r^2 + 3r - 14}{r - 2}$

65. $\dfrac{10a^3 + 9a^2 - 14a + 9}{5a - 3}$

66. $\dfrac{x^4 + 3x^3 - 5x^2 - 3x + 4}{x^2 - 1}$

67. $\dfrac{m^4 + 4m^3 - 5m^2 - 12m + 6}{m^2 - 3}$

[4.7] *Write each number in scientific notation.*

68. 48,000,000

69. 28,988,000,000

70. .0000000824

Write each number without exponents.

71. 2.4×10^4

72. 7.83×10^7

73. 8.97×10^{-7}

Perform each indicated operation and write the answer without exponents.

74. $(2 \times 10^{-3}) \times (4 \times 10^5)$

75. $\dfrac{8 \times 10^4}{2 \times 10^{-2}}$

76. $\dfrac{12 \times 10^{-5} \times 5 \times 10^4}{4 \times 10^3 \times 6 \times 10^{-2}}$

The quote is taken from the source cited. Write each number in scientific notation in the quote without exponents.

77. The muon, a close relative of the electron produced by the bombardment of cosmic rays against the upper atmosphere, has a half-life of 2 millionths of a second (2×10^{-6}s). (Excerpt from *Conceptual Physics,* 6th edition, by Paul G. Hewitt. Copyright © by Paul G. Hewitt. Published by HarperCollins College Publishers.)

78. There are 13 red balls and 39 black balls in a box. Mix them up and draw 13 out one at a time without returning any ball . . . the probability that the 13 drawings each will produce a red ball is . . . 1.6×10^{-12}. (Warren Weaver, *Lady Luck,* New York: Doubleday & Company, Inc., 1963, pp. 298–299.)

The quote is taken from the source cited. Write each number in scientific notation.

79. An electron and a positron attract each other in two ways: the electromagnetic attraction of their opposite electric charges, and the gravitational attraction of their two masses. The electromagnetic attraction is

4,200,000,000,000,000,000,000,000,000,000,000,000,000,000,000

times as strong as the gravitational. (Isaac Asimov, *Isaac Asimov's Book of Facts,* New York: Bell Publishing Company, 1981, p. 106.)

80. A Boeing 747 would not do well in a dogfight; it is too big to maneuver with the lively agility required of an Air Force fighter. It performs best when it flies straight and level, and this is true of the worldwide airline industry as a whole. That industry features annual revenues that approach \$200 billion. (T. A. Heppenheimer, *Turbulent Skies: The History of Commercial Aviation,* New York: John Wiley & Sons, 1995, p. 345.)

81. According to *The Wall Street Journal Almanac 1998,* the U.S. airline industry earned record profits in 1996 of \$2.82 billion and set records for both the number of passengers and the amount of cargo carried. Passenger traffic increased 7% to 578.4 billion revenue passenger miles, while cargo traffic rose 4.6% to 17.7 billion revenue ton miles.

Write each of the following numbers from the above paragraph using scientific notation.

(a) 2.82 billion **(b)** 578.4 billion **(c)** 17.7 billion

82. According to Campbell, Mitchell, and Reece in *Biology Concepts and Connections* (Benjamin Cummings, 1994, p. 230), "(t)he amount of DNA in a human cell is about 1000 times greater than the DNA in *E. coli.* Does this mean humans have 1000 times as many genes as the 2000 in *E. coli*? The answer is probably no; the human genome is thought to carry between 50,000 and 100,000 genes, which code for various proteins (as well as for tRNA and rRNA)."

Write each of the following numbers from the above quote using scientific notation.

(a) 1000 **(b)** 2000 **(c)** 50,000 **(d)** 100,000

RELATING CONCEPTS (EXERCISES 83–88)

In Exercises 83–88 we use the letters P and Q to denote the following polynomials:

$$P: \quad x - 3$$
$$Q: \quad x^3 + x^2 - 11x - 3.$$

Work these exercises in order.

83. (a) Evaluate P for $x = 5$.
(b) Evaluate Q for $x = 5$.
(c) Find the polynomial represented by $P + Q$.
(d) Evaluate the polynomial found in part (c) for $x = 5$ and verify that it is equal to the sum of the values you found in parts (a) and (b).

84. (a) Evaluate P for $x = 4$.
(b) Evaluate Q for $x = 4$.
(c) Find the polynomial represented by $P - Q$.
(d) Evaluate the polynomial found in part (c) for $x = 4$ and verify that it is equal to the difference between the values you found in parts (a) and (b).

85. (a) Evaluate P for $x = -3$.
(b) Evaluate Q for $x = -3$.
(c) Find the polynomial represented by $P \cdot Q$.
(d) Evaluate the polynomial found in part (c) for $x = -3$ and verify that it is equal to the product of the values you found in parts (a) and (b).

86. (a) Evaluate P for $x = 6$.
(b) Evaluate Q for $x = 6$.
(c) Find the polynomial represented by $\frac{Q}{P}$.
(d) Evaluate the polynomial found in part (c) for $x = 6$ and verify that it is equal to the quotient of the values you found in parts (a) and (b).

87. The concept illustrated in these exercises is this: If we evaluate P for a particular value of x and evaluate Q for the same value of x, and then perform an operation on P and Q, the resulting polynomial will give the appropriate value when evaluated for x. The values used in Exercises 83–86 were chosen arbitrarily. Repeat each exercise for a different value of x.

88. Refer to Exercise 86. Why could we *not* choose 3 as a replacement for x? Is there any other replacement for x that would not be valid?

Did you make the connection between operations with polynomials and operations with those same polynomials evaluated for particular values of the variables?

MIXED REVIEW EXERCISES

Perform the indicated operations. Write with positive exponents only. Assume all variables represent nonzero real numbers.

89. $5^0 + 7^0$

90. $\left(\frac{6r^2p}{5}\right)^3$

91. $(12a + 1)(12a - 1)$

92. 2^{-4}

93. $(8^{-3})^4$

94. $\frac{2p^3 - 6p^2 + 5p}{2p^2}$

95. $\frac{(2m^{-5})(3m^2)^{-1}}{m^{-2}(m^{-1})^2}$

96. $(3k - 6)(2k^2 + 4k + 1)$

97. $\frac{r^9 \cdot r^{-5}}{r^{-2} \cdot r^{-7}}$

98. $(2r + 5s)^2$

99. $(-5y^2 + 3y - 11) + (4y^2 - 7y + 15)$

100. $(2r + 5)(5r - 2)$

101. $\frac{2y^3 + 17y^2 + 37y + 7}{2y + 7}$

102. $(25x^2y^3 - 8xy^2 + 15x^3y) \div (10x^2y^3)$

103. $(6p^2 - p - 8) - (-4p^2 + 2p - 3)$

104. $\frac{5^8}{5^{19}}$

105. $(-7 + 2k)^2$

106. $\left(\frac{x}{y^{-3}}\right)^{-4}$

CHAPTER 4 TEST

For each polynomial, combine terms when possible and write the polynomial in descending powers of the variable. Give the degree of the simplified polynomial. Decide whether the simplified polynomial is a monomial, binomial, trinomial, or none of these.

1. $5x^2 + 8x - 12x^2$

2. $13n^3 - n^2 + n^4 + 3n^4 - 9n^2$

3. Use the table to complete a set of ordered pairs that lie on the graph of $y = 2x^2 - 4$. Then graph the equation.

x	y
-2	
-1	
0	
1	
2	

Perform the indicated operations.

4. $(2y^2 - 8y + 8) + (-3y^2 + 2y + 3) - (y^2 + 3y - 6)$

5. $(-9a^3b^2 + 13ab^5 + 5a^2b^2) - (6ab^5 + 12a^3b^2 + 10a^2b^2)$

6. Subtract.

$$\begin{array}{r} 9t^3 - 4t^2 + 2t + 2 \\ \underline{9t^3 + 8t^2 - 3t - 6} \end{array}$$

7. $3x^2(-9x^3 + 6x^2 - 2x + 1)$

8. $(t - 8)(t + 3)$

9. $(4x + 3y)(2x - y)$

10. $(5x - 2y)^2$

11. $(10v + 3w)(10v - 3w)$

12. $(2r - 3)(r^2 + 2r - 5)$

13. What polynomial expression represents the area of this square?

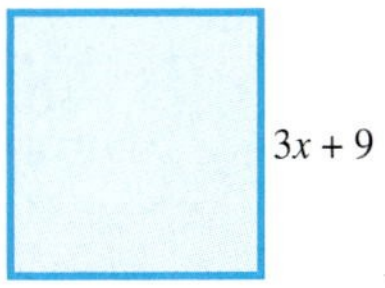

Evaluate each expression.

14. 5^{-4}

15. $(-3)^0 + 4^0$

16. $4^{-1} + 3^{-1}$

17. Use the rules for exponents to simplify $\dfrac{(3x^2y)^2(xy^3)^2}{(xy)^3}$. Assume x and y are nonzero.

Simplify, and write the answer using only positive exponents. Assume that variables represent nonzero numbers.

18. $\dfrac{8^{-1} \cdot 8^4}{8^{-2}}$

19. $\dfrac{(x^{-3})^{-2}(x^{-1}y)^2}{(xy^{-2})^2}$

20. Do you agree or disagree with the following statement?

3^{-4} represents a negative number.

Justify your answer.

Perform each division.

21. $\dfrac{8y^3 - 6y^2 + 4y + 10}{2y}$

22. $(-9x^2y^3 + 6x^4y^3 + 12xy^3) \div (3xy)$

23. $(3x^3 - x + 4) \div (x - 2)$

24. **(a)** Write 45,000,000,000 using scientific notation.
(b) Write 3.6×10^{-6} without using exponents.
(c) Write the quotient without using scientific notation: $\dfrac{9.5 \times 10^{-1}}{5 \times 10^3}$.

25. According to an article in *Aviation Week and Space Technology* (March 23, 1998), the MD-17 jet has the capacity to carry more than a 170,000-pound payload. Suppose we consider a group of 1000 such jets.
(a) Write the numbers 170,000 and 1000 using scientific notation.
(b) Multiply these two numbers to find the total payload that this group can carry. Express your answer using scientific notation.

CUMULATIVE REVIEW EXERCISES CHAPTERS 1–4

Write each fraction in lowest terms.

1. $\dfrac{28}{16}$

2. $\dfrac{55}{11}$

Perform each operation.

3. $\dfrac{2}{3} + \dfrac{1}{8}$

4. $\dfrac{7}{4} - \dfrac{9}{5}$

5. A contractor installs toolsheds. Each requires $1\frac{1}{4}$ cubic yards of concrete. How much concrete would be needed for 25 sheds?

6. A retailer has \$34,000 invested in her business. She finds that last year she earned 5.4% on this investment. How much did she earn?

7. List all positive integer factors of 45.

8. If $a < 0$ and $b > 0$, what is the sign of $b - a$?

Find the value of each expression if $x = -2$ and $y = 4$.

9. $\dfrac{4x - 2y}{x + y}$

10. $x^3 - 4xy$

Perform the indicated operations.

11. $\dfrac{(-13 + 15) - (3 + 2)}{6 - 12}$

12. $-7 - 3[2 + (5 - 8)]$

Decide which property justifies each statement.

13. $(9 + 2) + 3 = 9 + (2 + 3)$

14. $6(4 + 2) = 6(4) + 6(2)$

15. Use the bar graph to find a signed number that represents the change in the number of airline passenger deaths
(a) from 1994 to 1995
(b) from 1995 to 1996.

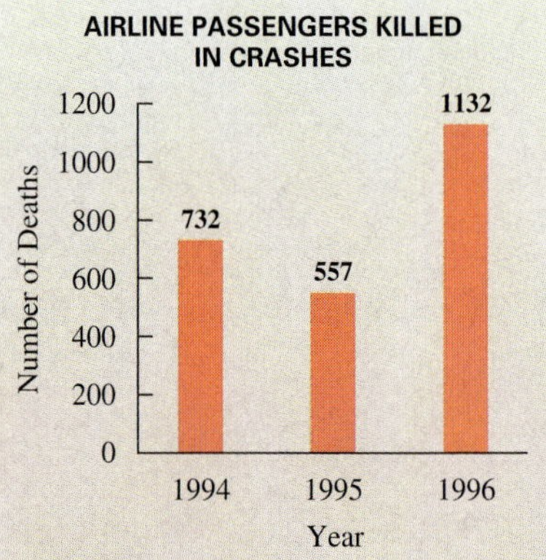

Source: International Civil Aviation Organization.

16. Simplify the expression $-3(2x^2 - 8x + 9) - (4x^2 + 3x + 2)$.

Solve each equation.

17. $2 - 3(t - 5) = 4 + t$

18. $2(5h + 1) = 10h + 4$

19. $d = rt$ for r

20. $\dfrac{x}{5} = \dfrac{x - 2}{7}$

21. $\dfrac{1}{3}p - \dfrac{1}{6}p = -2$

22. $.05x + .15(50 - x) = 5.50$

23. $4 - (3x + 12) = (2x - 9) - (5x - 1)$

Solve each problem.

24. A 1-ounce mouse takes about 16 times as many breaths as does a 3-ton elephant. If the two animals take a combined total of 170 breaths per minute, how many breaths does each take during that time period? (*Source:* Christopher McGowan, *Dinosaurs, Spitfires, and Sea Dragons,* Harvard University Press, 1991.)

25. If a number is subtracted from 8 and this difference is tripled, the result is 3 times the number. Find this number, and you will learn how many times a dolphin rests during a 24-hour period.

26. In 1995 and 1996, there were the same number of aircraft accidents involving passenger fatalities throughout worldwide scheduled air services. In 1994, there were 2 more than in each of the other years. How many accidents were there in each of these years if there was a total of 68 altogether? (*Source:* International Civil Aviation Organization.)

Solve each inequality.

27. $-8x \leq -80$ **28.** $-2(x + 4) > 3x + 6$ **29.** $-3 \leq 2x + 5 < 9$

Solve the problem.

30. One side of a triangle is twice as long as a second side. The third side of the triangle is 17 feet long. The perimeter of the triangle cannot be more than 50 feet. Find the longest possible values for the other two sides of the triangle.

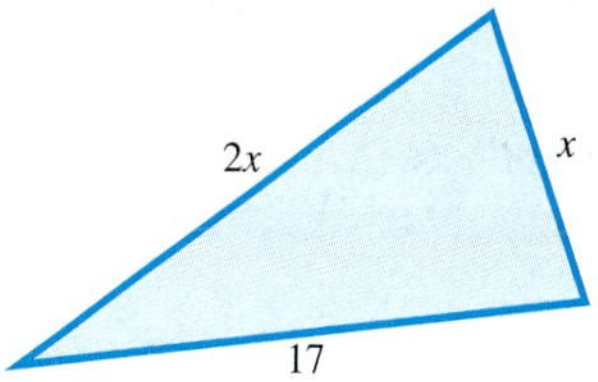

31. Complete the table of values for $-2x + 4y = 8$.

x	0		2		4
y		0		1	

32. Graph $y = -3x + 6$.

33. Graph $y = (x + 4)^2$, using the x-values -6, -5, -4, -3, and -2 to obtain a set of points.

Perform the indicated operations.

34. $(7x^3 - 12x^2 - 3x + 8) + (6x^2 + 4) - (-4x^3 + 8x^2 - 2x - 2)$

35. $6x^5(3x^2 - 9x + 10)$ **36.** $(7x + 4)(9x + 3)$

37. $(5x + 8)^2$ **38.** $\dfrac{14x^3 - 21x^2 + 7x}{7x}$

39. $\dfrac{y^3 - 3y^2 + 8y - 6}{y - 1}$

Evaluate each expression.

40. $4^{-1} + 3^0$ **41.** $2^{-4} \cdot 2^5$ **42.** $\dfrac{8^{-5} \cdot 8^7}{8^2}$

43. Write with positive exponents only: $\dfrac{(a^{-3}b^2)^2}{(2a^{-4}b^{-3})^{-1}}$.

44. Write in scientific notation: 34,500.

45. Write without exponents: 5.36×10^{-7}.

5 Factoring and Applications

Transportation

In the early days of the American frontier, it took four horses 75 days to haul one wagonload of goods a thousand miles. By the late 1800s, roads, canals, and railways connected "island" communities, greatly expediting the movement of people and goods. Whereas a stagecoach traveled 50 miles in a day, trains now covered 50 miles in an *hour*—over 700 miles per day. The vehicle that really changed twentieth-century American travel, however, was the automobile. Today emerging technologies are further transforming modern transportation systems. Just one example is the global network of computers moving millions of bits of data per second.

Much of this emerging technology depends on mathematics. Understanding mathematics makes it possible to model data using functions. In Chapter 4, we used polynomials of degree 2 to model data. Now, in this chapter, we introduce a method for solving polynomial equations of degree 2 called *quadratic equations,* and we give further examples of them as models. For instance, average fuel economy trends in miles per gallon for the automotive industry are closely approximated by the quadratic equation

$$y = -.04x^2 + .93x + 21,$$

where x represents the year. The years are coded so that $x = 0$ represents 1978, $x = 2$ represents 1980, $x = 4$ represents 1982, and so on. This equation was developed from the data in the table.

Fuel Economy

Year	1978	1980	1982	1984	1986	1988	1990	1992	1994	1996
Miles per Gallon	19.9	23.1	25.1	25.0	25.9	26.0	25.4	25.1	24.7	24.9

Source: National Highway Traffic Safety Administration.

Substituting 0 for x in the equation gives $y = 21$, the approximate average miles per gallon in 1978. For 1996, substituting $1996 - 1978 = 18$ for x in the equation gives $y = 24.78$ as the approximate average miles per gallon. Which of these is a closer approximation to the data in the table? We will return to this transportation model in Section 5.6, Example 4.

Visit our Web site at www.LialAlgebra.com

5.1 The Greatest Common Factor; Factoring by Grouping

OBJECTIVES

1. Find the greatest common factor of a list of terms.
2. Factor out the greatest common factor.
3. Factor by grouping.

FOR EXTRA HELP

SSG Sec. 5.1
SSM Sec. 5.1
Pass the Test Software
InterAct Math Tutorial Software
Video 8

Recall from Chapter 1 that to **factor** means to write a quantity as a product. That is, factoring is the opposite of multiplying. For example,

Multiplying	*Factoring*
$6 \cdot 2 = 12,$	$12 = 6 \cdot 2.$
↑ ↑ Factors ↑ Product	↑ Product ↑ ↑ Factors

Other factored forms of 12 are

$$(-6)(-2), \quad 3 \cdot 4, \quad (-3)(-4), \quad 12 \cdot 1, \quad \text{and} \quad (-12)(-1).$$

More than two factors may be used, so another factored form of 12 is $2 \cdot 2 \cdot 3$. The positive integer factors of 12 are

$$1, 2, 3, 4, 6, 12.$$

OBJECTIVE 1 Find the greatest common factor of a list of terms. An integer that is a factor of two or more integers is called a **common factor** of those integers. For example, 6 is a common factor of 18 and 24 since 6 is a factor of both 18 and 24. Other common factors of 18 and 24 are 1, 2, and 3. The **greatest common factor** of a list of integers is the largest common factor of those integers. Thus, 6 is the greatest common factor of 18 and 24, since it is the largest of the common factors of these numbers.

NOTE Factors of a number are also divisors of the number. The greatest common factor is actually the same as the greatest common divisor.

CONNECTIONS

There are many rules for deciding what numbers divide into a given number. Here are some especially useful divisibility rules for small numbers. It is surprising how many people do not know them.

A Whole Number Divisible by:	*Must Have the Following Property:*
2	Ends in 0, 2, 4, 6, or 8
3	Sum of its digits is divisible by 3
4	Last two digits form a number divisible by 4
5	Ends in 0 or 5
6	Divisible by both 2 and 3
8	Last three digits form a number divisible by 8
9	Sum of its digits is divisible by 9
10	Ends in 0

Recall from Chapter 1 that a prime number has only itself and 1 as factors. In Section 1.1 we factored numbers into prime factors. This is the first step in finding the greatest common factor of a list of numbers. We find the greatest common factor (GCF) of a list of numbers as follows.

Finding the Greatest Common Factor (GCF)

Step 1 **Factor.** Write each number in prime factored form.

Step 2 **List common factors.** List each prime number that is a factor of every number in the list.

Step 3 **Choose smallest exponents.** Use as exponents on the common prime factors the *smallest* exponent from the prime factored forms. (If a prime does not appear in one of the prime factored forms, it cannot appear in the greatest common factor.)

Step 4 **Multiply.** Multiply the primes from Step 3. If there are no primes left after Step 3, the greatest common factor is 1.

EXAMPLE 1 Finding the Greatest Common Factor for Numbers

Find the greatest common factor for each list of numbers.

(a) 30, 45

First write each number in prime factored form.

$$30 = 2 \cdot 3 \cdot 5$$
$$45 = 3 \cdot 3 \cdot 5$$

Now, take each prime the *least* number of times it appears in all the factored forms. There is no 2 in the prime factored form of 45, so there will be no 2 in the greatest common factor. The least number of times 3 appears in all the factored forms is 1, and the least number of times 5 appears is also 1. From this, the GCF is

$$3^1 \cdot 5^1 = 15.$$

(b) 72, 120, 432

Find the prime factored form of each number.

$$72 = 2 \cdot 2 \cdot 2 \cdot 3 \cdot 3$$
$$120 = 2 \cdot 2 \cdot 2 \cdot 3 \cdot 5$$
$$432 = 2 \cdot 2 \cdot 2 \cdot 2 \cdot 3 \cdot 3 \cdot 3$$

The least number of times 2 appears in all the factored forms is 3, and the least number of times 3 appears is 1. There is no 5 in the prime factored form of either 72 or 432, so the GCF is

$$2^3 \cdot 3 = 24.$$

(c) 10, 11, 14

Write the prime factored form of each number.

$$10 = 2 \cdot 5$$
$$11 = 11$$
$$14 = 2 \cdot 7$$

There are no primes common to all three numbers, so the GCF is 1.

The greatest common factor can also be found for a list of variable terms. For example, the terms x^4, x^5, x^6, and x^7 have x^4 as the greatest common factor because each of these terms can be written with x^4 as a factor.

$$x^4 = 1 \cdot x^4, \qquad x^5 = x \cdot x^4, \qquad x^6 = x^2 \cdot x^4, \qquad x^7 = x^3 \cdot x^4$$

NOTE The exponent on a variable in the GCF is the *smallest* exponent that appears in the factors.

EXAMPLE 2 Finding the Greatest Common Factor for Variable Terms

Find the greatest common factor for each list of terms.

(a) $21m^7, -18m^6, 45m^8, -24m^5$

$$\begin{aligned} 21m^7 &= 3 \cdot 7 \cdot m^7 \\ -18m^6 &= -1 \cdot 2 \cdot 3^2 \cdot m^6 \\ 45m^8 &= 3^2 \cdot 5 \cdot m^8 \\ -24m^5 &= -1 \cdot 2^3 \cdot 3 \cdot m^5 \end{aligned}$$

First, 3 is the greatest common factor of the coefficients 21, -18, 45, and -24. The smallest exponent on m is 5, so the GCF of the terms is $3m^5$.

(b) $x^4y^2, x^7y^5, x^3y^7, y^{15}$

$$\begin{aligned} x^4y^2 &= x^4 \cdot y^2 \\ x^7y^5 &= x^7 \cdot y^5 \\ x^3y^7 &= x^3 \cdot y^7 \\ y^{15} &= y^{15} \end{aligned}$$

There is no x in the last term, y^{15}, so x will not appear in the greatest common factor. There is a y in each term, however, and 2 is the smallest exponent on y. The GCF is y^2.

OBJECTIVE 2 Factor out the greatest common factor. We use the idea of a greatest common factor to write a polynomial (a sum) in factored form as a product. For example, the polynomial

$$3m + 12$$

has two terms, $3m$ and 12. The greatest common factor for these two terms is 3. We can write $3m + 12$ so that each term is a product with 3 as one factor.

$$3m + 12 = 3 \cdot m + 3 \cdot 4$$

Now use the distributive property.

$$3m + 12 = 3 \cdot m + 3 \cdot 4 = 3(m + 4)$$

The factored form of $3m + 12$ is $3(m + 4)$. This process is called **factoring out the greatest common factor.**

The polynomial $3m + 12$ is *not* in factored form when written as

$$3 \cdot m + 3 \cdot 4.$$

CAUTION The *terms* are factored, but the polynomial is not. The factored form of $3m + 12$ is the *product*

$$3(m + 4).$$

EXAMPLE 3 Factoring Out the Greatest Common Factor

Factor out the greatest common factor.

(a) $20m^5 + 10m^4 + 15m^3$

The GCF for the terms of this polynomial is $5m^3$.

$$\begin{aligned} 20m^5 + 10m^4 + 15m^3 &= (5m^3)(4m^2) + (5m^3)(2m) + (5m^3)3 \\ &= 5m^3(4m^2 + 2m + 3) \end{aligned}$$

Check by multiplying $5m^3$ and $4m^2 + 2m + 3$. You should get the original polynomial.

(b) $x^5 + x^3 = (x^3)x^2 + (x^3)1 = x^3(x^2 + 1)$

(c) $20m^7p^2 - 36m^3p^4 = 4m^3p^2(5m^4 - 9p^2)$

(d) $a(a + 3) + 4(a + 3)$

The binomial $a + 3$ is the greatest common factor here.

$$a(a + 3) + 4(a + 3) = (a + 3)(a + 4)$$

CAUTION Be sure to include the 1 in a problem like Example 3(b). Always check that the factored form can be multiplied out to give the original polynomial.

OBJECTIVE 3 Factor by grouping. Common factors are used in **factoring by grouping,** as explained in the next example.

EXAMPLE 4 Factoring by Grouping

Factor by grouping.

(a) $2x + 6 + ax + 3a$

The first two terms have a common factor of 2, and the last two terms have a common factor of a.

$$2x + 6 + ax + 3a = 2(x + 3) + a(x + 3)$$

The expression is still not in factored form because it is the *sum* of two terms. Now, however, $x + 3$ is a common factor and can be factored out.

$$2x + 6 + ax + 3a = 2(x + 3) + a(x + 3) = (x + 3)(2 + a)$$

The final result is in factored form because it is a *product.* Note that the goal in factoring by grouping is to get a common factor, $x + 3$ here, so that the last step is possible.

Same

(b) $m^2 + 6m + 2m + 12 = m(m + 6) + 2(m + 6)$
$= (m + 6)(m + 2)$

(c) $6xy - 21x - 8y + 28 = 3x(2y - 7) - 4(2y - 7) = (2y - 7)(3x - 4)$

Must be same

Since the quantities in parentheses in the second step must be the same, it was necessary here to factor out -4 rather than 4.

CAUTION Use negative signs carefully when grouping, as in Example 4(c). Otherwise, sign errors may result.

Use these steps when factoring four terms by grouping.

Factoring by Grouping

Step 1 **Group terms.** Collect the terms into two groups so that each group has a common factor.

Step 2 **Factor within groups.** Factor out the greatest common factor from each group.

Step 3 **Factor the entire polynomial.** Factor a common binomial factor from the results of Step 2.

Step 4 **If necessary, rearrange terms.** If Step 2 does not result in a common binomial factor, try a different grouping.

EXAMPLE 5 Rearranging Terms Before Factoring by Grouping

Factor by grouping.

(a) $10x^2 - 12y^2 + 15xy - 8xy$

Factoring out the common factor of 2 from the first two terms and the common factor of xy terms from the last two terms gives

$$10x^2 - 12y^2 + 15xy - 8xy = 2(5x^2 - 6y^2) + xy(15 - 8).$$

This did not lead to a common factor, so we try rearranging the terms. There is usually more than one way to do this. Let's try

$$10x^2 - 8xy - 12y^2 + 15xy,$$

grouping the first two terms and the last two terms as follows.

$$\begin{aligned} 10x^2 - 8xy - 12y^2 + 15xy &= 2x(5x - 4y) + 3y(-4y + 5x) \\ &= 2x(5x - 4y) + 3y(5x - 4y) \\ &= (5x - 4y)(2x + 3y) \end{aligned}$$

(b) $2xy + 12 - 3y - 8x$

We need to rearrange these terms to get two groups that each have a common factor. Trial and error suggests the following grouping.

$$\begin{aligned} 2xy + 12 - 3y - 8x &= (2xy - 3y) + (-8x + 12) \\ &= y(2x - 3) - 4(2x - 3) && \text{Factor each group.} \\ &= (2x - 3)(y - 4) && \text{Factor out the common binomial factor.} \end{aligned}$$

5.1 EXERCISES

1. Is 3 the greatest common factor of 18, 24, and 42? If not, what is?
2. Is pq the greatest common factor of pq^2, p^2, and p^2q^2? If not, what is?
3. Factoring is the opposite of what operation?
4. How can you check your answer when you factor a polynomial?
5. Give an example of three numbers whose greatest common factor is 5.
6. Explain how to find the greatest common factor of a list of terms. Use examples.

Find the greatest common factor for each list of terms. See Examples 1 and 2.

7. $16y$, 24
8. $18w$, 27
9. $30x^3$, $40x^6$, $50x^7$
10. $60z^4$, $70z^8$, $90z^9$
11. $12m^3n^2$, $18m^5n^4$, $36m^8n^3$
12. $25p^5r^7$, $30p^7r^8$, $50p^5r^3$
13. $-x^4y^3$, $-xy^2$
14. $-a^4b^5$, $-a^3b$
15. $42ab^3$, $-36a$, $90b$, $-48ab$
16. $45c^3d$, $75c$, $90d$, $-105cd$

An expression is factored when it is written as a product, not a sum. Which of the following are not *factored?*

17. $2k^2(5k)$
18. $2k^2(5k + 1)$
19. $2k^2 + (5k + 1)$
20. $(2k^2 + 1)(5k + 1)$
21. Is $-xy$ a common factor of $-x^4y^3$ and $-xy^2$? If so, what is the other factor that when multiplied by $-xy$ gives $-x^4y^3$?
22. Is $-a^5b^2$ a common factor of $-a^4b^5$ and $-a^3b$?

Complete each factoring.

23. $12 = 6(\quad)$
24. $18 = 9(\quad)$
25. $3x^2 = 3x(\quad)$
26. $8x^3 = 8x(\quad)$
27. $9m^4 = 3m^2(\quad)$
28. $12p^5 = 6p^3(\quad)$
29. $-8z^9 = -4z^5(\quad)$
30. $-15k^{11} = -5k^8(\quad)$
31. $6m^4n^5 = 3m^3n(\quad)$
32. $27a^3b^2 = 9a^2b(\quad)$
33. $-14x^4y^3 = 2xy(\quad)$
34. $-16m^3n^3 = 4mn^2(\quad)$

Factor out the greatest common factor. See Example 3.

35. $12y - 24$
36. $18p + 36$
37. $10a^2 - 20a$
38. $15x^3 - 30x^2$

39. $65y^{10} + 35y^6$
40. $100a^5 + 16a^3$
41. $11w^3 - 100$
42. $13z^5 - 80$
43. $8m^2n^3 + 24m^2n^2$
44. $19p^2y - 38p^2y^3$
45. $13y^8 + 26y^4 - 39y^2$
46. $5x^5 + 25x^4 - 20x^3$
47. $45q^4p^5 + 36qp^6 + 81q^2p^3$
48. $125a^3z^5 + 60a^4z^4 - 85a^5z^2$
49. $a^5 + 2a^3b^2 - 3a^5b^2 + 4a^4b^3$
50. $x^6 + 5x^4y^3 - 6xy^4 + 10xy$
51. $c(x + 2) - d(x + 2)$
52. $r(5 - x) + t(5 - x)$
53. $m(m + 2n) + n(m + 2n)$
54. $3p(1 - 4p) - 2q(1 - 4p)$

Students often have difficulty when factoring by grouping because they are not able to tell when the polynomial is indeed factored. For example,

$$5y(2x - 3) + 8t(2x - 3)$$

is not in factored form, because it is the *sum* of two terms, $5y(2x - 3)$ and $8t(2x - 3)$. However, because $2x - 3$ is a common factor of these two terms, the expression can now be factored as

$$(2x - 3)(5y + 8t).$$

The factored form is a *product* of two factors, $2x - 3$ and $5y + 8t$.

Determine whether each expression is in factored form or is not in factored form. If it is not in factored form, factor it if possible.

55. $8(7t + 4) + x(7t + 4)$
56. $3r(5x - 1) + 7(5x - 1)$
57. $(8 + x)(7t + 4)$
58. $(3r + 7)(5x - 1)$
59. $18x^2(y + 4) + 7(y + 4)$
60. $12k^3(s - 3) + 7(s + 3)$
61. Tell why it is not possible to factor the expression in Exercise 60.
62. Summarize the method of factoring a polynomial with four terms by grouping. Give an example.

Factor by grouping. See Examples 4 and 5.

63. $p^2 + 4p + 3p + 12$
64. $m^2 + 2m + 5m + 10$
65. $a^2 - 2a + 5a - 10$
66. $y^2 - 6y + 4y - 24$
67. $7z^2 + 14z - az - 2a$
68. $5m^2 + 15mp - 2mp - 6p^2$
69. $18r^2 + 12ry - 3xr - 2xy$
70. $8s^2 - 4st + 6sy - 3yt$
71. $3a^3 + 3ab^2 + 2a^2b + 2b^3$
72. $4x^3 + 3x^2y + 4xy^2 + 3y^3$
73. $1 - a + ab - b$
74. $6 - 3x - 2y + xy$
75. $16m^3 - 4m^2p^2 - 4mp + p^3$
76. $10t^3 - 2t^2s^2 - 5ts + s^3$
77. $5m + 15 - 2mp - 6p$
78. $y^2 - 3y - xy + 3x$
79. $18r^2 + 12ry - 3ry - 2y^2$
80. $3a^3 + 3ab^2 + 2a^2b + 2b^3$
81. $a^5 + 2a^5b - 3 - 6b$
82. $a^2b - 4a - ab^4 + 4b^3$

RELATING CONCEPTS (EXERCISES 83-88)

In most cases, the choice of which pairs of terms to group when factoring by grouping can be made in several ways.

Work Exercises 83–88 in order *to see how this applies to the polynomial in Example 5(b).*

83. Start with the polynomial in Example 5(b), $2xy + 12 - 3y - 8x$, and rearrange the terms as follows: $2xy - 8x + (-3y) + 12$. What properties from Section 1.7 allow us to do this?

EXAMPLE 2 Factoring a Trinomial with Two Negative Terms

Factor $p^2 - 2p - 15$.

Find two integers whose product is -15 and whose sum is -2. If these numbers do not come to mind right away, we can find them (if they exist) by listing all the pairs of integers whose product is -15. Because the last term, -15, is negative, we need pairs of integers with different signs.

$15, -1$	$15 + (-1) = 14$	
$5, -3$	$5 + (-3) = 2$	
$-15, 1$	$-15 + 1 = -14$	
$-5, 3$	$-5 + 3 = -2$	Sum is -2.

The necessary integers are -5 and 3, and

$$p^2 - 2p - 15 = (p - 5)(p + 3).$$

EXAMPLE 3 Deciding Whether a Polynomial Is Prime

(a) Factor $x^2 - 5x + 12$.

List all pairs of integers whose product is 12. Since the middle term is negative and the last term is positive, we need pairs with both numbers negative. Then examine the sums.

$-12, -1$	$-12 + (-1) = -13$
$-6, -2$	$-6 + (-2) = -8$
$-3, -4$	$-3 + (-4) = -7$

None of the pairs of integers has a sum of -5. Because of this, the trinomial $x^2 - 5x + 12$ *cannot be factored using only integer factors.* A polynomial that cannot be factored using only integer factors is called a **prime polynomial.**

(b) $k^2 - 8k + 11$

There is no pair of integers whose product is 11 and whose sum is -8, so $k^2 - 8k + 11$ is a prime polynomial.

The procedure for factoring a trinomial of the form $x^2 + bx + c$ is summarized here.

Factoring $x^2 + bx + c$

Find two integers whose product is c and whose sum is b.

1. Both integers must be positive if b and c are positive.
2. Both integers must be negative if c is positive and b is negative.
3. One integer must be positive and one must be negative if c is negative.

EXAMPLE 4 Factoring a Trinomial with Two Variables

Factor $z^2 - 2bz - 3b^2$.

To factor $z^2 - 2bz - 3b^2$, look for two expressions whose product is $-3b^2$ and whose sum is $-2b$. The expressions are $-3b$ and b, with

$$z^2 - 2bz - 3b^2 = (z - 3b)(z + b).$$

OBJECTIVE 2 Factor such polynomials after factoring out the greatest common factor. The trinomial in the next example does not have a coefficient of 1 for the squared term. (In fact, there is no squared term.) However, there may be a common factor.

EXAMPLE 5 Factoring a Trinomial with a Common Factor

Factor $4x^5 - 28x^4 + 40x^3$.

First, factor out the greatest common factor, $4x^3$.

$$4x^5 - 28x^4 + 40x^3 = 4x^3(x^2 - 7x + 10)$$

Now factor $x^2 - 7x + 10$. The integers -5 and -2 have a product of 10 and a sum of -7. The complete factored form is

$$4x^5 - 28x^4 + 40x^3 = 4x^3(x - 5)(x - 2).$$

CAUTION When factoring, always look for a common factor first. Remember to include the common factor as part of the answer. As a check, multiplying out the factored form should always give the original polynomial.

5.2 EXERCISES

1. In factoring a trinomial in x as $(x + a)(x + b)$, what must be true of a and b if the coefficient of the last term of the trinomial is negative?
2. In Exercise 1, what must be true of a and b if the coefficient of the last term is positive?
3. In your own words, explain the meaning of a *prime polynomial.*
4. A teacher asked a class to factor $m^3 + 3m^2 + 2m$. One student did not understand why her answer, $(m + 1)(m + 2)$ was incorrect. Explain her mistake, and give the correct answer.

In Exercises 5–8, list all pairs of integers with the given product. Then find the pair whose sum is given.

5. Product: 48 Sum: -19
6. Product: 48 Sum: 14
7. Product: -24 Sum: -5
8. Product: -36 Sum: -16
9. Which one of the following is the correct factored form of $x^2 - 12x + 32$?
   (a) $(x - 8)(x + 4)$ **(b)** $(x + 8)(x - 4)$
   (c) $(x - 8)(x - 4)$ **(d)** $(x + 8)(x + 4)$
10. Explain the steps you would use to factor $2x^3 + 8x^2 - 10x$.

Complete the following. See Examples 1–3.

11. $p^2 + 11p + 30 = (p + 5)(\quad)$
12. $x^2 + 10x + 21 = (x + 7)(\quad)$
13. $x^2 + 15x + 44 = (x + 4)(\quad)$
14. $r^2 + 15r + 56 = (r + 7)(\quad)$
15. $x^2 - 9x + 8 = (x - 1)(\quad)$
16. $t^2 - 14t + 24 = (t - 2)(\quad)$
17. $y^2 - 2y - 15 = (y + 3)(\quad)$
18. $t^2 - t - 42 = (t + 6)(\quad)$
19. $x^2 + 9x - 22 = (x - 2)(\quad)$
20. $x^2 + 6x - 27 = (x - 3)(\quad)$
21. $y^2 - 7y - 18 = (y + 2)(\quad)$
22. $y^2 - 2y - 24 = (y + 4)(\quad)$

RELATING CONCEPTS (EXERCISES 23–30)

To check your factoring of a trinomial, remember that when the factors are multiplied, the product must include *every* term of the original trinomial. In a trinomial such as $x^2 + x - 12$, students often factor so that the first and third terms are correct, but the middle term is incorrect, differing only in the sign.

***Work Exercises 23–30 in order** to see how this problem can be avoided.*

23. Add the pair of numbers: -3 and 4.

24. Add the pair of numbers: 3 and -4.

25. In Exercise 24, we added the *opposites* of the numbers in Exercise 23. How does the sum in Exercise 24 compare with the sum in Exercise 23?

26. Consider this factored form, given for $x^2 + x - 12$: $(x + 3)(x - 4)$. Multiply the two binomials. Is this the correct factored form? Why or why not?

27. Consider this factored form, given for $x^2 + x - 12$: $(x - 3)(x + 4)$. Multiply the two binomials. Is this the correct factored form? Why or why not?

28. In Exercises 26 and 27, one factored form is correct and one is not. For the one that is not, how does the sign of the middle term of the product compare with the signs of the second terms of the correct factored form?

29. Compare the two factored forms shown in Exercises 26 and 27, and use the result of Exercise 28 to complete the following statement: When I factor a trinomial into a product of binomials, and the middle term of the product is different only in sign, I should ______ in order to obtain the correct factored form.

30. Given that the factored form $(x + 5)(x - 3)$ is incorrect only in the sign of the middle term of the product for a particular trinomial, what would be the correct factored form?

Did you make the connection that if only the sign of the middle term is incorrect when you check the product of the binomial factors, then only the signs of the second terms in the binomials need to be changed?

Factor completely. If the polynomial cannot be factored, write prime. *See Examples 1–3.*

31. $y^2 + 9y + 8$

32. $a^2 + 9a + 20$

33. $b^2 + 8b + 15$

34. $x^2 + 6x + 8$

35. $m^2 + m - 20$

36. $p^2 + 4p - 5$

37. $y^2 - 8y + 15$

38. $y^2 - 6y + 8$

39. $x^2 + 4x + 5$

40. $t^2 + 11t + 12$

41. $t^2 - 8t + 16$

42. $s^2 - 10s + 25$

43. $r^2 - r - 30$

44. $q^2 - q - 42$

45. $n^2 - 12n - 35$

46. $x^2 - 4x + 12$

Factor completely. See Examples 4 and 5.

47. $r^2 + 3ra + 2a^2$

48. $x^2 + 5xa + 4a^2$

49. $t^2 - tz - 6z^2$

50. $a^2 - ab - 12b^2$

51. $x^2 + 4xy + 3y^2$

52. $p^2 + 9pq + 8q^2$

53. $v^2 - 11vw + 30w^2$

54. $v^2 - 11vx + 24x^2$

55. $4x^2 + 12x - 40$

56. $5y^2 - 5y - 30$

57. $2t^3 + 8t^2 + 6t$

58. $3t^3 + 27t^2 + 24t$

59. $2x^6 + 8x^5 - 42x^4$

60. $4y^5 + 12y^4 - 40y^3$

61. $m^3n - 10m^2n^2 + 24mn^3$

62. $y^3z + 3y^2z^2 - 54yz^3$

63. Use the FOIL method from Section 4.3 to show that $(2x + 4)(x - 3) = 2x^2 - 2x - 12$. If you are asked to completely factor $2x^2 - 2x - 12$, why would it be incorrect to give $(2x + 4)(x - 3)$ as your answer?

64. If you are asked to completely factor the polynomial $3x^2 + 9x - 12$, why would it be incorrect to give $(x - 1)(3x + 12)$ as your answer?

Use a combination of the factoring methods discussed in this section to factor each polynomial.

65. $a^5 + 3a^4b - 4a^3b^2$
66. $m^3n - 2m^2n^2 - 3mn^3$
67. $y^3z + y^2z^2 - 6yz^3$
68. $k^7 - 2k^6m - 15k^5m^2$
69. $z^{10} - 4z^9y - 21z^8y^2$
70. $x^9 + 5x^8w - 24x^7w^2$
71. $(a + b)x^2 + (a + b)x - 12(a + b)$
72. $(x + y)n^2 + (x + y)n + 16(x + y)$
73. $(2p + q)r^2 - 12(2p + q)r + 27(2p + q)$
74. $(3m - n)k^2 - 13(3m - n)k + 40(3m - n)$
75. What polynomial can be factored as $(a + 9)(a + 4)$?
76. What polynomial can be factored as $(y - 7)(y + 3)$?

5.3 More on Factoring Trinomials

OBJECTIVES

1. Factor trinomials by grouping when the coefficient of the squared term is not 1.
2. Factor trinomials using FOIL.

FOR EXTRA HELP

SSG Sec. 5.3
SSM Sec. 5.3
Pass the Test Software
InterAct Math Tutorial Software
Video 8

Trinomials such as $2x^2 + 7x + 6$, in which the coefficient of the squared term is *not* 1, can be factored with an extension of the method presented in the last section.

OBJECTIVE 1 **Factor trinomials by grouping when the coefficient of the squared term is not 1.** Recall that a trinomial such as $m^2 + 3m + 2$ is factored by finding two numbers whose product is 2 and whose sum is 3. To factor $2x^2 + 7x + 6$, we look for two integers whose product is $2 \cdot 6 = 12$ and whose sum is 7.

Sum is 7.

$$2x^2 + 7x + 6$$

Product is $2 \cdot 6 = 12$.

By considering the pairs of positive integers whose product is 12, the necessary integers are found to be 3 and 4. We use these integers to write the middle term, $7x$, as $7x = 3x + 4x$. With this, the trinomial $2x^2 + 7x + 6$ becomes

$$2x^2 + 7x + 6 = 2x^2 + \underbrace{3x + 4x}_{7x = 3x + 4x} + 6.$$

Factor the new polynomial by grouping as in Section 5.1.

$$\begin{aligned} 2x^2 + 3x + 4x + 6 &= x(2x + 3) + 2(2x + 3) \\ &= (2x + 3)(x + 2) \end{aligned}$$

The common factor of $2x + 3$ was factored out to get

$$2x^2 + 7x + 6 = (2x + 3)(x + 2).$$

Check by finding the product of $2x + 3$ and $x + 2$.

We could have written the middle term in the polynomial $2x^2 + 7x + 6$ as $7x = 4x + 3x$ to get

$$\begin{aligned} 2x^2 + 7x + 6 &= 2x^2 + 4x + 3x + 6 \\ &= 2x(x + 2) + 3(x + 2) \\ &= (x + 2)(2x + 3). \end{aligned}$$

Either result is correct.

EXAMPLE 1 Factoring Trinomials by Grouping

Factor the trinomial.

(a) $6r^2 + r - 1$

We must find two integers with a product of $6(-1) = -6$ and a sum of 1.

Sum is 1.

$$6r^2 + r - 1 = 6r^2 + 1r - 1$$

Product is $6(-1) = -6$.

The integers are -2 and 3. We write the middle term, $+r$, as $-2r + 3r$, so that

$$6r^2 + r - 1 = 6r^2 - 2r + 3r - 1.$$

Factor by grouping on the right-hand side.

$$\begin{aligned} 6r^2 + r - 1 &= 6r^2 - 2r + 3r - 1 \\ &= 2r(3r - 1) + 1(3r - 1) && \text{The binomials must be the same.} \\ &= (3r - 1)(2r + 1) \end{aligned}$$

(b) $12z^2 - 5z - 2$

Look for two integers whose product is $12(-2) = -24$ and whose sum is -5. The required integers are 3 and -8, and

$$\begin{aligned} 12z^2 - 5z - 2 &= 12z^2 + 3z - 8z - 2 && -5z = 3z - 8z \\ &= 3z(4z + 1) - 2(4z + 1) && \text{Group terms and factor each group.} \\ &= (4z + 1)(3z - 2). && \text{Factor out } 4z + 1. \end{aligned}$$

(c) $10m^2 + mn - 3n^2$

Two integers whose product is $10(-3) = -30$ and whose sum is 1 are -5 and 6. Rewrite the trinomial with four terms.

$$\begin{aligned} 10m^2 + mn - 3n^2 &= 10m^2 - 5mn + 6mn - 3n^2 && mn = -5mn + 6mn \\ &= 5m(2m - n) + 3n(2m - n) && \text{Group terms and factor each group.} \\ &= (2m - n)(5m + 3n) && \text{Factor out the common factor.} \end{aligned}$$

OBJECTIVE 2 Factor trinomials using FOIL. The rest of this section shows an alternative method of factoring trinomials in which the coefficient of the squared term is not 1. This method uses trial and error.

To factor $2x^2 + 7x + 6$ (the trinomial factored at the beginning of this section) by trial and error, we must use FOIL backwards. We want to write $2x^2 + 7x + 6$ as the product of two binomials.

$$2x^2 + 7x + 6 = (\qquad)(\qquad)$$

The product of the two first terms of the binomials is $2x^2$. The possible factors of $2x^2$ are $2x$ and x or $-2x$ and $-x$. Since all terms of the trinomial are positive, we consider only positive factors. Thus, we have

$$2x^2 + 7x + 6 = (2x \qquad)(x \qquad).$$

The product of the two last terms, 6, can be factored as $1 \cdot 6$, $6 \cdot 1$, $2 \cdot 3$, or $3 \cdot 2$. Try each pair to find the pair that gives the correct middle term.

$(2x + 1)(x + 6)$ Incorrect

x
$12x$
$13x$ Add.

$(2x + 6)(x + 1)$ Incorrect

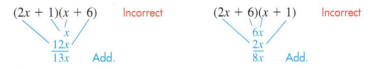

$6x$
$2x$
$8x$ Add.

Since $2x + 6 = 2(x + 3)$, the binomial $2x + 6$ has a common factor of 2, while $2x^2 + 7x + 6$ has no common factor other than 1. The product $(2x + 6)(x + 1)$ cannot be correct.

NOTE If the original polynomial has no common factor, then none of its binomial factors will either.

Now try the numbers 2 and 3 as factors of 6. Because of the common factor of 2 in $2x + 2$, $(2x + 2)(x + 3)$ will not work. Try $(2x + 3)(x + 2)$.

$(2x + 3)(x + 2) = 2x^2 + 7x + 6$ Correct

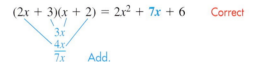

$3x$
$4x$
$7x$ Add.

Finally, we see that $2x^2 + 7x + 6$ factors as

$$2x^2 + 7x + 6 = (2x + 3)(x + 2).$$

Check by multiplying $2x + 3$ and $x + 2$.

EXAMPLE 2 Factoring a Trinomial with All Terms Positive Using FOIL

Factor $8p^2 + 14p + 5$.

The number 8 has several possible pairs of factors, but 5 has only 1 and 5 or -1 and -5. For this reason, it is easier to begin by considering the factors of 5. Ignore the negative factors since all coefficients in the trinomial are positive. If $8p^2 + 14p + 5$ can be factored, the factors will have the form

$$(\qquad + 5)(\qquad + 1).$$

The possible pairs of factors of $8p^2$ are $8p$ and p, or $4p$ and $2p$. Try various combinations, checking the middle term in each case.

$(8p + 5)(p + 1)$ Incorrect

$5p$
$8p$
$13p$ Add.

$(p + 5)(8p + 1)$ Incorrect

$40p$
p
$41p$ Add.

$(4p + 5)(2p + 1)$ Correct

$10p$
$4p$
$14p$ Add.

Since the sum $14p$ is the correct middle term, the trinomial $8p^2 + 14p + 5$ factors as $(4p + 5)(2p + 1)$.

EXAMPLE 3 **Factoring a Trinomial with a Negative Middle Term Using FOIL**

Factor $6x^2 - 11x + 3$.

Since 3 has only 1 and 3 or -1 and -3 as factors, it is better here to begin by factoring 3. The last term of the trinomial $6x^2 - 11x + 3$ is positive and the middle term has a negative coefficient, so consider only negative factors. We need negative factors because the *product* of two negative factors is positive and their *sum* is negative, as required. Use -3 and -1 as factors of 3:

$$(\quad - 3)(\quad - 1).$$

The factors of $6x^2$ may be either $6x$ and x, or $2x$ and $3x$. Try $2x$ and $3x$.

$(2x - 3)(3x - 1)$ Correct

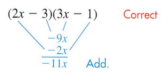

$-9x$
$-2x$
$-11x$ Add.

These factors give the correct middle term, so

$$6x^2 - 11x + 3 = (2x - 3)(3x - 1).$$

EXAMPLE 4 **Factoring a Trinomial with a Negative Last Term Using FOIL**

Factor $8x^2 + 6x - 9$.

The integer 8 has several possible pairs of factors, as does -9. Since the last term is negative, one positive factor and one negative factor of -9 are needed. Since the coefficient of the middle term is small, it is wise to avoid large factors such as 8 or 9. Let us try 4 and 2 as factors of 8, and 3 and -3 as factors of -9, and check the middle term.

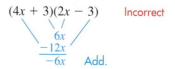

$(4x + 3)(2x - 3)$ Incorrect

$6x$
$-12x$
$-6x$ Add.

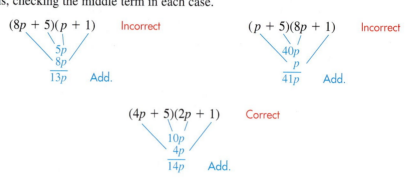

Now, try exchanging 3 and -3, since only the sign of the middle term is incorrect.

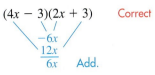

$$(4x - 3)(2x + 3) \qquad \text{Correct}$$

$$\begin{array}{r} -6x \\ 12x \\ \hline 6x \end{array} \quad \text{Add.}$$

This time we got the correct middle term, so

$$8x^2 + 6x - 9 = (4x - 3)(2x + 3).$$

EXAMPLE 5 Factoring a Trinomial with Two Variables

Factor $12a^2 - ab - 20b^2$.

There are several pairs of factors of $12a^2$, including $12a$ and a, $6a$ and $2a$, and $3a$ and $4a$, just as there are many pairs of factors of $-20b^2$, including $-20b$ and b, $10b$ and $-2b$, $-10b$ and $2b$, $4b$ and $-5b$, and $-4b$ and $5b$. Once again, since the desired middle term is small, avoid the larger factors. Try the factors $6a$ and $2a$ and $4b$ and $-5b$.

$$(6a + 4b)(2a - 5b)$$

This cannot be correct, as mentioned before, since $6a + 4b$ has a common factor while the given trinomial has none. Try $3a$ and $4a$ with $4b$ and $-5b$.

$$(3a + 4b)(4a - 5b) = 12a^2 + ab - 20b^2 \qquad \text{Incorrect}$$

Here the middle term has the wrong sign, so change the signs in the factors.

$$(3a - 4b)(4a + 5b) = 12a^2 - ab - 20b^2 \qquad \text{Correct}$$

EXAMPLE 6 Factoring a Trinomial with a Common Factor

Factor $28x^5 - 58x^4 - 30x^3$.

First factor out the greatest common factor, $2x^3$.

$$28x^5 - 58x^4 - 30x^3 = 2x^3(14x^2 - 29x - 15)$$

Now try to factor $14x^2 - 29x - 15$. Try $7x$ and $2x$ as factors of $14x^2$ and -3 and 5 as factors of -15.

$$(7x - 3)(2x + 5) = 14x^2 + 29x - 15 \qquad \text{Incorrect}$$

The middle term differs only in sign, so change the signs in the two factors.

$$(7x + 3)(2x - 5) = 14x^2 - 29x - 15 \qquad \text{Correct}$$

Finally, the factored form of $28x^5 - 58x^4 - 30x^3$ is

$$28x^5 - 58x^4 - 30x^3 = 2x^3(7x + 3)(2x - 5).$$

CAUTION Remember to include the common factor in the final result.

EXAMPLE 7 Factoring a Trinomial with a Negative Common Factor

Factor $-24a^3 - 42a^2 + 45a$.

The common factor could be $3a$ or $-3a$. If we factor out $-3a$, the first term of the trinomial factor will be positive, which makes it easier to factor.

$$-24a^3 - 42a^2 + 45a = -3a(8a^2 + 14a - 15) \quad \text{Factor out the greatest common factor.}$$
$$= -3a(4a - 3)(2a + 5) \quad \text{Use trial and error.}$$

5.3 EXERCISES

In Exercises 1–6 complete the steps in factoring the trinomial $2x^2 + x - 21$ by grouping.

1. Find the product of ______ and ______.
2. Find factors of ______ that have a sum of ______.
3. Write the middle term x as ______ + ______.
4. Factor the polynomial ______ by grouping as (______) + (______).
5. Factor out the common factor of ______.
6. The factored form is ______.

Fill in the blanks in Exercises 7–10.

7. $6a^2 + 7ab - 20b^2 = (3a \_\_\_\_\_\_)(\_\_\_\_\_\_ + 5b)$
8. $9m^2 - 3mn - 2n^2 = (3m \_\_\_\_\_\_)(\_\_\_\_\_\_ - 2n)$
9. $3x^2 - 9x - 30 = 3(x^2 \_\_\_\_\_\_) = 3(x \_\_\_\_\_\_)(x \_\_\_\_\_\_)$
10. $4z^3 - 10z^2 - 6z = \_\_\_\_\_\_(2z^2 \_\_\_\_\_\_) = 2z(\_\_\_\_\_\_ - 3)(2z \_\_\_\_\_\_)$

Decide which is the correct factored form of the given polynomial.

11. $4y^2 + 17y - 15$
    (a) $(y + 5)(4y - 3)$
    (b) $(2y - 5)(2y + 3)$
12. $12c^2 - 7c - 12$
    (a) $(6c - 2)(2c + 6)$
    (b) $(4c + 3)(3c - 4)$
13. $4k^2 + 13mk + 3m^2$
    (a) $(4k + m)(k + 3m)$
    (b) $(4k + 3m)(k + m)$
14. $2x^2 + 11x + 12$
    (a) $(2x + 3)(x + 4)$
    (b) $(2x + 4)(x + 3)$
15. For the polynomial $12x^2 + 7x - 12$, 2 is not a common factor. Explain why the binomial $2x - 6$, then, cannot be a factor of the polynomial.
16. Explain how the signs of the last terms of the two binomial factors of a trinomial are determined.

Factor completely. Use either method described in this section. See Examples 1–7.

17. $3a^2 + 10a + 7$
18. $7r^2 + 8r + 1$
19. $4r^2 + r - 3$
20. $4r^2 + 3r - 10$
21. $15m^2 + m - 2$
22. $6x^2 + x - 1$
23. $8m^2 - 10m - 3$
24. $12s^2 + 11s - 5$
25. $20x^2 + 11x - 3$
26. $20x^2 - 28x - 3$
27. $21m^2 + 13m + 2$
28. $38x^2 + 23x + 2$
29. $20y^2 + 39y - 11$
30. $10x^2 + 11x - 6$
31. $6b^2 + 7b + 2$
32. $6w^2 + 19w + 10$

33. $24x^2 - 42x + 9$

34. $48b^2 - 74b - 10$

35. $40m^2q + mq - 6q$

36. $15a^2b + 22ab + 8b$

37. $2m^3 + 2m^2 - 40m$

38. $3x^3 + 12x^2 - 36x$

39. $15n^4 - 39n^3 + 18n^2$

40. $24a^4 + 10a^3 - 4a^2$

41. $18x^5 + 15x^4 - 75x^3$

42. $32z^5 - 20z^4 - 12z^3$

43. $15x^2y^2 - 7xy^2 - 4y^2$

44. $14a^2b^3 + 15ab^3 - 9b^3$

45. $12p^2 + 7pq - 12q^2$

46. $6m^2 - 5mn - 6n^2$

47. $25a^2 + 25ab + 6b^2$

48. $6x^2 - 5xy - y^2$

49. $6a^2 - 7ab - 5b^2$

50. $25g^2 - 5gh - 2h^2$

51. $6m^6n + 7m^5n^2 + 2m^4n^3$

52. $12k^3q^4 - 4k^2q^5 - kq^6$

53. $5 - 6x + x^2$

54. $7 + 8x + x^2$

55. $16 + 16x + 3x^2$

56. $18 + 65x + 7x^2$

57. $-10x^3 + 5x^2 + 140x$

58. $-18k^3 - 48k^2 + 66k$

If a trinomial has a negative coefficient for the squared term, such as $-2x^2 + 11x - 12$, it is usually easier to factor by first factoring out the common factor -1:

$$\begin{aligned} -2x^2 + 11x - 12 &= -1(2x^2 - 11x + 12) \\ &= -1(2x - 3)(x - 4). \end{aligned}$$

Use this method to factor each trinomial. See Example 7.

59. $-x^2 - 4x + 21$

60. $-x^2 + x + 72$

61. $-3x^2 - x + 4$

62. $-5x^2 + 2x + 16$

63. $-2a^2 - 5ab - 2b^2$

64. $-3p^2 + 13pq - 4q^2$

65. The answer given in the back of the book for Exercise 59 is $-1(x + 7)(x - 3)$. Is $(x + 7)(3 - x)$ also a correct answer? Explain why or why not.

66. One answer for Exercise 60 is $-1(x + 8)(x - 9)$. Is $(-x - 8)(-x + 9)$ also a correct answer? Explain.

Factor each polynomial. Remember to factor out the greatest common factor as the first step.

67. $25q^2(m + 1)^3 - 5q(m + 1)^3 - 2(m + 1)^3$

68. $18x^2(y - 3)^2 - 21x(y - 3)^2 - 4(y - 3)^2$

69. $15x^2(r + 3)^3 - 34xy(r + 3)^3 - 16y^2(r + 3)^3$

70. $4t^2(k + 9)^7 + 20ts(k + 9)^7 + 25s^2(k + 9)^7$

Find all integers k so that the trinomial can be factored using the methods of this section. (Hint: *Try all possible factored forms with the given first and last terms. The coefficient of x will give the values of k.*)

71. $5x^2 + kx - 1$

72. $2c^2 + kc - 3$

73. $2m^2 + km + 5$

74. $3y^2 + ky + 3$

RELATING CONCEPTS (EXERCISES 75–82)

One of the most common questions that beginning algebra students ask is this: "If my answer doesn't look exactly like the one given in the back of the book, is it necessarily incorrect?" Often there are several different equivalent forms of an answer that are all correct.

Work Exercises 75–82 in order *to see how and why this is possible for factoring problems.*

75. Factor the integer 35 as the product of two prime numbers.

76. Factor the integer 35 as the product of the negatives of two prime numbers.

77. Verify the following factored form: $6x^2 - 11x + 4 = (3x - 4)(2x - 1)$.

78. Verify the following factored form: $6x^2 - 11x + 4 = (4 - 3x)(1 - 2x)$.

79. Compare the two valid factored forms in Exercises 77 and 78. How do the factors in each case compare?

80. Suppose you know that the correct factored form of a particular trinomial is $(7t - 3)(2t - 5)$. Based on your observations in Exercises 77–79, what is another valid factored form?

81. Look at your results in Exercises 75 and 76, and fill in the blanks: If an integer factors as the product of *a* and *b,* then it also factors as the product of _____ and _____.

82. Look at your results in Exercises 77 and 78 and fill in the blanks: If a trinomial factors as the product of the binomials *P* and *Q,* then it also factors as the product of the binomials _____ and _____.

Did you make the connection that the product of the negatives of two factors is the same as the product of the two factors?

5.4 Special Factoring Rules

OBJECTIVES

1. Factor the difference of two squares.
2. Factor a perfect square trinomial.
3. Factor the difference of two cubes.
4. Factor the sum of two cubes.

FOR EXTRA HELP

SSG Sec. 5.4
SSM Sec. 5.4
Pass the Test Software
InterAct Math Tutorial Software
Video 9

By reversing the rules for multiplying binomials that we learned in the last chapter, we get rules for factoring polynomials in certain forms.

OBJECTIVE 1 **Factor the difference of two squares.** Recall from the last chapter that

$$(x + y)(x - y) = x^2 - y^2.$$

Based on this product, we factor a **difference of two squares** as follows.

Difference of Two Squares

$$x^2 - y^2 = (x + y)(x - y)^*$$

*A pair of expressions like $x + y$ and $x - y$ are called *conjugates.*

EXAMPLE 1 Factoring a Difference of Squares

Factor each difference of two squares.

(a) $x^2 - 49 = x^2 - 7^2 = (x + 7)(x - 7)$

(b) $y^2 - m^2 = (y + m)(y - m)$

(c) $z^2 - \frac{9}{16} = z^2 - \left(\frac{3}{4}\right)^2 = \left(z + \frac{3}{4}\right)\left(z - \frac{3}{4}\right)$

(d) $x^2 - 8$
Because 8 is not the square of an integer, this is a prime polynomial.

(e) $p^2 + 16$
The polynomial is not the *difference* of two squares. Using FOIL,

$$(p + 4)(p - 4) = p^2 - 16$$
$$(p - 4)(p - 4) = p^2 - 8p + 16$$

and

$$(p + 4)(p + 4) = p^2 + 8p + 16$$

so $p^2 + 16$ is a prime polynomial.

CAUTION As Example 1(e) suggests, after any common factor is removed, the sum of two squares cannot be factored.

EXAMPLE 2 Factoring More Complex Differences of Squares

Factor completely.

(a) $9a^2 - 4b^2$
This is a difference of two squares because

$$9a^2 - 4b^2 = (3a)^2 - (2b)^2,$$

so $9a^2 - 4b^2 = (3a + 2b)(3a - 2b)$.

(b) $81y^2 - 36$
First factor out the common factor of 9.

$$\begin{aligned} 81y^2 - 36 &= 9(9y^2 - 4) \\ &= 9(3y + 2)(3y - 2) \end{aligned}$$

(c) $p^4 - 36 = (p^2)^2 - 6^2 = (p^2 + 6)(p^2 - 6)$
Neither $p^2 + 6$ nor $p^2 - 6$ can be factored further.

(d) $m^4 - 16 = (m^2)^2 - 4^2$

$$\begin{aligned} &= (m^2 + 4)(m^2 - 4) && \text{Difference of squares} \\ &= (m^2 + 4)(m + 2)(m - 2) && \text{Difference of squares} \end{aligned}$$

CAUTION Remember to factor again when any of the *factors* is a difference of squares, as in Example 2(d).

OBJECTIVE 2 **Factor a perfect square trinomial.** The expressions 144, $4x^2$, and $81m^6$ are called *perfect squares,* since

$$144 = 12^2, \qquad 4x^2 = (2x)^2, \qquad \text{and} \qquad 81m^6 = (9m^3)^2.$$

A **perfect square trinomial** is a trinomial that is the square of a binomial. For example, $x^2 + 8x + 16$ is a perfect square trinomial since it is the square of the binomial $x + 4$:

$$x^2 + 8x + 16 = (x + 4)^2.$$

For a trinomial to be a perfect square, two of its terms must be perfect squares. For this reason, $16x^2 + 4x + 15$ cannot be a perfect square trinomial since only the term $16x^2$ is a perfect square.

On the other hand, even though two of the terms are perfect squares, the trinomial may not be a perfect square trinomial. For example, $x^2 + 6x + 36$ has two perfect square terms, but it is not a perfect square trinomial. (Try to find a binomial that can be squared to give $x^2 + 6x + 36$; it cannot be done.)

We can multiply to see that the square of a binomial gives the following perfect square trinomials.

Perfect Square Trinomial

$$x^2 + 2xy + y^2 = (x + y)^2$$
$$x^2 - 2xy + y^2 = (x - y)^2$$

The middle term of a perfect square trinomial is always twice the product of the two terms in the squared binomial. (This was shown in Section 4.4.) Use this to check any attempt to factor a trinomial that appears to be a perfect square.

While perfect square trinomials can be factored using the procedures of Sections 5.2 and 5.3, it is usually more efficient to recognize the pattern and factor accordingly. The following example illustrates this.

EXAMPLE 3 **Factoring a Perfect Square Trinomial**

Factor the perfect square trinomial.

(a) $x^2 + 10x + 25$

The term x^2 is a perfect square and so is 25. Try to factor the trinomial as

$$x^2 + 10x + 25 = (x + 5)^2.$$

To check, take twice the product of the two terms in the squared binomial.

Twice → $2 \cdot x \cdot 5 = 10x$

First term of binomial ↑ ↑ Last term of binomial

Since $10x$ is the middle term of the trinomial, the trinomial is a perfect square and can be factored as $(x + 5)^2$.

(b) $x^2 - 22xz + 121z^2$

The first and last terms are perfect squares ($121 = 11^2$). Check to see whether the middle term of $x^2 - 22xz + 121z^2$ is twice the product of the first and last terms of the binomial $x - 11z$.

$$\text{Twice} \rightarrow 2 \cdot x \cdot 11z = 22xz$$

First term of binomial ↑ (under x); Last term of binomial ↑ (under $11z$)

Since twice the product of the first and last terms of the binomial is the middle term, $x^2 - 22xz + 121z^2$ is a perfect square trinomial and

$$x^2 - 22xz + 121z^2 = (x - 11z)^2.$$

(c) $9m^2 - 24m + 16 = (3m)^2 - 2(3m)(4) + 4^2 = (3m - 4)^2$

Twice ↑ (under 2); First term ↑ (under $3m$); Last term ↑ (under 4)

(d) $25y^2 + 20y + 16$

The first and last terms are perfect squares.

$$25y^2 = (5y)^2 \qquad \text{and} \qquad 16 = 4^2$$

Twice the product of the first and last terms of the binomial $5y + 4$ is

$$2 \cdot 5y \cdot 4 = 40y,$$

which is not the middle term of $25y^2 + 20y + 16$. This polynomial is not a perfect square. In fact, the polynomial cannot be factored even with the methods of Section 5.3; it is a prime polynomial.

NOTE: The sign of the second term in the squared binomial is always the same as the sign of the middle term in the trinomial. Also, the first and last terms of a perfect square trinomial must be positive, since they are squares. For example, the polynomial $x^2 - 2x - 1$ cannot be a perfect square because the last term is negative.

OBJECTIVE 3 **Factor the difference of two cubes.** The difference of two squares was factored above; we can also factor the **difference of two cubes.** Use the following pattern.

Difference of Two Cubes

$$x^3 - y^3 = (x - y)(x^2 + xy + y^2)$$

This pattern *should be memorized.* Multiply on the right to see that the pattern gives the correct factors.

$$\begin{array}{rrrr}
 & x^2 & + xy & + y^2 \\
 & & x & - y \\
\hline
 & - x^2y & - xy^2 & - y^3 \\
x^3 & + x^2y & + xy^2 & \\
\hline
x^3 & & & - y^3
\end{array}$$

Notice the pattern of the terms in the factored form of $x^3 - y^3$.

- $x^3 - y^3 =$ (a binomial factor)(a trinomial factor)
- The binomial factor has the difference of the cube roots of the given terms.
- The terms in the trinomial factor are all positive.
- What you write in the binomial factor determines the trinomial factor:

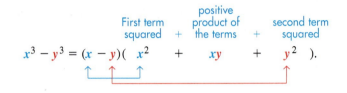

First term squared + positive product of the terms + second term squared

$$x^3 - y^3 = (x - y)(\;x^2 \;+\; xy \;+\; y^2\;).$$

CAUTION

The polynomial $x^3 - y^3$ is not equivalent to $(x - y)^3$, because $(x - y)^3$ can also be written as

$$\begin{aligned}(x - y)^3 &= (x - y)(x - y)(x - y)\\ &= (x - y)(x^2 - 2xy + y^2)\end{aligned}$$

but

$$x^3 - y^3 = (x - y)(x^2 + xy + y^2).$$

EXAMPLE 4 **Factoring Differences of Cubes**

Factor the following.

(a) $m^3 - 125$

Let $x = m$ and $y = 5$ in the pattern for the difference of two cubes.

$$\begin{aligned}x^3 - y^3 &= (x - y)(x^2 + xy + y^2)\\ m^3 - 125 &= m^3 - 5^3 = (m - 5)(m^2 + 5m + 5^2) \qquad \text{Let } x = m,\ y = 5.\\ &= (m - 5)(m^2 + 5m + 25)\end{aligned}$$

(b) $8p^3 - 27$

Since $8p^3 = (2p)^3$ and $27 = 3^3$, substitute into the rule using $2p$ for x and 3 for y.

$$\begin{aligned}8p^3 - 27 &= (2p)^3 - 3^3\\ &= (2p - 3)[(2p)^2 + (2p)3 + 3^2]\\ &= (2p - 3)(4p^2 + 6p + 9)\end{aligned}$$

(c) $$\begin{aligned}4m^3 - 32 &= 4(m^3 - 8)\\ &= 4(m^3 - 2^3)\\ &= 4(m - 2)(m^2 + 2m + 4)\end{aligned}$$

(d) $$\begin{aligned}125t^3 - 216s^6 &= (5t)^3 - (6s^2)^3\\ &= (5t - 6s^2)[(5t)^2 + (5t)(6s^2) + (6s^2)^2]\\ &= (5t - 6s^2)(25t^2 + 30ts^2 + 36s^4)\end{aligned}$$

CAUTION

A common error in factoring the difference of two cubes, such as $x^3 - y^3 = (x - y)(x^2 + xy + y^2)$, is to try to factor $x^2 + xy + y^2$. It is easy to confuse this factor with a perfect square trinomial, $x^2 + 2xy + y^2$. Because there is no 2 in $x^2 + xy + y^2$, it is very unusual to be able to further factor an expression of the form $x^2 + xy + y^2$.

OBJECTIVE 4 **Factor the sum of two cubes.** A sum of two squares, such as $m^2 + 25$, cannot be factored using real numbers, but the **sum of two cubes** can be factored by the following pattern, *which should be memorized.*

Sum of Two Cubes

$$x^3 + y^3 = (x + y)(x^2 - xy + y^2)$$

Compare the pattern for the *sum* of two cubes with the pattern for the *difference* of two cubes. The only difference between them is the positive and negative signs.

$$x^3 - y^3 = (x - y)(x^2 + xy + y^2)$$

Positive

Same sign

Opposite sign

$$x^3 + y^3 = (x + y)(x^2 - xy + y^2)$$

Positive

Same sign

Opposite sign

Observing these relationships should help you to remember these patterns.

EXAMPLE 5 **Factoring Sums of Cubes**

Factor.

(a) $k^3 + 27 = k^3 + 3^3$

$= (k + 3)(k^2 - 3k + 3^2)$

$= (k + 3)(k^2 - 3k + 9)$

(b) $8m^3 + 125 = (2m)^3 + 5^3$

$= (2m + 5)[(2m)^2 - (2m)(5) + 5^2]$

$= (2m + 5)(4m^2 - 10m + 25)$

(c) $1000a^6 + 27b^3 = (10a^2)^3 + (3b)^3$

$= (10a^2 + 3b)[(10a^2)^2 - (10a^2)(3b) + (3b)^2]$

$= (10a^2 + 3b)(100a^4 - 30a^2b + 9b^2)$

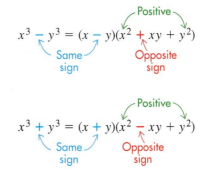

The methods of factoring discussed in this section are summarized here. All these rules should be memorized.

Special Factorizations

Difference of two squares	$x^2 - y^2 = (x + y)(x - y)$
Perfect square trinomials	$x^2 + 2xy + y^2 = (x + y)^2$
	$x^2 - 2xy + y^2 = (x - y)^2$
Difference of two cubes	$x^3 - y^3 = (x - y)(x^2 + xy + y^2)$
Sum of two cubes	$x^3 + y^3 = (x + y)(x^2 - xy + y^2)$

Remember the *sum* of two *squares* can be factored only if the terms have a common factor.

5.4 EXERCISES

1. To help you factor the difference of squares, complete the following list of squares.

$1^2 =$ _____ $2^2 =$ _____ $3^2 =$ _____ $4^2 =$ _____ $5^2 =$ _____

$6^2 =$ _____ $7^2 =$ _____ $8^2 =$ _____ $9^2 =$ _____ $10^2 =$ _____

$11^2 =$ _____ $12^2 =$ _____ $13^2 =$ _____ $14^2 =$ _____ $15^2 =$ _____

$16^2 =$ _____ $17^2 =$ _____ $18^2 =$ _____ $19^2 =$ _____ $20^2 =$ _____

2. The following powers of x are all perfect squares: $x^2, x^4, x^6, x^8, x^{10}$. Based on this observation, we may make a conjecture (an educated guess) that if the power of a variable is divisible by _____ (with 0 remainder), then we have a perfect square.

3. To help you factor the sum or difference of cubes, complete the following list of cubes.

$1^3 =$ _____ $2^3 =$ _____ $3^3 =$ _____ $4^3 =$ _____ $5^3 =$ _____

$6^3 =$ _____ $7^3 =$ _____ $8^3 =$ _____ $9^3 =$ _____ $10^3 =$ _____

4. The following powers of x are all perfect cubes: $x^3, x^6, x^9, x^{12}, x^{15}$. Based on this observation, we may make a conjecture that if the power of a variable is divisible by _____ (with 0 remainder), then we have a perfect cube.

5. Identify the monomial as a perfect square, a perfect cube, both of these, or neither of these.
(a) $64x^6y^{12}$ **(b)** $125t^6$ **(c)** $49x^{12}$ **(d)** $81r^{10}$

6. What must be true for x^n to be both a perfect square and a perfect cube?

Factor each binomial completely. Use your answers in Exercises 1 and 2 as necessary. See Examples 1 and 2.

7. $y^2 - 25$ **8.** $t^2 - 16$ **9.** $9r^2 - 4$ **10.** $4x^2 - 9$

11. $36m^2 - \frac{16}{25}$ **12.** $100b^2 - \frac{4}{49}$ **13.** $36x^2 - 16$ **14.** $32a^2 - 8$

15. $196p^2 - 225$ **16.** $361q^2 - 400$ **17.** $16r^2 - 25a^2$ **18.** $49m^2 - 100p^2$

19. $100x^2 + 49$ **20.** $81w^2 + 16$ **21.** $p^4 - 49$ **22.** $r^4 - 25$

23. $x^4 - 1$ **24.** $y^4 - 16$ **25.** $p^4 - 256$ **26.** $16k^4 - 1$

27. When a student was directed to factor $x^4 - 81$ completely, his teacher did not give him full credit for the answer $(x^2 + 9)(x^2 - 9)$. The student argued that since his answer does indeed give $x^4 - 81$ when multiplied out, he should be given full credit. Was the teacher justified in her grading of this item? Why or why not?

28. The binomial $4x^2 + 16$ is a sum of two squares that *can* be factored. How is this binomial factored? When can the sum of two squares be factored?

Factor each trinomial completely. It may be necessary to factor out the greatest common factor first. See Example 3.

29. $w^2 + 2w + 1$

30. $p^2 + 4p + 4$

31. $x^2 - 8x + 16$

32. $x^2 - 10x + 25$

33. $t^2 + t + \frac{1}{4}$

34. $m^2 + \frac{2}{3}m + \frac{1}{9}$

35. $x^2 - 1.0x + .25$

36. $y^2 - 1.4y + .49$

37. $2x^2 + 24x + 72$

38. $3y^2 - 48y + 192$

39. $16x^2 - 40x + 25$

40. $36y^2 - 60y + 25$

41. $49x^2 - 28xy + 4y^2$

42. $4z^2 - 12zw + 9w^2$

43. $64x^2 + 48xy + 9y^2$

44. $9t^2 + 24tr + 16r^2$

45. $-50h^2 + 40hy - 8y^2$

46. $-18x^2 - 48xy - 32y^2$

Factor each binomial completely. Use your answers in Exercises 3 and 4 as necessary. See Examples 4 and 5.

47. $a^3 + 1$

48. $m^3 + 8$

49. $a^3 - 1$

50. $m^3 - 8$

51. $p^3 + q^3$

52. $x^3 + z^3$

53. $27x^3 - 1$

54. $64y^3 - 27$

55. $8p^3 + 729q^3$

56. $64x^3 + 125y^3$

57. $y^3 - 8x^3$

58. $w^3 - 216z^3$

59. $27a^3 - 64b^3$

60. $125m^3 - 8p^3$

61. $125t^3 + 8s^3$

62. $27r^3 + 1000s^3$

63. State and give the name of each of the five special factoring rules. For each rule, give the numbers of three exercises from this exercise set that are examples.

RELATING CONCEPTS (EXERCISES 64–67)

We have seen that multiplication and factoring are reverse processes. We know that multiplication and division are also related: To check a division problem, we multiply the quotient by the divisor to get the dividend.

To see how factoring and division are related, ***work Exercises 64–67 in order.***

64. Factor $10x^2 + 11x - 6$.

65. Use long division to divide $10x^2 + 11x - 6$ by $2x + 3$.

66. Could we have predicted the result in Exercise 65 from the result in Exercise 64? Explain.

67. Divide $x^3 - 1$ by $x - 1$. Use your answer to factor $x^3 - 1$.

Did you make the connection that the quotient in a division problem is a factor of the dividend?

Extend the methods of factoring presented so far in this chapter to factor each polynomial completely.

68. $(m + n)^2 - (m - n)^2$

69. $(a - b)^3 - (a + b)^3$

70. $m^2 - p^2 + 2m + 2p$

71. $3r - 3k + 3r^2 - 3k^2$

How long would it take an object to fall from the top of the building described? Use a calculator as necessary, and round your answer to the nearest tenth of a second.

35. Navarre Building, New York City 512 feet

36. One Canada Square, London 800 feet

37. Central Plaza, Hong Kong 1028 feet

38. Las Vegas Stratosphere Tower, Las Vegas 1149 feet

Solve each problem. Refer to the discussion of consecutive integers and Examples 7 and 8 in Section 2.3.

39. The product of two consecutive integers is 11 more than their sum. Find the integers.

40. The product of two consecutive integers is 4 less than 4 times their sum. Find the integers.

41. Find three consecutive odd integers such that 3 times the sum of all three is 18 more than the product of the smaller two.

42. Find three consecutive odd integers such that the sum of all three is 42 less than the product of the larger two.

43. Find three consecutive even integers such that the sum of the squares of the smaller two is equal to the square of the largest.

44. Find three consecutive even integers such that the square of the sum of the smaller two is equal to twice the largest.

As we saw in Example 4, quadratic expressions are often used to approximate data. Exercises 45–47 further illustrate this kind of application.

45. The table shows the trend in transit ridership (in billions) from 1975 to 1995. The data is modeled fairly well by the quadratic expression $-.012x^2 + .381x + 5.96$. If y represents the number of passengers (in billions), then

$$y = -.012x^2 + .381x + 5.96.$$

Again, the years are coded, with $x = 5$ corresponding to 1975, $x = 10$ corresponding to 1980, and so on.

Transit Ridership Trend

Year	Number of Passengers (in billions)
1975	7.5
1980	8.5
1985	8.8
1990	9.0
1995	8.0

Source: American Public Transit Association.

(a) In what year shown in the table did the ridership peak?

(b) Use the equation to approximate the ridership y in 1990. (*Hint:* Let $x = 20$.) Does the equation give a good approximation of the value shown in the table?

(c) To what value does x correspond in 1995?

(d) Repeat part (b) for 1995.

(e) Which of the results in parts (b) and (d) is the better approximation of the actual data in the table?

46. Although motor vehicle accidents declined in New York City from 1990 to 1996, taxi accidents increased during that period, as indicated in the bar graph. The data in the graph is approximated by the quadratic equation

$$y = -143.3x^2 + 1823.3x + 6820.$$

Here, y represents the number of accidents in year x, where $x = 0$ corresponds to 1990, $x = 1$ corresponds to 1991, and so on.

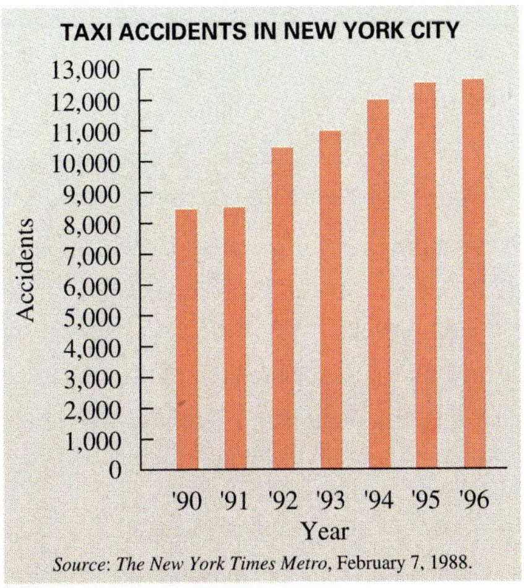

Source: The New York Times Metro, February 7, 1988.

(a) What value of x corresponds to 1998?

(b) According to the equation, if this trend continued, how many taxi accidents occurred in 1998?

47. (See Exercise 46.) The number of people injured in taxi accidents also increased over the years 1990 to 1996. The equation

$$y = -416.72x^2 + 4416.7x + 8500,$$

where y represents the number injured in year x, with $x = 0$ corresponding to 1990, $x = 1$ corresponding to 1991, and so on, approximates the data quite well.

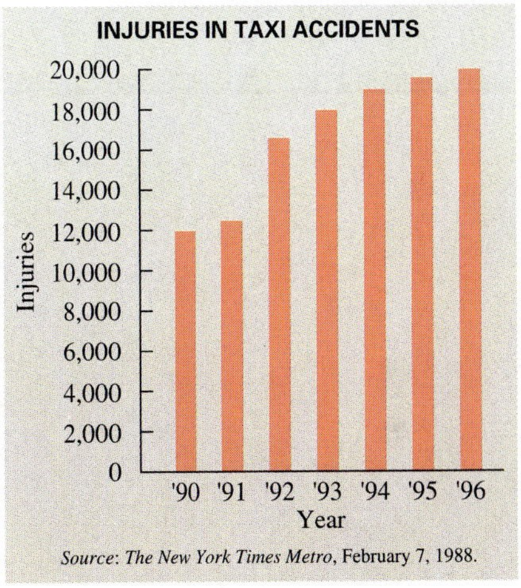

Source: The New York Times Metro, February 7, 1988.

(a) Use the equation to approximate the number injured in 1998.

(b) Find the ratio for 1998 (see Section 2.5) of the number of accidents (answer to Exercise 46(b)) to the number of people injured (answer to part (a) of this exercise). What does this tell us about the average number of people injured per accident?

48. A square piece of cardboard is to be formed into an open-topped box by cutting 3-inch squares from the corners and folding up the sides. See the figure. The volume V of the resulting box is given by the formula $V = 3(x - 6)^2$, where x is the original length of each side of the piece of cardboard. What is the volume of the box if the original length of each side of the piece of cardboard is 14 inches?

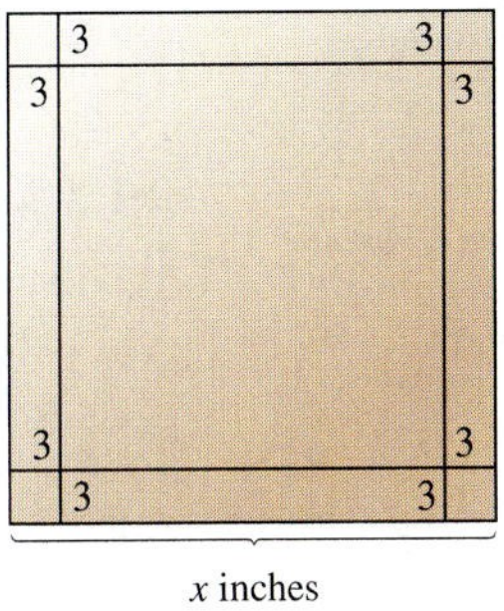

Original Piece of Cardboard

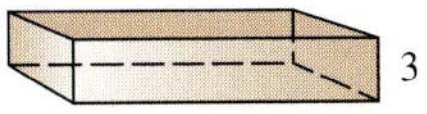

The Resulting Box

RELATING CONCEPTS (EXERCISES 49–52)

One of the many known proofs of the Pythagorean formula is based on the figures shown.

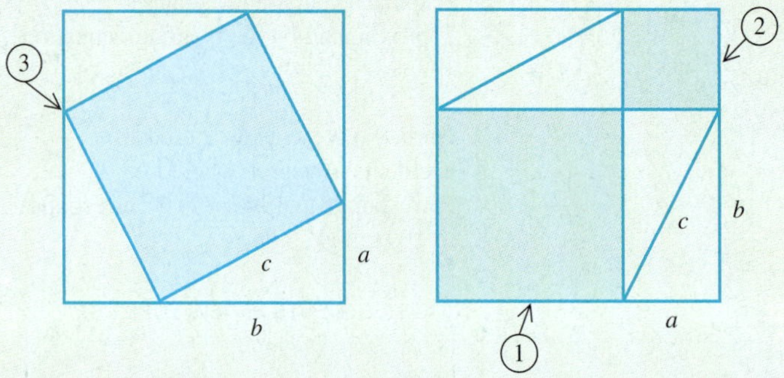

Figure A **Figure B**

Refer to the appropriate figure and ***answer Exercises 49–52 in order.***

49. What is an expression for the area of the dark square labeled ③ in Figure A?

50. The five regions in Figure A are equal in area to the six regions in Figure B. What is an expression for the area of the square labeled ① in Figure B?

51. What is an expression for the area of the square labeled ② in Figure B?

52. Represent this statement using algebraic expressions: The sum of the areas of the dark regions in Figure B is equal to the area of the dark region in Figure A. What does this equation represent?

Did you make the connection that the Pythagorean formula is the result of the fact that the dark areas in the two figures are equal?

5.7 Solving Quadratic Inequalities

OBJECTIVE

1 Solve quadratic inequalities and graph their solutions.

FOR EXTRA HELP

SSG Sec. 5.7
SSM Sec. 5.7
Pass the Test Software
InterAct Math Tutorial Software
Video 9

OBJECTIVE 1 Solve quadratic inequalities and graph their solutions. A **quadratic inequality** is an inequality that involves a second-degree polynomial. Examples of quadratic inequalities include

$$2x^2 + 3x - 5 < 0, \quad x^2 \leq 4, \quad \text{and} \quad x^2 + 5x + 6 > 0.$$

Examples 1 and 2 show how to solve such inequalities.

EXAMPLE 1 Solving a Quadratic Inequality Including Endpoints

Solve $x^2 - 3x - 10 \leq 0$.

To begin, we find the solution of the corresponding quadratic equation,

$$x^2 - 3x - 10 = 0.$$

Factor to get

$$(x - 5)(x + 2) = 0$$

from which

$$x - 5 = 0 \quad \text{or} \quad x + 2 = 0$$
$$x = 5 \quad \text{or} \quad x = -2.$$

Since 5 and -2 are the only values that satisfy $x^2 - 3x - 10 = 0$, all other values of x will make $x^2 - 3x - 10$ either less than 0 (< 0) or greater than 0 (> 0). The values $x = 5$ and $x = -2$ determine three regions on the number line, as shown in Figure 4. Region A includes all numbers less than -2, Region B includes the numbers between -2 and 5, and Region C includes all numbers greater than 5.

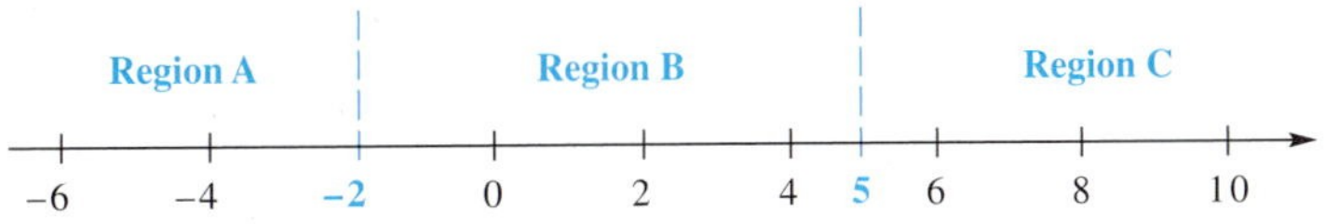

Figure 4

All values of x in a given region will cause $x^2 - 3x - 10$ to have the same sign (either positive or negative). Test one value of x from each region to see which regions satisfy $x^2 - 3x - 10 \le 0$. First, are the points in Region A part of the solution? As a trial value, choose any number less than -2, say -6.

$$x^2 - 3x - 10 \le 0 \qquad \text{Original inequality}$$
$$(-6)^2 - 3(-6) - 10 \le 0 \quad ? \qquad \text{Let } x = -6.$$
$$36 + 18 - 10 \le 0 \quad ? \qquad \text{Simplify.}$$
$$44 \le 0 \qquad \text{False}$$

Since $44 \le 0$ is false, the points in Region A do not belong to the solution.

What about Region B? Try the value $x = 0$.

$$0^2 - 3(0) - 10 \le 0 \quad ? \qquad \text{Let } x = 0.$$
$$-10 \le 0 \qquad \text{True}$$

Since $-10 \le 0$ is true, the points in Region B do belong to the solution.

Try $x = 6$ to check Region C.

$$6^2 - 3(6) - 10 \le 0 \quad ? \qquad \text{Let } x = 6.$$
$$36 - 18 - 10 \le 0 \quad ? \qquad \text{Simplify.}$$
$$8 \le 0 \qquad \text{False}$$

Since $8 \le 0$ is false, the points in Region C do not belong to the solution.

The points in Region B are the only ones that satisfy $x^2 - 3x - 10 \le 0$. As shown in Figure 5, the solution includes the points in Region B together with the endpoints -2 and 5. The solution set is written as the interval $[-2, 5]$.

Figure 5

To summarize, we use the following steps to solve a quadratic inequality.

Solving a Quadratic Inequality

Step 1 **Write an equation.** Change the inequality to an equation.

Step 2 **Solve.** Use the zero-factor property to solve the equation from Step 1.

Step 3 **Determine the regions.** Use the solutions of the equation in Step 2 to determine regions on the number line.

Step 4 **Test each region.** Choose a number from each region. Substitute the number into the original inequality. If the number satisfies the inequality, all numbers in that region satisfy the inequality.

Step 5 **Write the solution set.** Write the solution set in interval notation.

EXAMPLE 2 Solving a Quadratic Inequality Excluding Endpoints

Solve $-x^2 - 5x - 6 < 0$.

It will be easier to factor if we multiply both sides by -1. We must also remember to reverse the direction of the inequality symbol:

$$x^2 + 5x + 6 > 0.$$

Factoring the expression in the corresponding equation $x^2 + 5x + 6 = 0$, we get $(x + 2)(x + 3) = 0$. The solutions of the equation are -2 and -3. These points determine three regions on the number line. See Figure 6. This time, these points will not belong to the solution because only values of x that make $x^2 + 5x + 6$ *greater than* 0 are solutions.

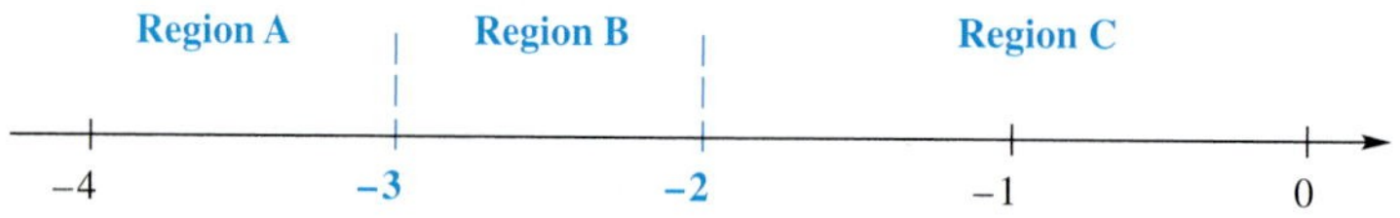

Figure 6

Do the points in Region A belong to the solution? Decide by selecting any number in Region A, such as -4. Does -4 satisfy the inequality?

$$-x^2 - 5x - 6 < 0 \qquad \text{Original inequality}$$

$$-(-4)^2 - 5(-4) - 6 < 0 \quad ? \qquad \text{Let } x = -4.$$

$$-16 + 20 - 6 < 0 \quad ? \qquad \text{Simplify.}$$

$$-2 < 0 \qquad \text{True}$$

Since $-2 < 0$ is true, all the points in Region A belong to the solution set of the inequality.

For Region B, choose a number between -3 and -2, say $-2\frac{1}{2}$, or $-\frac{5}{2}$.

$$-\left(-\frac{5}{2}\right)^2 - 5\left(-\frac{5}{2}\right) - 6 < 0 \quad ? \qquad \text{Let } x = -\tfrac{5}{2}.$$

$$-\frac{25}{4} + \frac{25}{2} - 6 < 0 \quad ? \qquad \text{Simplify.}$$

$$\frac{1}{4} < 0 \qquad \text{False}$$

Since $\frac{1}{4} < 0$ is false, no point in Region B belongs to the solution set.

For Region C, try the number 0.

$$-0^2 - 5(0) - 6 < 0 \quad ? \qquad \text{Let } x = 0.$$
$$-6 < 0 \qquad \text{True}$$

Since $-6 < 0$ is true, the points in Region C belong to the solution set.

The solutions include all values of x less than -3, together with all values of x greater than -2, as shown in the graph in Figure 7. Using inequalities, the solutions may be written as $x < -3$ *or* $x > -2$. Any number that satisfies *either* of these inequalities will be a solution, because *or* means "either one or the other or both." In interval notation, we write the solution set using the **union symbol** $\cup$ for *or*:

$$(-\infty, -3) \cup (-2, \infty).$$

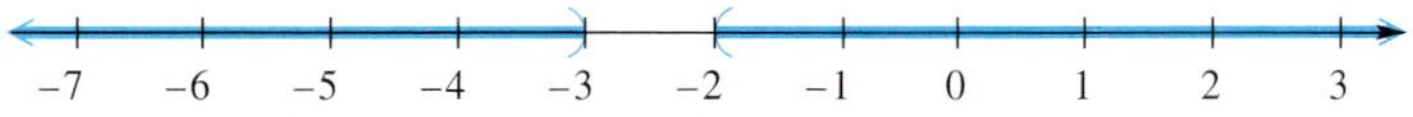

Figure 7

CAUTION There is no shortcut way to write the solution set $x < -3$ or $x > -2$.

CONNECTIONS

The number of items that must be sold for a company's revenue to equal its cost to produce those items is called the *break-even point.* If revenue, R, and cost, C, are given by expressions in x, where x is the number of items produced and sold, then the solution of the equation $R - C = 0$ gives the break-even point. For the company to make a profit, $R - C$ must be greater than 0, and the company would want to solve the inequality $R - C > 0$.

FOR DISCUSSION OR WRITING

Explain why $R - C > 0$ gives values of x that produce a profit. Suppose $R - C = x^2 - x - 12$. What values of x produce a profit? Why is only one part of the solution of the corresponding inequality valid here?

5.7 EXERCISES

1. Which of these inequalities is not a quadratic inequality?
   (a) $5x^2 - x + 7 < 0$ **(b)** $x^2 + 4 > 0$ **(c)** $3x + 5 \geq 0$
   (d) $(x - 1)(x + 2) \leq 0$

2. To solve a quadratic inequality, we use the _____ of the corresponding equation to determine _____ on the number line.

To prepare for using test values in the various regions, answer true or false depending on whether the given value of x satisfies or does not satisfy the inequality.

3. $(2x + 1)(x - 5) \geq 0$
   (a) $x = -\dfrac{1}{2}$ **(b)** $x = 5$ **(c)** $x = 0$ **(d)** $x = -6$

4. $(x - 6)(3x + 4) < 0$
   (a) $x = -3$ **(b)** $x = 6$ **(c)** $x = 0$ **(d)** $x = -\dfrac{4}{3}$

Solve each inequality and graph the solution set. See Examples 1 and 2.

5. $(a + 3)(a - 3) < 0$

6. $(b - 2)(b + 2) > 0$

7. $(a + 6)(a - 7) \geq 0$

8. $(z - 5)(z - 4) \leq 0$

9. $m^2 + 5m + 6 > 0$

10. $y^2 - 3y + 2 < 0$

11. $z^2 - 4z - 5 \leq 0$

12. $3p^2 - 5p - 2 \leq 0$

13. $5m^2 + 3m - 2 < 0$

14. $2k^2 + 7k - 4 > 0$

15. $6r^2 - 5r < 4$

16. $6r^2 + 7r > 3$

17. $q^2 - 7q < -6$

18. $2k^2 - 7k \leq 15$

19. $6m^2 + m - 1 > 0$

20. $30r^2 + 3r - 6 \leq 0$

21. $12p^2 + 11p + 2 < 0$

22. $a^2 - 16 < 0$

23. $9m^2 - 36 > 0$

24. $r^2 - 100 \geq 0$

25. $r^2 > 16$

26. $m^2 \geq 25$

TECHNOLOGY INSIGHTS (EXERCISES 27-30)

The calculator screens show the x-intercepts of the graph of the quadratic equation corresponding to each quadratic inequality. When the graph is above the x-axis, the values of y (which represents the quadratic expression) are positive. When the graph is below the x-axis, the values of y are negative. Therefore, the quadratic expression is positive for those x-values that correspond to positive y-values and negative for those x-values that correspond to negative y-values.

Use the graphs and the x-intercepts to determine the intervals that satisfy each inequality.

27. $x^2 + .7x - 1.2 \leq 0$

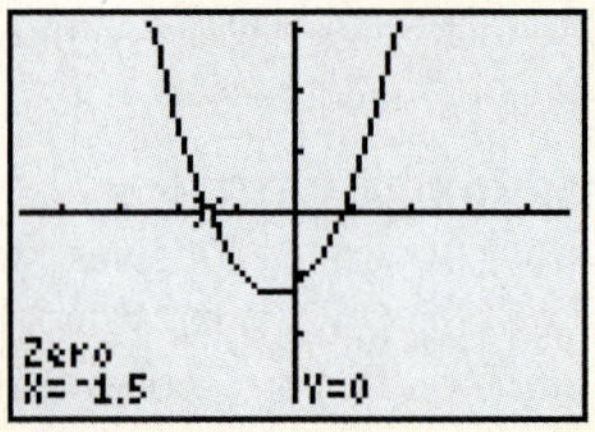

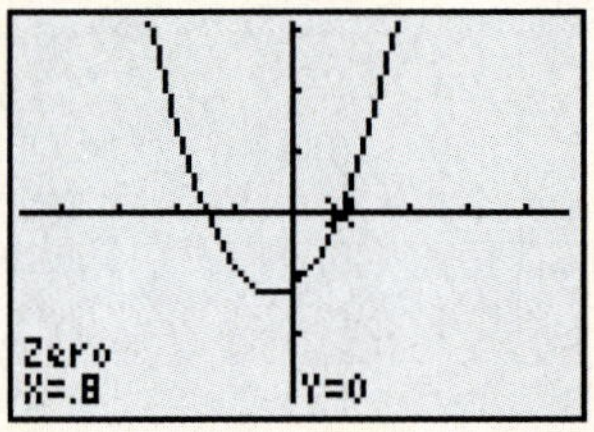

28. $3x^2 + 4x \geq 0$

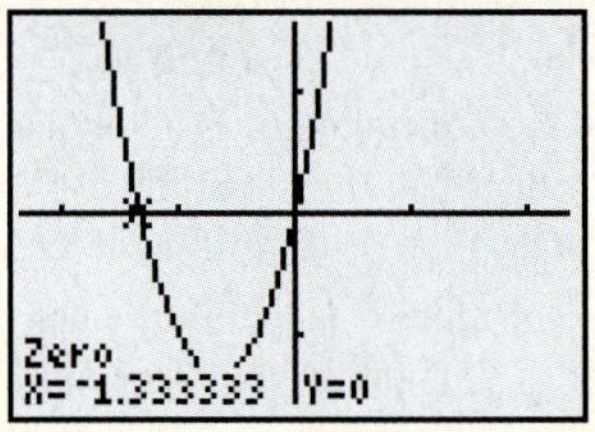

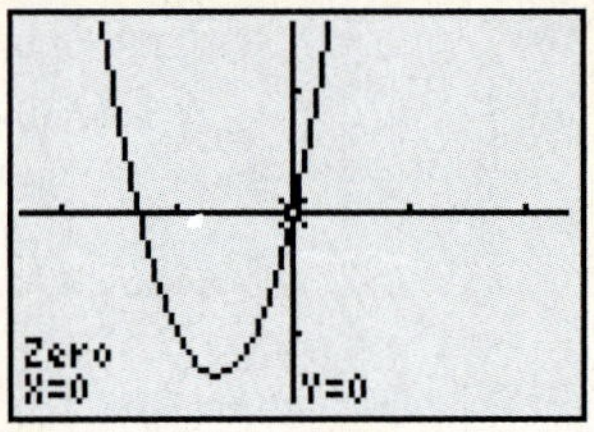

TECHNOLOGY INSIGHTS (EXERCISES 27-30) (CONTINUED)

29. $4x^2 - 20x + 21 \geq 0$

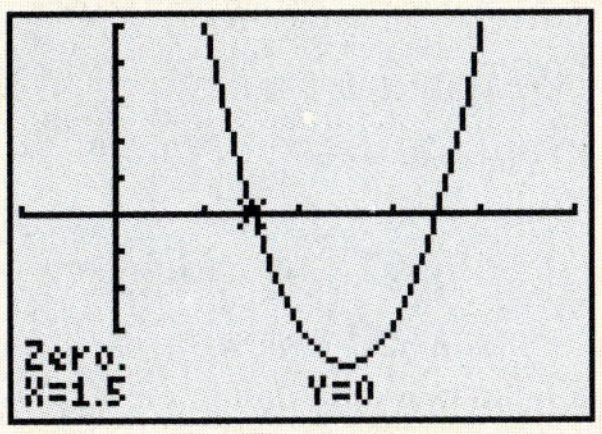

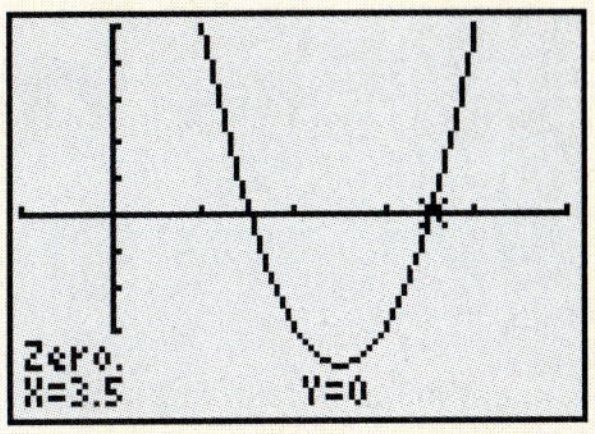

30. $6x^2 + 13x - 8 \leq 0$

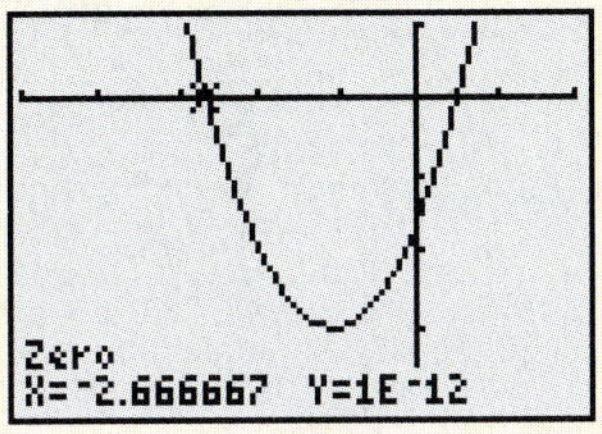

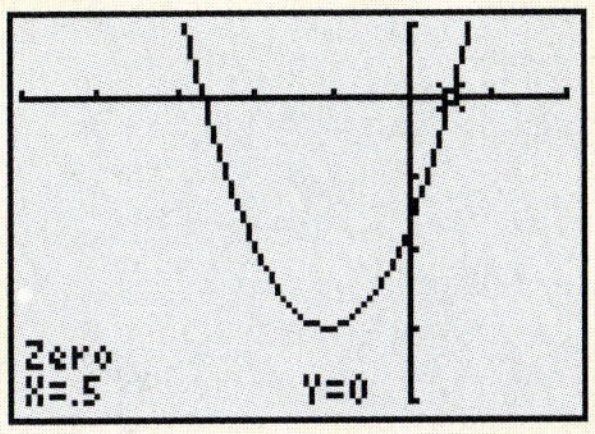

*The given inequality is not quadratic, but it may be solved in a similar manner. Solve and graph each inequality. (*Hint: *Because these inequalities correspond to equations with three solutions, they determine four regions on a number line.)*

31. $(a + 2)(3a - 1)(a - 4) \geq 0$

32. $(2p - 7)(p - 1)(p + 3) \leq 0$

33. $(r - 2)(r^2 - 3r - 4) < 0$

34. $(m + 5)(m^2 - m - 6) > 0$

An object is propelled upward from ground level with an initial velocity of 256 feet per second. After t seconds, its height h is given by the equation $h = -16t^2 + 256t$, where h is in feet.

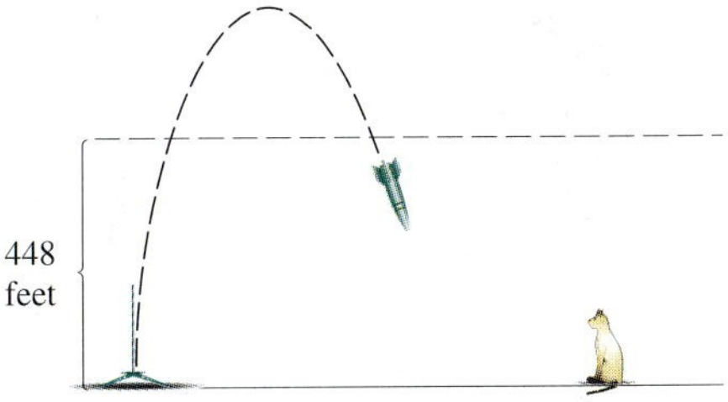

35. The object reaches a height of 448 feet when $h = 448$ in the equation. Use the method of Section 5.5 to solve the resulting equation and determine the times at which the object reaches this height.

36. The object is more than 448 feet above the ground when $h > 448$. This is given by the inequality $-16t^2 + 256t > 448$. Use the method of this section to find the time interval over which the object is more than 448 feet above the ground.

37. By solving $-16t^2 + 256t < 448$ and realizing that t must be greater than or equal to 0, we can determine the time intervals over which the object is less than 448 feet above the ground. Use the method of this section to find this time interval. (*Note:* We must also have $t \leq 16$.)

38. Explain the restrictions on t described in Exercise 37.

CHAPTER 5 GROUP ACTIVITY

Transportation for Tomorrow

Objective: Use quadratic equations to find vehicle speed and stopping distance given acceleration rate.

In 1998 major automotive manufacturers began to produce electric cars. In this activity the speed and stopping distances of three different electric vehicles with given acceleration rates will be examined. Acceleration is the rate of change of speed with respect to time and is given in feet per second per second. Complete the table below for Cars 1, 2, and 3 to find the time, speed, and distance for each vehicle. (Round to the nearest whole number.)

Car	*Acceleration Time*	*Speed*	*Stopping Distance*
1			
2			
3			

A. Car 1: The Toyota Rav4EV has an acceleration constant of $a = 4.89$ feet per second per second.*

1. Determine how long it would take for this vehicle to go $\frac{3}{10}$ mile (1584 feet) by solving the formula $d = at^2$, where d represents distance in feet, a is the acceleration constant, and t is time in seconds.
2. Now find the speed r in miles per hour the car has achieved using the distance formula $d = rt$. (Convert time to hours, distance to miles.)
3. Next find how many feet it will take the car to stop on a dry road at that speed (in miles per hour) using the following quadratic equation: $d = .045r^2 + 1.1r$.

B. Determine the same information for two other vehicles with the following acceleration constants.

Car 2: $a = 3.96$ feet per second per second

Car 3: $a = 5.48$ feet per second per second

**Source:* Ecomall.com.

CHAPTER 5 SUMMARY

KEY TERMS

5.1 factor, common factor, greatest common factor, factoring by grouping

5.2 factored form, factoring, prime polynomial

5.4 perfect square trinomial

5.5 quadratic equation, standard form

5.6 hypotenuse, legs

5.7 quadratic inequality

TEST YOUR WORD POWER

See how well you have learned the vocabulary in this chapter. Answers, with examples, are given at the bottom of the page.

1. Factoring is

(a) a method of multiplying polynomials

(b) the process of writing a polynomial as a product

(c) the answer in a multiplication problem

(d) a way to add the terms of a polynomial.

2. A polynomial is in **factored form** when

(a) it is prime

(b) it is written as a sum

(c) the squared term has a coefficient of 1

(d) it is written as a product.

3. A **perfect square trinomial** is a trinomial

(a) that can be factored as the square of a binomial

(b) that cannot be factored

(c) that is multiplied by a binomial

(d) where all terms are perfect squares.

4. A **quadratic equation** is a polynomial equation of

(a) degree one

(b) degree two

(c) degree three

(d) degree four.

5. A **hypotenuse** is

(a) either of the two shorter sides of a triangle

(b) the shortest side of a triangle

(c) the side opposite the right angle in a triangle

(d) the longest side in any triangle.

QUICK REVIEW

CONCEPTS	EXAMPLES
5.1 THE GREATEST COMMON FACTOR; FACTORING BY GROUPING	
Finding the Greatest Common Factor (GCF) **1.** Include the largest numerical factor of every term. **2.** Include each variable that is a factor of every term raised to the smallest exponent that appears in a term.	Find the greatest common factor of $4x^2y, \quad -6x^2y^3, \quad 2xy^2.$ $4x^2y = 2^2 \cdot x^2 \cdot y$ $-6x^2y^3 = -1 \cdot 2 \cdot 3 \cdot x^2 \cdot y^3$ $2xy^2 = 2 \cdot x \cdot y^2$ The greatest common factor is $2xy$.
Factoring by Grouping *Step 1* Group the terms so that each group has a common factor.	Factor $3x^2 + 5x - 24xy - 40y$. $(3x^2 + 5x) + (-24xy - 40y)$
Step 2 Factor out the greatest common factor in each group.	$x(3x + 5) - 8y(3x + 5)$
Step 3 Factor a common binomial factor from the result of Step 2.	$(3x + 5)(x - 8y)$
Step 4 If Step 3 cannot be performed, try a different grouping.	
5.2 FACTORING TRINOMIALS	
To factor $x^2 + bx + c$, find m and n such that $mn = c$ and $m + n = b$. $mn = c$ ↓ $x^2 + bx + c$ ↑ $m + n = b$ Then $x^2 + bx + c = (x + m)(x + n).$	Factor $x^2 + 6x + 8$. $mn = 8$ ↓ $x^2 + 6x + 8$ ↑ $m + n = 6$ $m = 2$ and $n = 4$ $x^2 + 6x + 8 = (x + 2)(x + 4)$

Answers to Test Your Word Power

1. (b) *Example:* $x^2 - 5x - 14 = (x - 7)(x + 2)$ **2.** (d) *Example:* The factored form of $x^2 - 5x - 14$ is $(x - 7)(x + 2)$. **3.** (a) *Example:* $a^2 + 2a + 1$ is a perfect square trinomial; its factored form is $(a + 1)^2$. **4.** (b) *Examples:* $y^2 - 3y + 2 = 0$, $x^2 - 9 = 0$, $2m^2 = 6m + 8$
5. (c) *Example:* In Figure 3 of Section 5.6, the hypotenuse is the side labeled $x + 2$.

CONCEPTS	EXAMPLES

5.3 MORE ON FACTORING TRINOMIALS

To factor $ax^2 + bx + c$:

By Grouping
Find m and n:

$$mn = ac$$
$$ax^2 + bx + c$$
$$m + n = b$$

Factor $3x^2 + 14x - 5$.

$$-15$$

$mn = -15$, $m + n = 14$

By Trial and Error
Use FOIL backwards.

By trial and error or by grouping.

$$3x^2 + 14x - 5 = (3x - 1)(x + 5)$$

5.4 SPECIAL FACTORING RULES

Difference of Squares
$x^2 - y^2 = (x + y)(x - y)$

Factor.

$$4x^2 - 9 = (2x + 3)(2x - 3)$$

Perfect Square Trinomial
$x^2 + 2xy + y^2 = (x + y)^2$
$x^2 - 2xy + y^2 = (x - y)^2$

$$9x^2 + 6x + 1 = (3x + 1)^2$$
$$4x^2 - 20x + 25 = (2x - 5)^2$$

Difference of Cubes
$x^3 - y^3 = (x - y)(x^2 + xy + y^2)$

$$m^3 - 8 = m^3 - 2^3 = (m - 2)(m^2 + 2m + 4)$$

Sum of Cubes
$x^3 + y^3 = (x + y)(x^2 - xy + y^2)$

$$27 + z^3 = 3^3 + z^3 = (3 + z)(9 - 3z + z^2)$$

5.5 SOLVING QUADRATIC EQUATIONS BY FACTORING

Zero-Factor Property
If a and b are real numbers and if $ab = 0$, then $a = 0$ or $b = 0$.

If $(x - 2)(x + 3) = 0$, then $x - 2 = 0$ or $x + 3 = 0$.

Solving a Quadratic Equation by Factoring

Solve $2x^2 = 7x + 15$.

Step 1 Write in standard form.

$$2x^2 - 7x - 15 = 0$$

Step 2 Factor.

$$(2x + 3)(x - 5) = 0$$

Step 3 Use the zero-factor property.

$$2x + 3 = 0 \quad \text{or} \quad x - 5 = 0$$
$$2x = -3 \qquad x = 5$$
$$x = -\frac{3}{2}$$

Step 4 Check.

Both solutions satisfy the original equation; the solution set is $\left\{-\frac{3}{2}, 5\right\}$.

5.6 APPLICATIONS OF QUADRATIC EQUATIONS

Pythagorean Formula
In a right triangle, the square of the hypotenuse equals the sum of the squares of the legs.

$$a^2 + b^2 = c^2$$

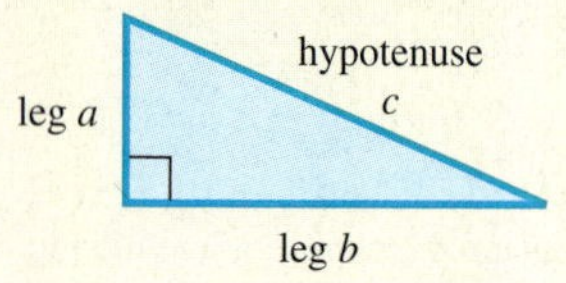

In a right triangle, one leg measures 2 feet longer than the other. The hypotenuse measures 4 feet longer than the shorter leg. Find the lengths of the three sides of the triangle.

Let x = the length of the shorter leg. Then

$$x^2 + (x + 2)^2 = (x + 4)^2.$$

Verify that the solutions of this equation are -2 and 6. Discard -2 as a solution. Check that the sides are 6, $6 + 2 = 8$, and $6 + 4 = 10$ feet in length.

CONCEPTS	EXAMPLES
5.7 SOLVING QUADRATIC INEQUALITIES	
Solving a Quadratic Inequality	Solve $2x^2 + 5x - 3 < 0$.
Step 1 Change the inequality to an equation.	$2x^2 + 5x - 3 = 0$
Step 2 Use the zero-factor property to solve the equation from Step 1.	$(2x - 1)(x + 3) = 0$ $2x - 1 = 0$ or $x + 3 = 0$ $2x = 1$ or $x = -3$ $x = \frac{1}{2}$
Step 3 Use the solutions of the equation in Step 2 to determine regions on the number line.	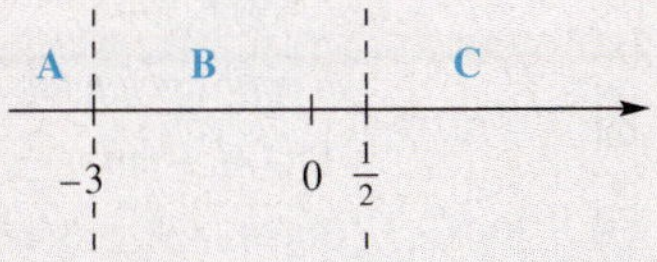
Step 4 Choose a number from each region. Substitute the number into the original inequality. If the number satisfies the inequality, all numbers in that region satisfy the inequality.	Choose -4 from A: $2(-4)^2 + 5(-4) - 3 < 0$? $9 < 0$ False Choose 0 from B: $2(0)^2 + 5(0) - 3 < 0$? $-3 < 0$ True Choose 1 from C: $2(1)^2 + 5(1) - 3 < 0$? $4 < 0$ False Only the numbers from Region B satisfy the inequality.
Step 5 Write the solution set in interval notation.	The solution set is written $\left(-3, \frac{1}{2}\right)$.

CHAPTER 5 REVIEW EXERCISES

[5.1] *Factor out the greatest common factor or factor by grouping.*

1. $7t + 14$
2. $60z^3 + 30z$
3. $2xy - 8y + 3x - 12$
4. $6y^2 + 9y + 4y + 6$

[5.2] *Factor completely.*

5. $x^2 + 5x + 6$
6. $y^2 - 13y + 40$
7. $q^2 + 6q - 27$
8. $r^2 - r - 56$
9. $r^2 - 4rs - 96s^2$
10. $p^2 + 2pq - 120q^2$
11. $8p^3 - 24p^2 - 80p$
12. $3x^4 + 30x^3 + 48x^2$
13. $p^7 - p^6q - 2p^5q^2$
14. $3r^5 - 6r^4s - 45r^3s^2$

[5.3]

15. To begin factoring $6r^2 - 5r - 6$, what are the possible first terms of the two binomial factors if we consider only positive integer coefficients?
16. What is the first step you would use to factor $2z^3 + 9z^2 - 5z$?

Factor completely.

17. $2k^2 - 5k + 2$

18. $3r^2 + 11r - 4$

19. $6r^2 - 5r - 6$

20. $10z^2 - 3z - 1$

21. $8v^2 + 17v - 21$

22. $24x^5 - 20x^4 + 4x^3$

23. $-6x^2 + 3x + 30$

24. $10r^3s + 17r^2s^2 + 6rs^3$

[5.4]

25. Which one of the following is the difference of two squares?
(a) $32x^2 - 1$ **(b)** $4x^2y^2 - 25z^2$ **(c)** $x^2 + 36$ **(d)** $25y^3 - 1$

26. Which one of the following is a perfect square trinomial?
(a) $x^2 + x + 1$ **(b)** $y^2 - 4y + 9$ **(c)** $4x^2 + 10x + 25$ **(d)** $x^2 - 20x + 100$

Factor completely.

27. $n^2 - 49$

28. $25b^2 - 121$

29. $49y^2 - 25w^2$

30. $144p^2 - 36q^2$

31. $x^2 + 100$

32. $r^2 - 12r + 36$

33. $9t^2 - 42t + 49$

34. $m^3 + 1000$

35. $125k^3 + 64x^3$

36. $343x^3 - 64$

[5.5] *Solve each equation and check your solutions.*

37. $z^2 + 4z + 3 = 0$

38. $2m^2 - 10m + 8 = 0$

39. $x(x - 8) = -15$

40. $3z^2 - 11z - 20 = 0$

41. $81t^2 - 64 = 0$

42. $y^2 = 8y$

43. $3n(n - 5) = 18$

44. $t^2 - 14t + 49 = 0$

45. $t^2 = 12(t - 3)$

46. $(5z + 2)(z^2 + 3z + 2) = 0$

[5.6] *Solve each problem.*

47. The length of a rectangle is 6 meters more than the width. The area is 40 square meters. Find the length and width of the rectangle.

48. The length of a rectangle is three times the width. If the width were increased by 3 meters while the length remained the same, the new rectangle would have an area of 30 square meters. Find the length and width of the original rectangle.

49. The volume of a box is to be 120 cubic meters. The width of the box is to be 4 meters, and the height 1 meter less than the length. Find the length and height of the box.

50. The sides of a right triangle have lengths (in feet) that are consecutive integers. What are the lengths of the sides?

If an object is propelled straight up from ground level with an initial velocity of 128 feet per second, its height h in feet after t seconds is $h = 128t - 16t^2$.

Find the height of the object after the following periods of time.

51. 1 second

52. 2 seconds

53. 4 seconds

54. When does the object described above return to the ground?

55. A 9-inch by 12-inch picture is to be placed on a cardboard mat so that there is an equal border around the picture. The area of the finished mat and picture is to be 208 square inches. How wide will the border be?

56. A box is made from a 12-centimeter by 10-centimeter piece of cardboard by cutting equal-sized squares from each corner and folding up the sides. The area of the bottom of the box is to be 48 square centimeters. Find the length of a side of the cutout squares.

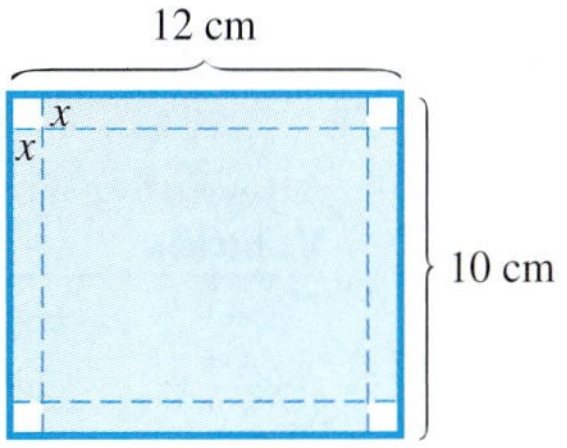

57. In 1994, Greyhound Lines, Inc. was near its second bankruptcy in three years. Since then, a new CEO, Craig R. Lentzsch, appears to have turned the company around. The bar graph and table of values show the operating income for 1994, when Lentzsch took over, 1995, and 1996. (*Source:* Company Reports, Rothschild Inc.)

Year	*Operating Income (in millions of dollars)*
1994	−65
1995	9.4
1996	37

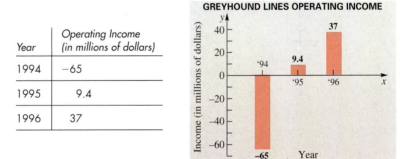

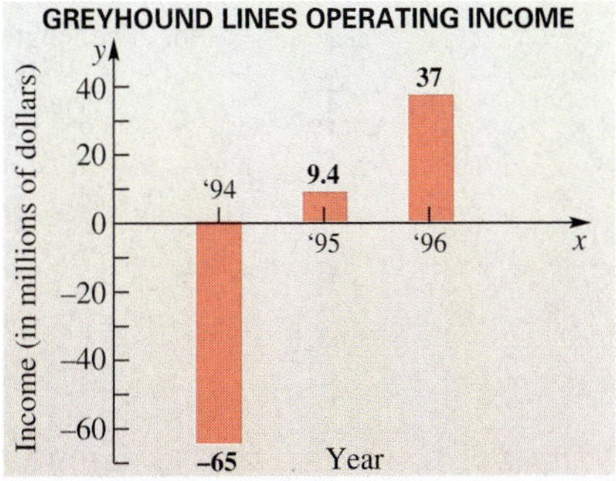

Using the data, we constructed the quadratic equation

$$y = -23.4x^2 + 285x - 831,$$

which gives the operating income y (in millions of dollars) in year x. We use $x = 4$ to represent 1994, $x = 5$ to represent 1995, and so on.

(a) Use the equation to predict the operating income in 1997.

(b) Use the equation to predict the operating income in 1998.

(c) Comment on the validity of the answers for parts (a) and (b).

[5.7] *Solve each inequality.*

58. $(q + 5)(q - 3) > 0$

59. $(2r - 1)(r + 4) \geq 0$

60. $m^2 - 5m + 6 \leq 0$

61. $2x^2 + 5x - 12 \geq 0$

62. $2p^2 + 5p - 12 < 0$

63. Suppose you know that the solution set of a quadratic inequality involving the $<$ symbol is $(-5, 7)$. If the symbol is changed to $>$, what is the solution set of the new inequality?

MIXED REVIEW EXERCISES

Factor completely.

64. $z^2 - 11zx + 10x^2$

65. $3k^2 + 11k + 10$

66. $15m^2 + 20mp - 12mp - 16p^2$

67. $y^4 - 625$

68. $6m^3 - 21m^2 - 45m$

69. $24ab^3c^2 - 56a^2bc^3 + 72a^2b^2c$

70. $25a^2 + 15ab + 9b^2$

71. $12x^2yz^3 + 12xy^2z - 30x^3y^2z^4$

72. $2a^5 - 8a^4 - 24a^3$

73. $12r^2 + 18rq - 10rq - 15q^2$

74. $1000a^3 + 27$

75. $49t^2 + 56t + 16$

Solve.

76. $t(t - 7) = 0$

77. $x(x + 3) = 10$

78. $4x^2 - x - 3 \leq 0$

79. The numbers of alternative-fueled vehicles, in thousands, in use for the years 1995–1997 are given in the table.

Alternative-Fueled Vehicles

Year	Number (in thousands)
1995	333
1996	357
1997	386

Source: Energy Information Administration, Alternatives to Traditional Fuels, 1993.

Using statistical methods, we constructed the quadratic equation

$$y = 2.5x^2 - 453.5x + 20{,}850$$

to model the number of vehicles y in year x. Here we used $x = 95$ for 1995, $x = 96$ for 1996, and so on. Because only three years of data were used to determine the model, we must be particularly careful about using it to estimate for years before 1995 or after 1997.

(a) What prediction for 1998 is given by the equation?

(b) Why might the prediction for 1998 be unreliable?

80. The sum of two consecutive even integers is 34 less than their product. Find the integers.

81. The floor plan for a house is a rectangle with length 7 meters more than its width. The area is 170 square meters. Find the width and length of the house.

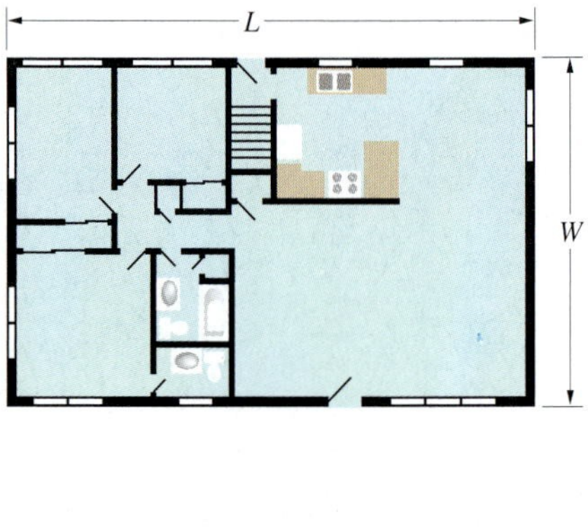

82. The triangular sail of a schooner has an area of 30 square meters. The height of the sail is 4 meters more than the base. Find the base of the sail.

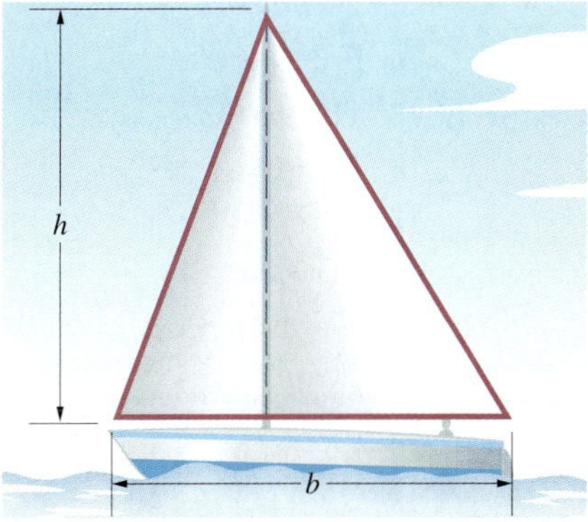

83. Two cars left an intersection at the same time. One traveled north. The other traveled 14 miles farther, but to the east. How far apart were they then, if the distance between them was 4 miles more than the distance traveled east?

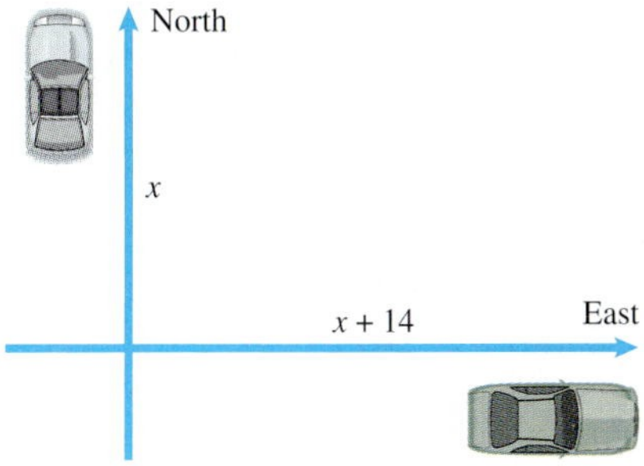

84. A ladder is leaning against a building. The distance from the bottom of the ladder to the building is 4 feet less than the length of the ladder. How high up the side of the building is the top of the ladder if that distance is 2 feet less than the length of the ladder?

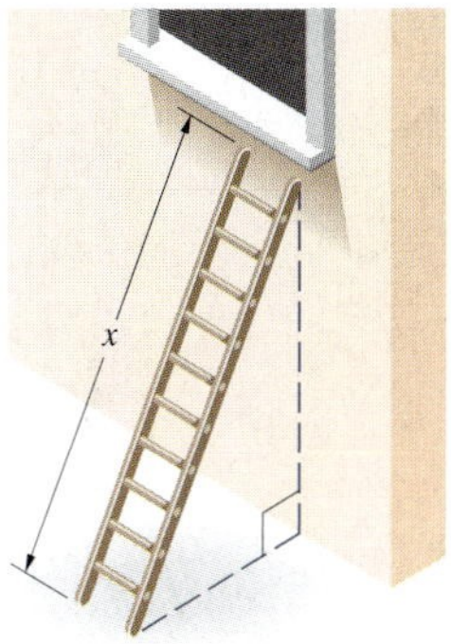

85. A bicyclist heading east and a motorist traveling south left an intersection at the same time. When the motorist had gone 17 miles farther than the bicyclist, the distance between them was 1 mile more than the distance traveled by the motorist. How far apart were they then? (*Hint:* Draw a sketch.)

86. Although $(2x + 8)(3x - 4) = 6x^2 + 16x - 32$ is a true statement, the polynomial is not factored completely. Explain why and give the completely factored form.

CHAPTER 5 TEST

1. Which one of the following is the correct, completely factored form of $2x^2 - 2x - 24$?
(a) $(2x + 6)(x - 4)$ **(b)** $(x + 3)(2x - 8)$
(c) $2(x + 4)(x - 3)$ **(d)** $2(x + 3)(x - 4)$

Factor each polynomial completely. If it cannot be factored, write prime.

2. $2m^3n^2 + 3m^3n - 5m^2n^2$

3. $x^2 - 5x - 24$

4. $2x^2 + x - 3$

5. $10z^2 - 17z + 3$

6. $t^2 + 2t + 3$

7. $x^2 + 36$

8. $12 - 6a + 2b - ab$

9. $9y^2 - 64$

10. $4x^2 - 28xy + 49y^2$

11. $-2x^2 - 4x - 2$

12. $6t^4 + 3t^3 - 108t^2$

13. $r^3 - 125$

14. $8k^3 + 64$

15. Why is $(p + 3)(p + 3)$ not the correct factored form of $p^2 + 9$?

Solve each equation.

16. $2r^2 - 13r + 6 = 0$

17. $25x^2 - 4 = 0$

18. $x(x - 20) = -100$

19. $t^3 = 9t$

Solve each problem.

20. If an object is propelled from ground level at an initial velocity of 96 feet per second, after t seconds its height h in feet is given by the formula $h = -16t^2 + 96t$. After how many seconds will its height h be 108 feet?

21. A carpenter needs to cut a brace to support a wall stud. See the figure. The brace should be 7 feet less than three times the length of the stud. The brace will be fastened on the floor 1 foot less than twice the length of the stud away from the stud.

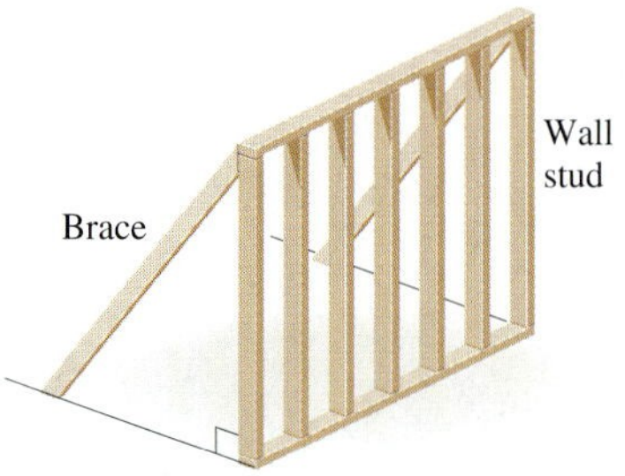

(a) Let x represent the length of the stud. Write an expression for the length of the brace.

(b) Write an expression for the distance from the wall to where the brace is fastened.

(c) How long should the brace be? (*Hint:* Use the Pythagorean formula.)

22. Assume the data given in Exercise 79 of the Chapter 5 Review Exercises represent the number of alternative-fueled vehicles at the beginning of each of the years shown. Then 95.5 would represent the number of such vehicles halfway through 1995 (July 1). Use the equation that models the data,

$$y = 2.5x^2 - 453.5x + 20{,}850,$$

to approximate the numbers of alternative-fueled vehicles on July 1 of 1995 and 1996. (Round your answers to the nearest whole number.)

Solve each inequality and graph the solutions.

23. $(3x - 1)(2x + 5) < 0$

24. $x^2 - 2x - 24 \geq 0$

25. Why isn't "$x = \frac{2}{3}$" the correct response to "Solve the equation $x^2 = \frac{4}{9}$"?

CUMULATIVE REVIEW EXERCISES CHAPTERS 1–5

Solve each equation.

1. $3x + 2(x - 4) = 4(x - 2)$

2. $.3x + .9x = .06$

3. $\frac{2}{3}y - \frac{1}{2}(y - 4) = 3$

4. Solve for P: $A = P + Prt$.

Solve each problem.

5. Find the measures of the marked angles.

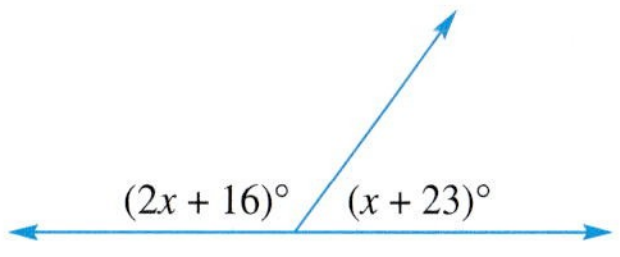

6. In a recent year, the United States exported to the Bahamas \$315 million more in goods than it imported. Together, the two amounts totaled \$1167 million. How much were the exports and how much were the imports?

7. In a mixture of concrete, there are 3 pounds of cement mix for every 1 pound of gravel. If the mixture contains a total of 140 pounds of these two ingredients, how many pounds of gravel are there?

8. Fill in each blank with *positive* or *negative.* The point with coordinates (a, b) is in
(a) quadrant II if a is _____ and b is _____.
(b) quadrant III if a is _____ and b is _____.

9. The sales of a small company are a function of the number of years x it has been in business. Sales (in thousands) are given by $y = 12x + 3$. Write ordered pairs for this function for the second and fifth years it has been in business.

10. What are the x- and y-intercepts of the graph of the sales equation $y = 12x + 3$?

11. Graph the linear equation $y = 12x + 3$ for $x \geq 0$.

12. The points on the graph show the number of U.S. radio stations in the years 1990–1995, along with the graph of a linear equation that models the data. Use the ordered pairs shown on the graph to find the slope of the line. Interpret the slope.

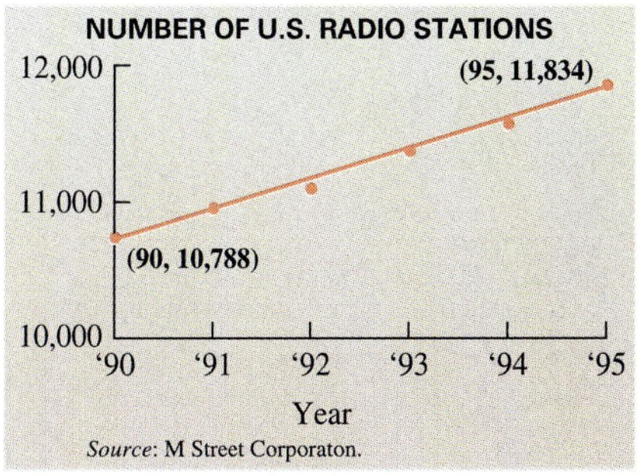

Evaluate each expression.

13. $2^{-3} \cdot 2^5$

14. $\left(\frac{3}{4}\right)^{-2}$

15. $\left(\frac{4^{-3} \cdot 4^4}{4^5}\right)^{-1}$

16. Simplify $\frac{(p^2)^3 p^{-4}}{(p^{-3})^{-1} p}$ and write the answer using only positive exponents. Assume $p \neq 0$.

Perform the indicated operations.

17. $(2k^2 + 4k) - (5k^2 - 2) - (k^2 + 8k - 6)$

18. $3m^3(2m^5 - 5m^3 + m)$

19. $(y^2 + 3y + 5)(3y - 1)$

20. $(2p + 3q)(2p - 3q)$

21. $\dfrac{8x^4 + 12x^3 - 6x^2 + 20x}{2x}$

22. $(12p^3 + 2p^2 - 12p + 5) \div (2p - 2)$

Factor completely.

23. $2a^2 + 7a - 4$

24. $10m^2 + 19m + 6$

25. $15x^2 - xy - 6y^2$

26. $9x^2 + 6x + 1$

27. $-32t^2 - 112tz - 98z^2$

28. $25r^2 - 81t^2$

29. $100x^2 + 25$

30. $2pq + 6p^3q + 8p^2q$

31. $2ax - 2bx + ay - by$

Solve each equation.

32. $(2p - 3)(p + 2)(p - 6) = 0$

33. $6m^2 + m - 2 = 0$

34. Solve the inequality and graph the solution set: $2x^2 + x - 6 \geq 0$.

Solve each problem.

35. The difference between the squares of two consecutive even integers is 28 less than the square of the smaller integer. Find the two integers.

36. The length of the hypotenuse of a right triangle is twice the length of the shorter leg, plus 3 meters. The longer leg is 7 meters longer than the shorter leg. Find the lengths of the sides.

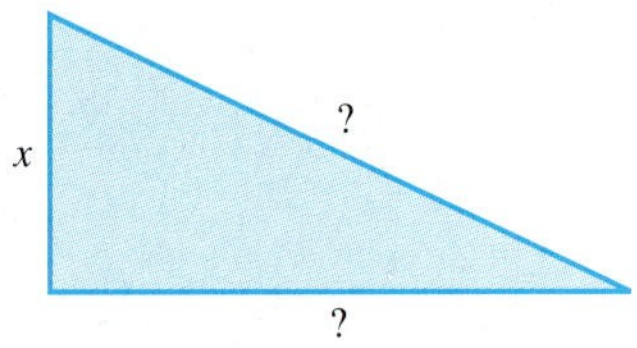

Rational Expressions 6

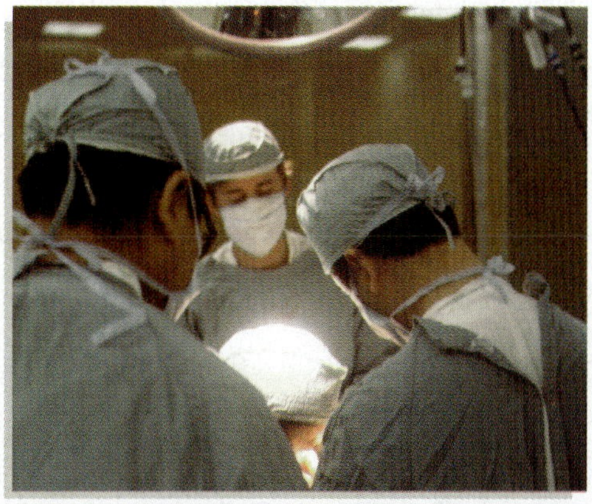

One of the reasons that people are living longer today than in the past is the increased efficiency of health care. Organ transplants allow at-risk patients to survive longer than they would have a decade ago. Heart, liver, kidney, lung, pancreas, and cornea transplants are now available. The table shows the number of heart transplants in the United States for each year from 1990 through 1994.

Year	1990	1991	1992	1993	1994
Number of Heart Transplants	2108	2125	2171	2297	2340

Source: U.S. Department of Health and Human Services, Public Health Service, Division of Organ Transplantation, and United Network for Organ Sharing.

In 1995, the number of heart transplants was still on the rise, and compared to ten years before, in 1985, the increase was drastic. In those two years, the total number of transplants was 3080, and the ratio of the number in 1995 to the number in 1985 was approximately 33 to 10. How many transplants were performed in each of these two years? To answer this question, write an equation. If we let x represent the number of transplants in 1995, then $3080 - x$ represents the number of transplants in 1985. Writing this information as the ratio

$$\frac{\text{Number of transplants in 1995}}{\text{Number of transplants in 1985}},$$

we have

$$\frac{x}{3080 - x} = \frac{33}{10}.$$

The fractions on each side of this equation are *rational expressions,* the topic of this chapter. The equation is an example of a *rational equation.* In Section 6.6, we show how to solve this type of equation. You are asked to solve this problem in Exercise 71 of Section 6.7.

Health Care

6.1 The Fundamental Property of Rational Expressions

OBJECTIVES

1. Find the values for which a rational expression is undefined.
2. Find the numerical value of a rational expression.
3. Write rational expressions in lowest terms.
4. Recognize equivalent forms of rational expressions.

FOR EXTRA HELP

SSG Sec. 6.1
SSM Sec. 6.1
Pass the Test Software
InterAct Math Tutorial Software
Video 10

The quotient of two integers (with divisor not zero) is called a rational number. In the same way, the quotient of two polynomials with divisor not equal to zero is called a *rational expression.* The techniques of factoring, studied in Chapter 5, are essential in working with rational expressions.

Rational Expression

A **rational expression** is an expression of the form

$$\frac{P}{Q}$$

where P and Q are polynomials, with $Q \neq 0$.

Examples of rational expressions include $\dfrac{-6x}{x^3+8}$, $\dfrac{9x}{y+3}$, and $\dfrac{2m^3}{9}$.

OBJECTIVE 1 **Find the values for which a rational expression is undefined.** A fraction with denominator 0 is *not* a rational expression, since division by 0 is not possible. For this reason, be careful when substituting a number for a variable in the denominator of a rational expression. For example, in

$$\frac{8x^2}{x-3}$$

x can take on any value except 3. When $x = 3$, the denominator becomes $3 - 3 = 0$, making the expression undefined.

NOTE The numerator of a rational expression may be *any* number.

To determine the values for which a rational expression is undefined, use the following procedure.

Determining When a Rational Expression Is Undefined

Step 1 Set the denominator of the rational expression equal to 0.

Step 2 Solve this equation.

Step 3 The solutions of the equation are the values that make the rational expression undefined.

This procedure is illustrated in Example 1.

EXAMPLE 1 **Finding Values That Make Rational Expressions Undefined**

Find any values for which the following rational expressions are undefined.

(a) $\dfrac{p+5}{3p+2}$

Remember that the *numerator* may be any number; we must find any value of p that makes the *denominator* equal to 0 since division by 0 is undefined.

Step 1 Set the denominator equal to 0.

$$3p + 2 = 0$$

Step 2 Solve this equation.

$$3p = -2$$

$$p = -\frac{2}{3}$$

Step 3 Since $p = -\frac{2}{3}$ will make the denominator 0, the given expression is undefined for $-\frac{2}{3}$.

(b) $\dfrac{9m^2}{m^2 - 5m + 6}$

Find the numbers that make the denominator 0 by solving the equation

$$m^2 - 5m + 6 = 0.$$

$$(m - 2)(m - 3) = 0 \qquad \text{Factor.}$$

$$m - 2 = 0 \quad \text{or} \quad m - 3 = 0 \qquad \text{Zero-factor property}$$

$$m = 2 \quad \text{or} \quad m = 3 \qquad \text{Solve.}$$

The original expression is undefined for 2 and for 3.

(c) $\dfrac{2r}{r^2 + 1}$

This denominator cannot equal 0 for any value of r, since r^2 is always greater than or equal to 0 and adding 1 makes the sum greater than 0. Thus, there are no values for which this rational expression is undefined.

OBJECTIVE 2 Find the numerical value of a rational expression.

EXAMPLE 2 Evaluating a Rational Expression

Find the numerical value of $\dfrac{3x + 6}{2x - 4}$ for the given values of x.

(a) $x = 1$

Find the value of the rational expression by substituting 1 for x.

$$\frac{3x + 6}{2x - 4} = \frac{3(1) + 6}{2(1) - 4} \qquad \text{Let } x = 1.$$

$$= \frac{9}{-2}$$

$$= -\frac{9}{2}$$

(b) $x = 2$

Substituting 2 for x makes the denominator 0, so the rational expression is undefined when $x = 2$.

EXAMPLE 4 **Writing a Fraction with a Given Denominator**

Rewrite each expression with the indicated denominator.

(a) $\dfrac{3}{8} = \dfrac{}{40}$ **(b)** $\dfrac{9k}{25} = \dfrac{}{50k}$

For each example, first factor the denominator on the right. Then compare the denominator on the left with the one on the right to decide what factors are missing.

$$\frac{3}{8} = \frac{}{5 \cdot 8}$$

A factor of 5 is missing. Multiply by $\frac{5}{5}$ to get a denominator of 40.

$$\frac{3}{8} = \frac{3}{8} \cdot \frac{5}{5} = \frac{15}{40}$$

$\frac{5}{5} = 1$

$$\frac{9k}{25} = \frac{}{25 \cdot 2k}$$

Factors of 2 and k are missing. Get a denominator of $50k$ by multiplying by $\frac{2k}{2k}$.

$$\frac{9k}{25} = \frac{9k}{25} \cdot \frac{2k}{2k} = \frac{18k^2}{50k}$$

$\frac{2k}{2k} = 1$

Notice the use of the multiplicative identity property in each part of this example.

EXAMPLE 5 **Writing a Fraction with a Given Denominator**

Rewrite the following rational expression with the indicated denominator.

$$\frac{12p}{p^2 + 8p} = \frac{}{p^3 + 4p^2 - 32p}$$

Factor $p^2 + 8p$ as $p(p + 8)$. Compare with the denominator on the right which factors as $p(p + 8)(p - 4)$. The factor $p - 4$ is missing, so multiply $\dfrac{12p}{p(p + 8)}$ by $\dfrac{p - 4}{p - 4}$.

$$\frac{12p}{p^2 + 8p} = \frac{12p}{p(p + 8)} \cdot \frac{p - 4}{p - 4} \quad \text{Multiplicative identity property}$$

$$= \frac{12p(p - 4)}{p(p + 8)(p - 4)} \quad \text{Multiplication of rational expressions}$$

$$= \frac{12p^2 - 48p}{p^3 + 4p^2 - 32p} \quad \text{Multiply the factors.}$$

In the next section we add and subtract rational expressions, which sometimes requires the steps illustrated in Examples 4 and 5. While it is beneficial to leave the denominator in factored form, we multiplied the factors in the denominator in Example 5 to give the answer in the same form as the original problem.

6.3 EXERCISES

Choose the correct response in Exercises 1–4.

1. Suppose that the greatest common factor of a and b is 1. Then the least common denominator for $\frac{1}{a}$ and $\frac{1}{b}$ is

(a) a. **(b)** b. **(c)** ab. **(d)** 1.

2. If a is a factor of b, then the least common denominator for $\frac{1}{a}$ and $\frac{1}{b}$ is

(a) a. **(b)** b. **(c)** ab. **(d)** 1.

3. The least common denominator for $\frac{11}{20}$ and $\frac{1}{2}$ is

(a) 40. **(b)** 2. **(c)** 20. **(d)** none of these.

4. Suppose that we wish to write the fraction $\frac{1}{(x-4)^2(y-3)}$ with denominator $(x-4)^3(y-3)^2$. We must multiply both the numerator and the denominator by

(a) $(x-4)(y-3)$. **(b)** $(x-4)^2$. **(c)** $x-4$. **(d)** $(x-4)^2(y-3)$.

Find the LCD for the fractions in each list. See Examples 1–3.

5. $\frac{-7}{15}, \frac{21}{20}$

6. $\frac{9}{10}, \frac{12}{25}$

7. $\frac{17}{100}, \frac{23}{120}, \frac{43}{180}$

8. $\frac{17}{250}, \frac{-21}{300}, \frac{127}{360}$

9. $\frac{9}{x^2}, \frac{8}{x^5}$

10. $\frac{12}{m^7}, \frac{13}{m^8}$

11. $\frac{-2}{5p}, \frac{15}{6p}$

12. $\frac{14}{15k}, \frac{9}{4k}$

13. $\frac{17}{15y^2}, \frac{55}{36y^4}$

14. $\frac{4}{25m^3}, \frac{-9}{10m^4}$

15. $\frac{13}{5a^2b^3}, \frac{29}{15a^5b}$

16. $\frac{-7}{3r^4s^5}, \frac{-22}{9r^6s^8}$

17. $\frac{7}{6p}, \frac{15}{4p-8}$

18. $\frac{7}{8k}, \frac{-23}{12k-24}$

RELATING CONCEPTS (EXERCISES 19–22)

Suppose we want to find the LCD for the two common fractions

$$\frac{1}{24} \quad \text{and} \quad \frac{1}{20}.$$

In their prime factored forms, the denominators are

$$24 = 2^3 \cdot 3$$

and

$$20 = 2^2 \cdot 5.$$

Refer to this information as necessary and ***work Exercises 19–22 in order.***

19. What is the prime factored form of the LCD of the two fractions?

20. Suppose that two algebraic fractions have denominators $(t+4)^3(t-3)$ and $(t+4)^2(t+8)$. What is the factored form of the LCD of these?

21. What is the similarity between your answers in Exercises 19 and 20?

22. Comment on the following statement: The method for finding the LCD for two algebraic fractions is the same as the method for finding the LCD for two common fractions.

Did you make the connection between finding the LCD for common fractions and for algebraic fractions?

Find the LCD for the fractions in each list. See Examples 1–3.

23. $\dfrac{37}{6r-12}, \dfrac{25}{9r-18}$

24. $\dfrac{-14}{5p-30}, \dfrac{5}{6p-36}$

25. $\dfrac{5}{12p+60}, \dfrac{17}{p^2+5p}, \dfrac{16}{p^2+10p+25}$

26. $\dfrac{13}{r^2+7r}, \dfrac{-3}{5r+35}, \dfrac{-7}{r^2+14r+49}$

27. $\dfrac{3}{8y+16}, \dfrac{22}{y^2+3y+2}$

28. $\dfrac{-2}{9m-18}, \dfrac{-9}{m^2-7m+10}$

29. $\dfrac{12}{m-3}, \dfrac{-4}{3-m}$

30. $\dfrac{-17}{8-a}, \dfrac{2}{a-8}$

31. $\dfrac{29}{p-q}, \dfrac{18}{q-p}$

32. $\dfrac{16}{z-x}, \dfrac{8}{x-z}$

33. $\dfrac{6}{a^2+6a}, \dfrac{-5}{a^2+3a-18}$

34. $\dfrac{8}{y^2-5y}, \dfrac{-2}{y^2-2y-15}$

35. $\dfrac{-5}{k^2+2k-35}, \dfrac{-8}{k^2+3k-40}, \dfrac{9}{k^2-2k-15}$

36. $\dfrac{19}{z^2+4z-12}, \dfrac{16}{z^2+z-30}, \dfrac{6}{z^2+2z-24}$

37. Suppose that $(2x-5)^2$ is the LCD for two fractions. Is $(5-2x)^2$ also acceptable as an LCD? Why?

38. Suppose that $(4t-3)(5t-6)$ is the LCD for two fractions. Is $(3-4t)(6-5t)$ also acceptable as an LCD? Why?

RELATING CONCEPTS (EXERCISES 39–44)

Work Exercises 39–44 in order.

39. Suppose that you want to write $\frac{3}{4}$ as an equivalent fraction with denominator 28. By what number must you multiply both the numerator and the denominator?

40. If you write $\frac{3}{4}$ as an equivalent fraction with denominator 28, by what number are you actually multiplying the fraction?

41. What property of multiplication is being used when we write a common fraction as an equivalent one with a larger denominator? (See Section 1.7.)

42. Suppose that you want to write $\dfrac{2x+5}{x-4}$ as an equivalent fraction with denominator $7x-28$. By what number must you multiply both the numerator and the denominator?

43. If you write $\dfrac{2x+5}{x-4}$ as an equivalent fraction with denominator $7x-28$, by what number are you actually multiplying the fraction?

44. Repeat Exercise 41, changing "a common" to "an algebraic."

Did you make the connection between writing common fractions with larger denominators and writing algebraic fractions with more denominator factors?

Write each rational expression on the left with the indicated denominator. See Examples 4 and 5.

45. $\dfrac{15m^2}{8k} = \dfrac{\quad}{32k^4}$

46. $\dfrac{5t^2}{3y} = \dfrac{\quad}{9y^2}$

47. $\dfrac{19z}{2z-6} = \dfrac{\quad}{6z-18}$

48. $\dfrac{2r}{5r-5} = \dfrac{\quad}{15r-15}$

49. $\dfrac{-2a}{9a-18} = \dfrac{\quad}{18a-36}$

50. $\dfrac{-5y}{6y+18} = \dfrac{\quad}{24y+72}$

51. $\dfrac{6}{k^2-4k} = \dfrac{\quad}{k(k-4)(k+1)}$

52. $\dfrac{15}{m^2-9m} = \dfrac{\quad}{m(m-9)(m+8)}$

53. $\dfrac{36r}{r^2-r-6} = \dfrac{\quad}{(r-3)(r+2)(r+1)}$

54. $\dfrac{4m}{m^2-8m+15} = \dfrac{\quad}{(m-5)(m-3)(m+2)}$

55. $\dfrac{a+2b}{2a^2+ab-b^2} = \dfrac{\quad}{2a^3b+a^2b^2-ab^3}$

56. $\dfrac{m-4}{6m^2+7m-3} = \dfrac{\quad}{12m^3+14m^2-6m}$

57. $\dfrac{4r-t}{r^2+rt+t^2} = \dfrac{\quad}{t^3-r^3}$

58. $\dfrac{3x-1}{x^2+2x+4} = \dfrac{\quad}{x^3-8}$

59. $\dfrac{2(z-y)}{y^2+yz+z^2} = \dfrac{\quad}{y^4-z^3y}$

60. $\dfrac{2p+3q}{p^2+2pq+q^2} = \dfrac{\quad}{(p+q)(p^3+q^3)}$

61. Write an explanation of how to find the least common denominator for a group of denominators. Give an example.

62. Write an explanation of how to write a rational expression as an equivalent rational expression with a given denominator. Give an example.

6.4 Addition and Subtraction of Rational Expressions

OBJECTIVES

1. Add rational expressions having the same denominator.
2. Add rational expressions having different denominators.
3. Subtract rational expressions.

FOR EXTRA HELP

SSG Sec. 6.4
SSM Sec. 6.4
Pass the Test Software
InterAct Math Tutorial Software
Video 10

To add and subtract rational expressions, we use our previous work on finding least common denominators and writing fractions with the LCD.

OBJECTIVE 1 Add rational expressions having the same denominator. We find the sum of two rational expressions with a procedure similar to the one used for adding two fractions.

Adding Rational Expressions

If $\frac{P}{Q}$ and $\frac{R}{Q}$ are rational expressions, then

$$\frac{P}{Q} + \frac{R}{Q} = \frac{P+R}{Q}.$$

Again, the first example shows how addition of rational expressions compares with that of rational numbers.

Add or subtract. Write the answer in lowest terms. See Examples 1 and 6.

9. $\frac{4}{m}+\frac{7}{m}$

10. $\frac{5}{p}+\frac{11}{p}$

11. $\frac{a+b}{2}-\frac{a-b}{2}$

12. $\frac{x-y}{2}-\frac{x+y}{2}$

13. $\frac{x^2}{x+5}+\frac{5x}{x+5}$

14. $\frac{t^2}{t-3}+\frac{-3t}{t-3}$

15. $\frac{y^2-3y}{y+3}+\frac{-18}{y+3}$

16. $\frac{r^2-8r}{r-5}+\frac{15}{r-5}$

17. Explain with an example how to add or subtract rational expressions with the same denominators.

18. Explain with an example how to add or subtract rational expressions with different denominators.

Add or subtract. Write the answer in lowest terms. See Examples 2, 3, 4, and 7.

19. $\frac{z}{5}+\frac{1}{3}$

20. $\frac{p}{8}+\frac{3}{5}$

21. $\frac{5}{7}-\frac{r}{2}$

22. $\frac{10}{9}-\frac{z}{3}$

23. $-\frac{3}{4}-\frac{1}{2x}$

24. $-\frac{5}{8}-\frac{3}{2a}$

25. $\frac{x+1}{6}+\frac{3x+3}{9}$

26. $\frac{2x-6}{4}+\frac{x+5}{6}$

27. $\frac{x+3}{3x}+\frac{2x+2}{4x}$

28. $\frac{x+2}{5x}+\frac{6x+3}{3x}$

29. $\frac{7}{3p^2}-\frac{2}{p}$

30. $\frac{12}{5m^2}-\frac{5}{m}$

31. $\frac{x}{x-2}+\frac{4}{x+2}-\frac{8}{x^2-4}$

32. $\frac{2x}{x-1}+\frac{3}{x+1}-\frac{4}{x^2-1}$

33. $\frac{t}{t+2}+\frac{5-t}{t}-\frac{4}{t^2+2t}$

34. $\frac{2p}{p-3}+\frac{2+p}{p}-\frac{-6}{p^2-3p}$

35. What are the two possible LCDs that could be used for the sum

$$\frac{10}{m-2}+\frac{5}{2-m}?$$

36. If one form of the correct answer to a sum or difference of rational expressions is $\frac{4}{k-3}$, what would an alternate form of the answer be if the denominator is $3-k$?

Add or subtract. Write the answer in lowest terms. See Examples 5 and 8.

37. $\frac{4}{x-5}+\frac{6}{5-x}$

38. $\frac{10}{m-2}+\frac{5}{2-m}$

39. $\frac{-1}{1-y}+\frac{3-4y}{y-1}$

40. $\frac{-4}{p-3}-\frac{p+1}{3-p}$

41. $\frac{2}{x-y^2}+\frac{7}{y^2-x}$

42. $\frac{-8}{p-q^2}+\frac{3}{q^2-p}$

43. $\frac{x}{5x-3y}-\frac{y}{3y-5x}$

44. $\frac{t}{8t-9s}-\frac{s}{9s-8t}$

45. $\frac{3}{4p-5}+\frac{9}{5-4p}$

46. $\frac{8}{3-7y}-\frac{2}{7y-3}$

In these subtraction problems, the rational expression that follows the subtraction sign has a numerator with more than one term. Be very careful with signs and find each difference.

47. $\frac{2m}{m-n}-\frac{5m+n}{2m-2n}$

48. $\frac{5p}{p-q}-\frac{3p+1}{4p-4q}$

49. $\frac{5}{x^2-9}-\frac{x+2}{x^2+4x+3}$

50. $\frac{1}{a^2-1}-\frac{a-1}{a^2+3a-4}$

51. $\dfrac{2q+1}{3q^2+10q-8} - \dfrac{3q+5}{2q^2+5q-12}$

52. $\dfrac{4y-1}{2y^2+5y-3} - \dfrac{y+3}{6y^2+y-2}$

Perform the indicated operations. See Examples 1–9.

53. $\dfrac{4}{r^2-r} + \dfrac{6}{r^2+2r} - \dfrac{1}{r^2+r-2}$

54. $\dfrac{6}{k^2+3k} - \dfrac{1}{k^2-k} + \dfrac{2}{k^2+2k-3}$

55. $\dfrac{x+3y}{x^2+2xy+y^2} + \dfrac{x-y}{x^2+4xy+3y^2}$

56. $\dfrac{m}{m^2-1} + \dfrac{m-1}{m^2+2m+1}$

57. $\dfrac{r+y}{18r^2+12ry-3ry-2y^2} + \dfrac{3r-y}{36r^2-y^2}$

58. $\dfrac{2x-z}{2x^2-4xz+5xz-10z^2} - \dfrac{x+z}{x^2-4z^2}$

Perform the indicated operations. Remember the order of operations.

59. $\left(\dfrac{-k}{2k^2-5k-3} + \dfrac{3k-2}{2k^2-k-1}\right)\dfrac{2k+1}{k-1}$

60. $\left(\dfrac{3p+1}{2p^2+p-6} - \dfrac{5p}{3p^2-p}\right)\dfrac{2p-3}{p+2}$

61. $\dfrac{k^2+4k+16}{k+4}\left(\dfrac{-5}{16-k^2} + \dfrac{2k+3}{k^3-64}\right)$

62. $\dfrac{m-5}{2m+5}\left(\dfrac{-3m}{m^2-25} - \dfrac{m+4}{125-m^3}\right)$

63. Refer to the rectangle in the figure.
(a) Find an expression that represents its perimeter. Give the simplified form.
(b) Find an expression that represents its area. Give the simplified form.

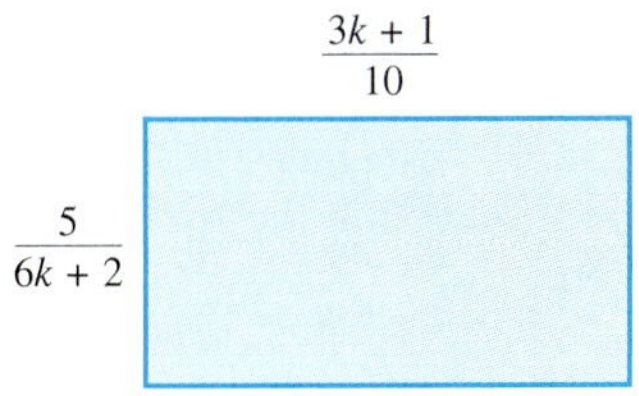

64. Refer to the triangle in the figure. Find an expression that represents its perimeter.

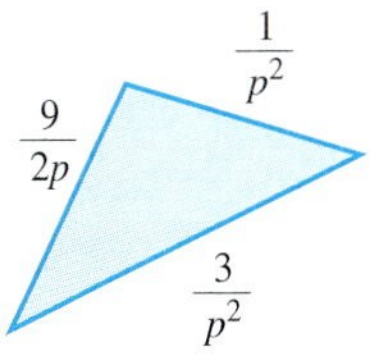

6.5 Complex Fractions

OBJECTIVES

1. Simplify a complex fraction by writing it as a division problem (Method 1).
2. Simplify a complex fraction by multiplying by the least common denominator (Method 2).

FOR EXTRA HELP

SSG Sec. 6.5
SSM Sec. 6.5
Pass the Test Software
InterAct Math Tutorial Software
Video 11

The quotient of two mixed numbers in arithmetic, such as $2\frac{1}{2} \div 3\frac{1}{4}$, can be written as a fraction:

$$2\frac{1}{2} \div 3\frac{1}{4} = \frac{2\frac{1}{2}}{3\frac{1}{4}} = \frac{2+\frac{1}{2}}{3+\frac{1}{4}}.$$

The last expression is the quotient of expressions that involve fractions. In algebra, some rational expressions also have fractions in the numerator, or denominator, or both.

Complex Fraction

A rational expression with fractions in the numerator, denominator, or both, is called a **complex fraction.**

Examples of complex fractions include

$$\frac{2+\frac{1}{2}}{3+\frac{1}{4}}, \quad \frac{\frac{3x^2-5x}{6x^2}}{2x-\frac{1}{x}}, \quad \text{and} \quad \frac{3+x}{5-\frac{2}{x}}.$$

The parts of a complex fraction are named as follows.

$$\frac{\frac{2}{p}-\frac{1}{q}}{\frac{3}{p}+\frac{5}{q}}$$

← Numerator of complex fraction
← Main fraction bar
← Denominator of complex fraction

OBJECTIVE 1 Simplify a complex fraction by writing it as a division problem (Method 1). Since the main fraction bar represents division in a complex fraction, one method of simplifying a complex fraction involves division.

Method 1

To simplify a complex fraction:

Step 1 Write both the numerator and denominator as single fractions.

Step 2 Change the complex fraction to a division problem.

Step 3 Perform the indicated division.

Once again, in this section the first example shows complex fractions from both arithmetic and algebra.

EXAMPLE 1 Simplifying Complex Fractions by Method 1

Simplify each complex fraction.

(a) $\dfrac{\frac{2}{3}+\frac{5}{9}}{\frac{1}{4}+\frac{1}{12}}$ **(b)** $\dfrac{6+\frac{3}{x}}{\frac{x}{4}+\frac{1}{8}}$

Step 1 First, write each numerator as a single fraction.

$$\frac{2}{3}+\frac{5}{9}=\frac{2(3)}{3(3)}+\frac{5}{9}=\frac{6}{9}+\frac{5}{9}=\frac{11}{9}$$

$$6+\frac{3}{x}=\frac{6}{1}+\frac{3}{x}=\frac{6x}{x}+\frac{3}{x}=\frac{6x+3}{x}$$

Do the same thing with each denominator.

$$\frac{1}{4}+\frac{1}{12}=\frac{1(3)}{4(3)}+\frac{1}{12}=\frac{3}{12}+\frac{1}{12}=\frac{4}{12}$$

$$\frac{x}{4}+\frac{1}{8}=\frac{x(2)}{4(2)}+\frac{1}{8}=\frac{2x}{8}+\frac{1}{8}=\frac{2x+1}{8}$$

Step 2 The original complex fraction can now be written as follows.

$$\frac{\frac{11}{9}}{\frac{4}{12}} \qquad \Bigg| \qquad \frac{\frac{6x+3}{x}}{\frac{2x+1}{8}}$$

Step 3 Now use the rule for division and the fundamental property.

$$\frac{11}{9} \div \frac{4}{12} = \frac{11}{9} \cdot \frac{12}{4} = \frac{11 \cdot 3 \cdot 4}{3 \cdot 3 \cdot 4} = \frac{11}{3}$$

$$\frac{6x+3}{x} \div \frac{2x+1}{8} = \frac{6x+3}{x} \cdot \frac{8}{2x+1} = \frac{3(2x+1)}{x} \cdot \frac{8}{2x+1} = \frac{24}{x}$$

EXAMPLE 2 Simplifying a Complex Fraction by Method 1

Simplify the complex fraction.

$$\frac{\frac{xp}{q^3}}{\frac{p^2}{qx^2}}$$

Here the numerator and denominator are already single fractions, so use the division rule and then the fundamental property.

$$\frac{xp}{q^3} \div \frac{p^2}{qx^2} = \frac{xp}{q^3} \cdot \frac{qx^2}{p^2} = \frac{x^3}{q^2p}$$

EXAMPLE 3 Simplifying a Complex Fraction by Method 1

Simplify the complex fraction.

$$\frac{\frac{3}{x+2} - 4}{\frac{2}{x+2} + 1} = \frac{\frac{3}{x+2} - \frac{4(x+2)}{x+2}}{\frac{2}{x+2} + \frac{1(x+2)}{x+2}}$$ Write both second terms with a denominator of $x + 2$.

$$= \frac{\frac{3 - 4(x+2)}{x+2}}{\frac{2 + 1(x+2)}{x+2}}$$ Subtract in the numerator. Add in the denominator.

$$= \frac{\frac{3 - 4x - 8}{x+2}}{\frac{2 + x + 2}{x+2}}$$ Distributive property

$$\frac{\frac{3-4x-8}{x+2}}{\frac{2+x+2}{x+2}} = \frac{\frac{-5-4x}{x+2}}{\frac{4+x}{x+2}} \qquad \text{Combine terms.}$$

$$= \frac{-5-4x}{x+2} \cdot \frac{x+2}{4+x} \qquad \text{Multiply by the reciprocal.}$$

$$= \frac{-5-4x}{4+x} \qquad \text{Lowest terms}$$

OBJECTIVE 2 Simplify a complex fraction by multiplying by the least common denominator (Method 2). As an alternative method, a complex fraction may be simplified by a method that uses the fundamental property of rational expressions. Since any expression can be multiplied by a form of 1 to get an equivalent expression, we may multiply both the numerator and the denominator of a complex fraction by the same nonzero expression to get an equivalent complex fraction. If we choose the expression to be the LCD of all the fractions within the complex fraction, the complex fraction will be simplified. This is Method 2.

Method 2

To simplify a complex fraction:

Step 1 Find the LCD of all fractions within the complex fraction.

Step 2 Multiply both the numerator and the denominator of the complex fraction by this LCD using the distributive property as necessary. Write in lowest terms.

In the next example, Method 2 is used to simplify the complex fractions from Example 1.

EXAMPLE 4 Simplifying Complex Fractions by Method 2

Simplify each complex fraction.

(a) $\dfrac{\frac{2}{3}+\frac{5}{9}}{\frac{1}{4}+\frac{1}{12}}$ **(b)** $\dfrac{6+\frac{3}{x}}{\frac{x}{4}+\frac{1}{8}}$

Step 1 Find the LCD for all denominators in the complex fraction.

The LCD for 3, 9, 4, and 12 is 36. | The LCD for x, 4, and 8 is $8x$.

Step 2 Multiply numerator and denominator of the complex fraction by the LCD.

$$\frac{\frac{2}{3}+\frac{5}{9}}{\frac{1}{4}+\frac{1}{12}} = \frac{36\left(\frac{2}{3}+\frac{5}{9}\right)}{36\left(\frac{1}{4}+\frac{1}{12}\right)} \qquad \frac{6+\frac{3}{x}}{\frac{x}{4}+\frac{1}{8}} = \frac{8x\left(6+\frac{3}{x}\right)}{8x\left(\frac{x}{4}+\frac{1}{8}\right)}$$

$$= \frac{36\left(\frac{2}{3}\right) + 36\left(\frac{5}{9}\right)}{36\left(\frac{1}{4}\right) + 36\left(\frac{1}{12}\right)} \qquad = \frac{8x(6) + 8x\left(\frac{3}{x}\right)}{8x\left(\frac{x}{4}\right) + 8x\left(\frac{1}{8}\right)}$$ Distributive property

$$= \frac{24 + 20}{9 + 3} \qquad = \frac{48x + 24}{2x^2 + x}$$

$$= \frac{44}{12} = \frac{4 \cdot 11}{4 \cdot 3} \qquad = \frac{24(2x + 1)}{x(2x + 1)}$$ Factor.

$$= \frac{11}{3} \qquad = \frac{24}{x}$$ Lowest terms

EXAMPLE 5 Simplifying a Complex Fraction by Method 2

Simplify the complex fraction.

$$\frac{\frac{3}{5m} - \frac{2}{m^2}}{\frac{9}{2m} + \frac{3}{4m^2}}$$

The LCD for $5m$, m^2, $2m$, and $4m^2$ is $20m^2$. Multiply numerator and denominator by $20m^2$.

$$\frac{\frac{3}{5m} - \frac{2}{m^2}}{\frac{9}{2m} + \frac{3}{4m^2}} = \frac{20m^2\left(\frac{3}{5m} - \frac{2}{m^2}\right)}{20m^2\left(\frac{9}{2m} + \frac{3}{4m^2}\right)}$$

$$= \frac{20m^2\left(\frac{3}{5m}\right) - 20m^2\left(\frac{2}{m^2}\right)}{20m^2\left(\frac{9}{2m}\right) + 20m^2\left(\frac{3}{4m^2}\right)}$$ Distributive property

$$= \frac{12m - 40}{90m + 15}$$

Either of the two methods shown in this section can be used to simplify a complex fraction. You may want to choose one method and stick with it to eliminate confusion. However, some students prefer to use Method 1 for problems like Example 2, which is the quotient of two fractions. They prefer Method 2 for problems like Examples 1, 3, 4, and 5, which have sums or differences in the numerators or denominators or both.

CONNECTIONS

Some numbers can be expressed as *continued fractions,* which are infinite complex fractions. For example, the irrational number $\sqrt{2}$ can be expressed as follows.

$$\sqrt{2} = 1 + \cfrac{1}{2 + \cfrac{1}{2 + \cfrac{1}{2 + \cfrac{1}{2 + \ddots}}}}$$

Better and better approximations of $\sqrt{2}$ can be found by using more and more terms of the fraction. We give the first three approximations here.

$$1 + \frac{1}{2} = 1.5$$

$$1 + \cfrac{1}{2 + \cfrac{1}{2}} = 1 + \cfrac{1}{\cfrac{5}{2}} = 1 + \frac{2}{5} = \frac{7}{5} = 1.4$$

$$1 + \cfrac{1}{2 + \cfrac{1}{2 + \cfrac{1}{2}}} = 1 + \cfrac{1}{2 + \cfrac{1}{\cfrac{5}{2}}} = 1 + \cfrac{1}{2 + \cfrac{2}{5}} = 1 + \cfrac{1}{\cfrac{12}{5}}$$

$$= 1 + \frac{5}{12} = \frac{17}{12} = 1.41\overline{6}$$

A calculator gives $\sqrt{2} \approx 1.414213562$ to nine decimal places.

FOR DISCUSSION OR WRITING

Give the next approximation of $\sqrt{2}$ and compare it to the calculator value shown above. How many places after the decimal agree?

6.5 EXERCISES

Note: In many problems involving complex fractions, several different equivalent forms of the answer exist. If your answer does not look exactly like the one given in the back of the book, check to see if your answer is an equivalent form.

1. Consider the complex fraction $\dfrac{\frac{1}{2} - \frac{1}{3}}{\frac{5}{6} - \frac{1}{12}}$. Answer each of the following, outlining Method 1 for simplifying this complex fraction.

(a) To combine the terms in the numerator, we must find the LCD of $\frac{1}{2}$ and $\frac{1}{3}$. What is this LCD? Determine the simplified form of the numerator of the complex fraction.

(b) To combine the terms in the denominator, we must find the LCD of $\frac{5}{6}$ and $\frac{1}{12}$. What is this LCD? Determine the simplified form of the denominator of the complex fraction.

(c) Now use the results from parts (a) and (b) to write the complex fraction as a division problem using the symbol $\div$.

(d) Perform the operation from part (c) to obtain the final simplification.

2. Consider the same complex fraction given in Exercise 1, $\dfrac{\frac{1}{2}-\frac{1}{3}}{\frac{5}{6}-\frac{1}{12}}$. Answer each of the following, outlining Method 2 for simplifying this complex fraction.

(a) We must determine the LCD of all the fractions within the complex fraction. What is this LCD?

(b) Multiply every term in the complex fraction by the LCD found in part (a), but do not combine the terms in the numerator and the denominator yet.

(c) Combine the terms from part (b) to obtain the simplified form of the complex fraction.

3. Which one of the following complex fractions is equivalent to $\dfrac{3-\frac{1}{2}}{2-\frac{1}{4}}$? Answer this question without showing any work, and explain the reason for the equivalence.

(a) $\dfrac{3+\frac{1}{2}}{2+\frac{1}{4}}$ **(b)** $\dfrac{-3+\frac{1}{2}}{2-\frac{1}{4}}$ **(c)** $\dfrac{-3-\frac{1}{2}}{-2-\frac{1}{4}}$ **(d)** $\dfrac{-3+\frac{1}{2}}{-2+\frac{1}{4}}$

4. Only one of the choices below is equal to $\dfrac{\frac{1}{2}+\frac{1}{4}}{\frac{1}{3}+\frac{1}{12}}$. Which one is it?

Answer this question without showing any work, and explain the reason for the equivalence.

(a) $\frac{9}{5}$ **(b)** $-\frac{9}{5}$ **(c)** $-\frac{5}{9}$ **(d)** -12

5. In your own words, describe Method 1 for simplifying complex fractions. Illustrate with the example $\dfrac{\frac{1}{2}}{\frac{2}{3}}$.

6. In your own words, describe Method 2 for simplifying complex fractions. Illustrate with the example $\dfrac{\frac{1}{2}}{\frac{2}{3}}$.

Simplify each complex fraction. Use either method. See Examples 1–5.

7. $\dfrac{-\frac{4}{3}}{\frac{2}{9}}$ **8.** $\dfrac{-\frac{5}{6}}{\frac{5}{4}}$ **9.** $\dfrac{\frac{x}{y^2}}{\frac{x^2}{y}}$ **10.** $\dfrac{\frac{p^4}{r}}{\frac{p^2}{r^2}}$

11. $\dfrac{\frac{4a^4b^3}{3a}}{\frac{2ab^4}{b^2}}$ **12.** $\dfrac{\frac{2r^4t^2}{3t}}{\frac{5r^2t^5}{3r}}$ **13.** $\dfrac{\frac{m+2}{3}}{\frac{m-4}{m}}$ **14.** $\dfrac{\frac{q-5}{q}}{\frac{q+5}{3}}$

15. $\dfrac{\dfrac{2}{x} - 3}{\dfrac{2 - 3x}{2}}$

16. $\dfrac{6 + \dfrac{2}{r}}{\dfrac{3r + 1}{4}}$

17. $\dfrac{\dfrac{1}{x} + x}{\dfrac{x^2 + 1}{8}}$

18. $\dfrac{\dfrac{3}{m} - m}{\dfrac{3 - m^2}{4}}$

19. $\dfrac{a - \dfrac{5}{a}}{a + \dfrac{1}{a}}$

20. $\dfrac{q + \dfrac{1}{q}}{q + \dfrac{4}{q}}$

21. $\dfrac{\dfrac{5}{8} + \dfrac{2}{3}}{\dfrac{7}{3} - \dfrac{1}{4}}$

22. $\dfrac{\dfrac{6}{5} - \dfrac{1}{9}}{\dfrac{2}{5} + \dfrac{5}{3}}$

23. $\dfrac{\dfrac{1}{x^2} + \dfrac{1}{y^2}}{\dfrac{1}{x} - \dfrac{1}{y}}$

24. $\dfrac{\dfrac{1}{a^2} - \dfrac{1}{b^2}}{\dfrac{1}{a} - \dfrac{1}{b}}$

25. $\dfrac{\dfrac{2}{p^2} - \dfrac{3}{5p}}{\dfrac{4}{p} + \dfrac{1}{4p}}$

26. $\dfrac{\dfrac{2}{m^2} - \dfrac{3}{m}}{\dfrac{2}{5m^2} + \dfrac{1}{3m}}$

27. $\dfrac{\dfrac{5}{x^2y} - \dfrac{2}{xy^2}}{\dfrac{3}{x^2y^2} + \dfrac{4}{xy}}$

28. $\dfrac{\dfrac{1}{m^3p} + \dfrac{2}{mp^2}}{\dfrac{4}{mp} + \dfrac{1}{m^2p}}$

29. $\dfrac{\dfrac{1}{4} - \dfrac{1}{a^2}}{\dfrac{1}{2} + \dfrac{1}{a}}$

30. $\dfrac{\dfrac{1}{9} - \dfrac{1}{m^2}}{\dfrac{1}{3} + \dfrac{1}{m}}$

31. $\dfrac{\dfrac{1}{z + 5}}{\dfrac{4}{z^2 - 25}}$

32. $\dfrac{\dfrac{1}{a + 1}}{\dfrac{2}{a^2 - 1}}$

33. $\dfrac{\dfrac{1}{m + 1} - 1}{\dfrac{1}{m + 1} + 1}$

34. $\dfrac{\dfrac{2}{x - 1} + 2}{\dfrac{2}{x - 1} - 2}$

35. $\dfrac{\dfrac{1}{m - 1} + \dfrac{2}{m + 2}}{\dfrac{2}{m + 2} - \dfrac{1}{m - 3}}$

36. $\dfrac{\dfrac{5}{r + 3} - \dfrac{1}{r - 1}}{\dfrac{2}{r + 2} + \dfrac{3}{r + 3}}$

37. In a fraction, what operation does the fraction bar represent?

38. What property of real numbers justifies Method 2 of simplifying complex fractions?

RELATING CONCEPTS (EXERCISES 39–42)

In order to find the average of two numbers, we add them and divide by 2. Suppose that we wish to find the average of $\frac{3}{8}$ and $\frac{5}{6}$.

Work Exercises 39–42 in order, *to see how a complex fraction occurs in a problem like this.*

39. Write in symbols: the sum of $\dfrac{3}{8}$ and $\dfrac{5}{6}$, divided by 2. Your result should be a complex fraction.

40. Simplify the complex fraction from Exercise 39 using Method 1.

41. Simplify the complex fraction from Exercise 39 using Method 2.

42. Your answers in Exercises 40 and 41 should be the same. Which method did you prefer? Why?

Did you make the connection between finding the average of two fractions and simplifying a complex fraction?

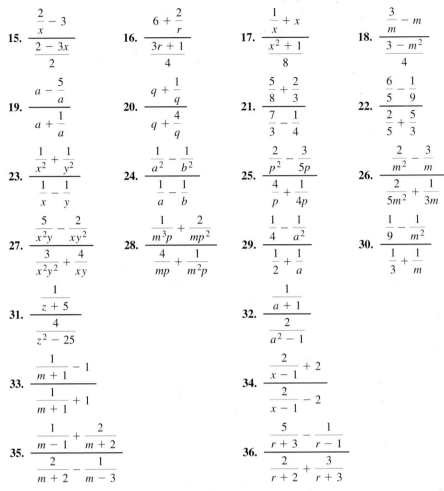

The expressions in Exercises 43–48 are called continued fractions. *Simplify these continued fractions by starting at "the bottom" and working upward.*

43. $1+\dfrac{1}{1+\dfrac{1}{1+1}}$

44. $5+\dfrac{5}{5+\dfrac{5}{5+5}}$

45. $7-\dfrac{3}{5+\dfrac{2}{4-2}}$

46. $3-\dfrac{2}{4+\dfrac{2}{4-2}}$

47. $r+\dfrac{r}{4-\dfrac{2}{6+2}}$

48. $\dfrac{2q}{7}-\dfrac{q}{6+\dfrac{8}{4+4}}$

6.6 Solving Equations Involving Rational Expressions

OBJECTIVES

1. Distinguish between expressions with rational coefficients and equations with terms that are rational expressions.
2. Solve equations with rational expressions.
3. Solve a formula for a specified variable.

FOR EXTRA HELP

SSG Sec. 6.6
SSM Sec. 6.6
Pass the Test Software
InterAct Math Tutorial Software
Video 11

In Section 2.2 we solved equations with fractions as coefficients. By using the multiplication property of equality, we cleared the fractions by multiplying by the LCD. We continue this work here.

OBJECTIVE 1 Distinguish between expressions with rational coefficients and equations with terms that are rational expressions. Before solving equations with rational expressions, you must understand the difference between *sums* and *differences* of terms with rational coefficients, and *equations* with terms that are rational expressions. Sums and differences are *simplified,* while equations are *solved.*

EXAMPLE 1 Distinguishing between Expressions and Equations

Identify each of the following as an expression or an equation. If it is an expression, simplify it. If it is an equation, solve it.

(a) $\dfrac{3}{4}x-\dfrac{2}{3}x$

This is a difference of two terms, so it is an expression. (There is no equals sign.) Simplify by finding the LCD, writing each coefficient with this LCD, and combining like terms.

$$\frac{3}{4}x-\frac{2}{3}x=\frac{9}{12}x-\frac{8}{12}x \quad \text{Get a common denominator.}$$

$$=\frac{1}{12}x \quad \text{Combine like terms.}$$

(b) $\dfrac{3}{4}x-\dfrac{2}{3}x=\dfrac{1}{2}$

Because of the equals sign, this is an equation to be solved. We proceed as in Section 2.2, using the multiplication property of equality to clear fractions. The LCD is 12.

$$\frac{3}{4}x-\frac{2}{3}x=\frac{1}{2}$$

$$12\left(\frac{3}{4}x-\frac{2}{3}x\right)=12\left(\frac{1}{2}\right) \quad \text{Multiply by 12.}$$

$$12\left(\frac{3}{4}x\right)-12\left(\frac{2}{3}x\right)=12\left(\frac{1}{2}\right) \quad \text{Distributive property}$$

$$9x - 8x = 6 \quad \text{Multiply.}$$
$$x = 6 \quad \text{Combine like terms.}$$

The solution, 6, should be checked in the original equation. Because it leads to a true statement, the solution set is $\{6\}$.

The ideas of Example 1 can be summarized as follows.

When adding or subtracting, the LCD must be kept throughout the simplification. When solving an equation, the LCD is used to multiply both sides so that denominators are eliminated.

OBJECTIVE 2 **Solve equations with rational expressions.** The next few examples illustrate clearing an equation of fractions in order to solve it.

EXAMPLE 2 **Solving an Equation Involving Rational Expressions**

Solve $\frac{p}{2} - \frac{p-1}{3} = 1$.

Multiply both sides by the LCD, 6.

$$6\left(\frac{p}{2} - \frac{p-1}{3}\right) = 6 \cdot 1$$
$$6\left(\frac{p}{2}\right) - 6\left(\frac{p-1}{3}\right) = 6 \quad \text{Distributive property}$$
$$3p - 2(p - 1) = 6$$

Be careful to put parentheses around $p - 1$; otherwise you may get an incorrect solution. Continue simplifying and solve the equation.

$$3p - 2p + 2 = 6 \quad \text{Distributive property}$$
$$p + 2 = 6 \quad \text{Combine like terms.}$$
$$p = 4 \quad \text{Subtract 2.}$$

Check to see that $\{4\}$ is the solution set by replacing p with 4 in the original equation.

CAUTION The most common error in equations like the one found in Example 2 occurs when parentheses are not used for the numerator $p - 1$. If parentheses are not used, a sign error may result.

The equations in Examples 1(b) and 2 did not have variables in denominators. When solving equations that have a variable in a denominator, remember that the number 0 cannot be a denominator. Therefore, the solution cannot be a number that will make the denominator equal 0. Example 3 illustrates this.

EXAMPLE 3 **Solving an Equation Involving Rational Expressions with No Solution**

Solve $\frac{x}{x-2} = \frac{2}{x-2} + 2$.

Multiply both sides by the LCD, $x - 2$.

$$(x-2)\left(\frac{x}{x-2}\right) = (x-2)\left(\frac{2}{x-2}\right) + (x-2)(2)$$

$x = 2 + 2x - 4$	Distributive property
$x = -2 + 2x$	Combine terms.
$-x = -2$	Subtract $2x$.
$x = 2$	Multiply by -1.

The proposed solution is 2. If we substitute 2 into the original equation, we get

$$\frac{2}{2-2} = \frac{2}{2-2} + 2 \qquad ?$$

$$\frac{2}{0} = \frac{2}{0} + 2. \qquad ?$$

Notice that 2 makes both denominators equal 0. Because 0 cannot be the denominator of a fraction, the solution set is $\emptyset$.

While it is always a good idea to check solutions to guard against arithmetic and algebraic errors, it is *essential* to check proposed solutions when variables appear in denominators in the original equation. Some students like to determine which numbers cannot be solutions *before* solving the equation.

Solving Equations with Rational Expressions

Step 1 **Multiply by the LCD.** Multiply both sides of the equation by the least common denominator. (This clears the equation of fractions.)

Step 2 **Solve.** Solve the resulting equation.

Step 3 **Check.** Check each proposed solution by substituting it in the original equation. Reject any that cause a denominator to equal 0.

EXAMPLE 4 **Solving an Equation Involving Rational Expressions**

Solve $\frac{2}{x^2 - x} = \frac{1}{x^2 - 1}$.

Step 1 Begin by finding the LCD. Since $x^2 - x$ can be factored as $x(x - 1)$, and $x^2 - 1$ can be factored as $(x + 1)(x - 1)$, the LCD is $x(x + 1)(x - 1)$.

$$\frac{2}{x(x-1)} = \frac{1}{(x+1)(x-1)} \qquad \text{Factor the denominators.}$$

Step 2 Notice that 0, -1, and 1 cannot be solutions of this equation. Multiply both sides of the equation by $x(x+1)(x-1)$.

$$x(x+1)(x-1)\frac{2}{x(x-1)} = x(x+1)(x-1)\frac{1}{(x+1)(x-1)}$$

$$2(x+1) = x$$

$$2x + 2 = x \qquad \text{Distributive property}$$

$$2 = -x \qquad \text{Subtract } 2x.$$

$$x = -2 \qquad \text{Multiply by } -1.$$

Step 3 The proposed solution is -2, which does not make any denominator equal 0. A check will verify that no arithmetic or algebraic errors have been made. Thus, the solution set is $\{-2\}$.

EXAMPLE 5 Solving an Equation Involving Rational Expressions

Solve $\frac{1}{x-1} + \frac{1}{2} = \frac{2}{x^2-1}$.

Factor the denominator on the right.

$$\frac{1}{x-1} + \frac{1}{2} = \frac{2}{(x+1)(x-1)}$$

Notice that 1 and -1 cannot be solutions of this equation. Multiply both sides of the equation by the LCD, $2(x+1)(x-1)$.

$$2(x+1)(x-1)\left(\frac{1}{x-1} + \frac{1}{2}\right) = 2(x+1)(x-1)\frac{2}{(x+1)(x-1)}$$

$$2(x+1)(x-1)\frac{1}{x-1} + 2(x+1)(x-1)\frac{1}{2} = 2(x+1)(x-1)\frac{2}{(x+1)(x-1)}$$

$$2(x+1) + (x+1)(x-1) = 4$$

$$2x + 2 + x^2 - 1 = 4 \qquad \text{Distributive property}$$

$$x^2 + 2x + 1 = 4$$

$$x^2 + 2x - 3 = 0 \qquad \text{Get 0 on the right side.}$$

Factoring gives

$$(x+3)(x-1) = 0.$$

$$x + 3 = 0 \quad \text{or} \quad x - 1 = 0 \qquad \text{Zero-factor property}$$

$$x = -3 \quad \text{or} \quad x = 1$$

-3 and 1 are proposed solutions. However, as noted above, 1 makes an original denominator equal 0, so 1 is not a solution. Substituting -3 for x gives a true statement, so $\{-3\}$ is the solution set.

EXAMPLE 6 Solving an Equation Involving Rational Expressions

Solve $\frac{1}{k^2+4k+3} + \frac{1}{2k+2} = \frac{3}{4k+12}$.

Factoring each denominator gives the equation

$$\frac{1}{(k+1)(k+3)} + \frac{1}{2(k+1)} = \frac{3}{4(k+3)}.$$

The LCD is $4(k+1)(k+3)$, indicating that -1 and -3 cannot be solutions of the equation. Multiply both sides by this LCD.

$$4(k+1)(k+3)\left(\frac{1}{(k+1)(k+3)} + \frac{1}{2(k+1)}\right) = 4(k+1)(k+3)\frac{3}{4(k+3)}$$

$$4(k+1)(k+3)\frac{1}{(k+1)(k+3)} + 2 \cdot 2(k+1)(k+3)\frac{1}{2(k+1)} = 4(k+1)(k+3)\frac{3}{4(k+3)}$$

$$4 + 2(k+3) = 3(k+1)$$

$$4 + 2k + 6 = 3k + 3 \qquad \text{Distributive property}$$

$$2k + 10 = 3k + 3$$

$$10 - 3 = 3k - 2k$$

$$7 = k$$

The proposed solution, 7, does not make an original denominator equal 0. A check shows that the algebra is correct, so $\{7\}$ is the solution set.

OBJECTIVE 3 Solve a formula for a specified variable. Solving a formula for a specified variable was discussed in Chapter 2. In the next example, this procedure is applied to a formula involving fractions.

EXAMPLE 7 Solving for a Specified Variable

Solve the formula $\frac{1}{a} = \frac{1}{b} + \frac{1}{c}$ for c.

The LCD of all the fractions in the equation is abc, so multiply both sides by abc.

$$abc\left(\frac{1}{a}\right) = abc\left(\frac{1}{b} + \frac{1}{c}\right)$$

$$abc\left(\frac{1}{a}\right) = abc\left(\frac{1}{b}\right) + abc\left(\frac{1}{c}\right) \qquad \text{Distributive property}$$

$$bc = ac + ab$$

Since we are solving for c, get all terms with c on one side of the equation. Do this by subtracting ac from both sides.

$$bc - ac = ab \qquad \text{Subtract } ac.$$

Factor out the common factor c on the left.

$$c(b - a) = ab \qquad \text{Factor out } c.$$

Finally, divide both sides by the coefficient of c, which is $b - a$.

$$c = \frac{ab}{b - a}$$

RELATING CONCEPTS (EXERCISES 79–84)

In Section 6.4 we saw how, after adding or subtracting two rational expressions, the result can often be simplified to lowest terms. In this section we learned that multiplying both sides of an equation by the LCD may lead to a solution that must be rejected because it causes a denominator to equal 0.

Work Exercises 79–84 in order, to see a relationship between these two concepts.

79. Solve the equation $\dfrac{x^2}{x-3} + \dfrac{2x-15}{x-3} = 0$ by multiplying both sides by $x - 3$. What is the solution? What is the number that must be rejected as a solution?

80. Combine the two rational expressions on the left side of the given equation in Exercise 79 and reduce to lowest terms.

 (a) If you set the simplified form equal to 0 and solve, how does your solution compare to the actual solution in Exercise 79?

 (b) If you set the common factor that you divided out equal to 0 and solve, how does your solution compare to the rejected solution in Exercise 79?

81. Consider the equation $\dfrac{1}{x-1} + \dfrac{1}{2} - \dfrac{2}{x^2-1} = 0$, which is equivalent to the equation solved in Example 5. According to Example 5, what is the solution? What is the number that must be rejected as a solution?

82. Combine the three rational expressions on the left side of the given equation in Exercise 81 and reduce to lowest terms. Repeat parts (a) and (b) of Exercise 80, comparing this time to the equation in Exercise 81.

83. Write a short paragraph summarizing what you have learned in working Exercises 79–82.

84. Devise a procedure whereby you could solve an equation involving rational expressions which would not lead to values that must be rejected.

Did you make the connection between rational expressions that can be simplified and rational equations with rejected solutions?

TECHNOLOGY INSIGHTS (EXERCISES 85–90)

The first line in the calculator screen on the left shows that 6 has been stored into memory location X. The expression $\dfrac{1}{X+4}$ is then calculated to be $\dfrac{1}{10}$, as expected.

However, if the calculator is directed to evaluate $\dfrac{1}{X-6}$, an error message occurs, as shown in the screen on the right. The value 6 leads to division by 0, which is not allowed.

```
6→X
                 6
1/(X+4)▶Frac
              1/10
1/(X-6)▶Frac
```

```
ERR:DIVIDE BY 0
1▪Quit
2:Goto
```

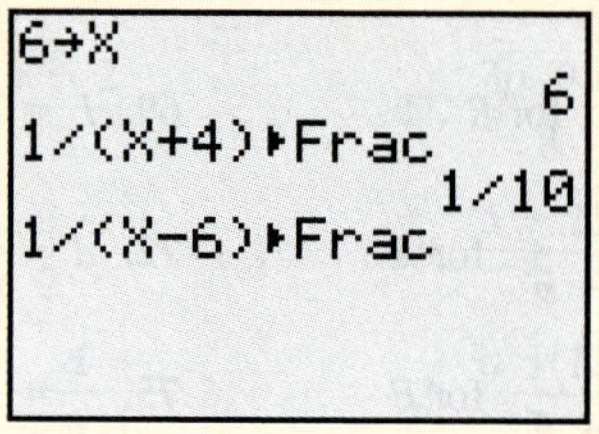

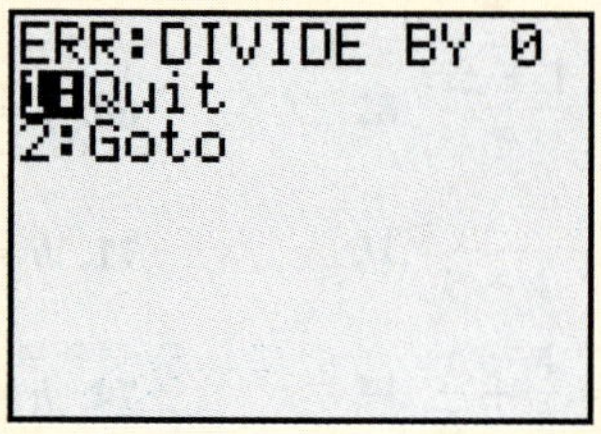

TECHNOLOGY INSIGHTS (EXERCISES 85–90) (CONTINUED)

For the following screens, either evaluate the expression or tell whether an error message would occur for the expression entered.

85.
```
5→X
                5
-3/(X+5)▸Frac
```

86.
```
8→X
                8
-4/(X+8)▸Frac
```

87.
```
12→X
               12
-3/(12-X)
```

88.
```
16→X
               16
-8/(16-X)
```

89.
```
3→X
                3
1/((X+4)*(X-3))▸
Frac
```

90.
```
5→X
                5
1/((X-5)*(X-3))▸
Frac
```

SUMMARY Exercises on Operations and Equations with Rational Expressions

We have performed the four operations of arithmetic with rational expressions and solved equations with rational expressions. The exercises in this summary include a mixed variety of problems of these types. To work them, recall the procedures explained in the earlier sections of this chapter. They are summarized here.

Multiplication of Rational Expressions	Multiply numerators and multiply denominators. Use the fundamental property to express in lowest terms.
Division of Rational Expressions	First, change the second fraction to its reciprocal; then multiply as described above.
Addition of Rational Expressions	Find the least common denominator (LCD) if necessary. Write all rational expressions with this LCD. Add numerators, and keep the same denominator. Express in lowest terms.

15. $\dfrac{1}{m^2 + 5m + 6} + \dfrac{2}{m^2 + 4m + 3}$

16. $\dfrac{2k^2 - 3k}{20k^2 - 5k} \div \dfrac{2k^2 - 5k + 3}{4k^2 + 11k - 3}$

17. $\dfrac{2}{x + 1} + \dfrac{5}{x - 1} = \dfrac{10}{x^2 - 1}$

18. $\dfrac{3}{x + 3} + \dfrac{4}{x + 6} = \dfrac{9}{x^2 + 9x + 18}$

19. $\dfrac{4t^2 - t}{6t^2 + 10t} \div \dfrac{8t^2 + 2t - 1}{3t^2 + 11t + 10}$

20. $\dfrac{x}{x - 2} + \dfrac{3}{x + 2} = \dfrac{8}{x^2 - 4}$

6.7 Applications of Rational Expressions

OBJECTIVES

1 Solve problems about numbers and data.

2 Solve problems about distance.

3 Solve problems about work.

4 Solve problems about variation.

FOR EXTRA HELP

SSG Sec. 6.7
SSM Sec. 6.7
Pass the Test Software
InterAct Math Tutorial Software
Video 11

Every time we learn to solve a new type of equation, we are able to apply our knowledge to solving new types of applications. In Section 6.6 we solved equations involving rational expressions; now we can solve applications that involve this type of equation. The six-step problem solving method of Chapter 2 still applies.

OBJECTIVE 1 Solve problems about numbers and data. We begin with an example about an unknown number.

EXAMPLE 1 Solving a Problem about an Unknown Number

If the same number is added to both the numerator and the denominator of the fraction $\frac{2}{5}$, the result is $\frac{2}{3}$. Find the number.

Step 1 Let $x =$ the number added to the numerator and the denominator.

Step 2 Then

$$\frac{2 + x}{5 + x}$$

represents the result of adding the same number to both the numerator and denominator.

Step 3 Since this result is $\frac{2}{3}$, the equation is

$$\frac{2 + x}{5 + x} = \frac{2}{3}.$$

Step 4 Solve this equation by multiplying both sides by the LCD, $3(5 + x)$.

$$3(5 + x)\frac{2 + x}{5 + x} = 3(5 + x)\frac{2}{3}$$

$$3(2 + x) = 2(5 + x)$$

$$6 + 3x = 10 + 2x \qquad \text{Distributive property}$$

$$x = 4 \qquad \text{Subtract } 2x\text{; subtract 6.}$$

Step 5 The number is 4.

Step 6 Check the solution in the words of the original problem. If 4 is added to both the numerator and denominator of $\frac{2}{5}$, the result is $\frac{6}{9} = \frac{2}{3}$, as required.

EXAMPLE 2 Examining the Leading Causes of Death in the United States

In 1996, approximately 734,000 Americans died of heart disease, the leading cause of death in the United States. Cancer was not far behind. However, there was a large drop between cancer and the third leading cause, stroke. Approximately 380,000 more deaths were caused by cancer than stroke, and the ratio of cancer deaths to stroke deaths was 3.4 to 1. How many deaths were caused by cancer, and how many were caused by stroke?

Let x represent the number of deaths caused by stroke. Then $x + 380{,}000$ deaths were caused by cancer. Since the ratio of cancer deaths to stroke deaths was 3.4 to 1 (or simply 3.4), we can write the following equation.

$$\frac{x + 380{,}000}{x} = 3.4$$

Now solve the equation.

$$x + 380{,}000 = 3.4x \qquad \text{Multiply by } x.$$

$$380{,}000 = 2.4x \qquad \text{Subtract } x.$$

$$x \approx 158{,}000 \qquad \text{Divide by 2.4.}$$

Since x represents the number of stroke deaths, there were about 158,000 deaths due to stroke, and about $x + 380{,}000 = 538{,}000$ deaths due to cancer. A check shows that these numbers satisfy the conditions of the problem. (*Note:* These are only approximations based on actual data from the Centers for Disease Control and Prevention and the National Center for Health Statistics.)

OBJECTIVE 2 **Solve problems about distance.** Recall from Chapter 2 the following formulas relating distance, rate, and time.

Distance, Rate, and Time Relationship

$$d = rt \qquad r = \frac{d}{t} \qquad t = \frac{d}{r}$$

You may wish to refer to Example 6 in Section 2.6, which illustrates the basic use of these formulas.

PROBLEM SOLVING

In Section 2.6 we solved applications involving distance, rate, and time. Recall the importance of setting up a chart to organize the information given in these problems.

EXAMPLE 3 Solving a Problem about Distance, Rate, and Time

The Big Muddy River has a current of 3 miles per hour. A motorboat takes the same amount of time to go 12 miles downstream as it takes to go 8 miles upstream. What is the speed of the boat in still water?

This problem requires the distance formula, $d = rt$. Let x = the speed of the boat in still water. Since the current pushes the boat when the boat is going downstream, the speed of the boat downstream will be the sum of the speed of the boat and the speed of the current, or $x + 3$ miles per hour. Also, the boat's speed going upstream is $x - 3$ miles per hour. See Figure 1.

OBJECTIVE 3 Solve problems about work. Suppose that you can mow your lawn in 4 hours. After 1 hour, you will have mowed $\frac{1}{4}$ of the lawn. After 2 hours, you will have mowed $\frac{2}{4}$ or $\frac{1}{2}$ of the lawn, and so on. This idea is generalized as follows.

Rate of Work

If a job can be completed in t units of time, then the rate of work is

$$\frac{1}{t} \text{ job per unit of time.}$$

PROBLEM SOLVING

The relationship between problems involving work and problems involving distance is a very close one. Recall that the formula $d = rt$ says that distance traveled is equal to rate of travel multiplied by time traveled. Similarly, the fractional part of a job accomplished is equal to the rate of the work multiplied by the time worked. In the lawn-mowing example, after 3 hours, the fractional part of the job done is

$$\underbrace{\frac{1}{4}}_{\text{Rate of work}} \cdot \underbrace{3}_{\text{Time worked}} = \underbrace{\frac{3}{4}}_{\text{Fractional part of job done}}.$$

After 4 hours, $(\frac{1}{4})(4) = 1$ whole job has been done.

These ideas are used in solving problems about the length of time needed to do a job.

EXAMPLE 4 Solving a Problem about Work

With a riding lawn mower, Mateo, the groundskeeper in a large park, can cut the lawn in 8 hours. With a small mower, his assistant Chet needs 14 hours to cut the same lawn. If both Mateo and Chet work on the lawn, how long will it take to cut it?

Let x = the number of hours it will take for Mateo and Chet to mow the lawn, working together.

Certainly, x will be less than 8, since Mateo alone can mow the lawn in 8 hours. Begin by making a chart. Based on the previous discussion, Mateo's rate alone is $\frac{1}{8}$ job per hour, and Chet's rate is $\frac{1}{14}$ job per hour.

	Rate	*Time Working Together*	*Fractional Part of the Job Done When Working Together*
Mateo	$\frac{1}{8}$	x	$\frac{1}{8}x$
Chet	$\frac{1}{14}$	x	$\frac{1}{14}x$

Sum is 1 whole job.

Since together Mateo and Chet complete 1 whole job, we must add their individual fractional parts and set the sum equal to 1.

$$\begin{array}{ccccc} \text{Fractional part done by Mateo} & + & \text{Fractional part done by Chet} & = & 1 \text{ whole job} \\ \frac{1}{8}x & + & \frac{1}{14}x & = & 1 \end{array}$$

$$\begin{aligned} 56\left(\frac{1}{8}x + \frac{1}{14}x\right) &= 56(1) && \text{Multiply by LCD, 56.} \\ 56\left(\frac{1}{8}x\right) + 56\left(\frac{1}{14}x\right) &= 56(1) && \text{Distributive property} \\ 7x + 4x &= 56 \\ 11x &= 56 && \text{Combine like terms.} \\ x &= \frac{56}{11} && \text{Divide by 11.} \end{aligned}$$

Working together, Mateo and Chet can mow the lawn in $\frac{56}{11}$ hours, or $5\frac{1}{11}$ hours. (Is this answer reasonable?)

NOTE

An alternative approach in work problems is to consider the part of the job that can be done in 1 hour. For instance, in Example 4 Mateo can do the entire job in 8 hours, and Chet can do it in 14. Thus, their work rates, as we saw in Example 4, are $\frac{1}{8}$ and $\frac{1}{14}$, respectively. Since it takes them x hours to complete the job when working together, in one hour they can mow $\frac{1}{x}$ of the lawn. The amount mowed by Mateo in one hour plus the amount mowed by Chet in one hour must equal the amount they can do together. This leads to the equation

$$\underset{\text{Amount by Mateo}}{\frac{1}{8}} + \underset{\text{Amount by Chet}}{\frac{1}{14}} = \underset{\text{Amount together}}{\frac{1}{x}}.$$

Compare this with the equation in Example 4. Multiplying both sides by $56x$ leads to

$$7x + 4x = 56,$$

the same equation found in the third step in the example. The same solution results.

OBJECTIVE 4 Solve problems about variation. In Section 2.5 we discussed direct variation. (See Example 6 in that section.) Recall that y *varies directly* as x if there exists a constant (a numerical value) k such that $y = kx$. This value is called the **constant of variation.** For example, the circumference C of a circle varies directly as its diameter d, because $C = \pi d$. Here, the constant of variation is π.

Some types of variation involve rational expressions. Suppose that a rectangle has area 48 square units and length 24 units. Then its width is 2 units. However, if the area stays the same (48 square units) and the length decreases to 12 units, the width increases to 4 units. See Figure 2.

As length decreases, width increases. This is an example of **inverse variation,** since $W = \dfrac{A}{L}$.

4. The numerator of a certain fraction is 4 times the denominator. If 6 is added to both the numerator and the denominator, the resulting fraction is equivalent to 2. What was the original fraction?

5. A quantity, its $\frac{2}{3}$, its $\frac{1}{2}$, and its $\frac{1}{7}$, added together, become 33. What is the quantity? (From the *Rhind Mathematical Papyrus.*)

6. A quantity, its $\frac{3}{4}$, its $\frac{1}{2}$, and its $\frac{1}{3}$, added together, become 93. What is the quantity?

Solve each problem using the six-step method. See Example 2.

7. (All enrollments are in thousands.) The Medicare enrollment in 1990 was 624 more than in 1989. The ratio of these enrollments was $\frac{877}{861}$. What was the enrollment in each of these years? (*Source:* U.S. Health Care Financing Administration.)

8. In 1985, during the early years of the appearance of AIDS, there were about 3760 more cases reported than in 1984. The ratio of these numbers of cases was $\frac{205}{111}$. How many cases were reported in each of these years? (*Source:* Centers for Disease Control.)

9. In 1995, the number of female physicians in the United States was approximately $\frac{1}{4}$ the number of male physicians. The total number of physicians was approximately 720,000. How many female and how many male physicians were there? (*Source:* American Medical Association.)

10. Two of the largest national managed care firms (as ranked by total HMO enrollments) in 1996 were United Health Care Corporation and Kaiser Foundation Health Plans, Inc. The number of plans provided by United was $\frac{10}{3}$ that of Kaiser. Together, the number of plans was 52. How many plans were provided by each company? (*Source:* InterStudy Publications.)

Solve each problem. See Example 6 in Section 2.6. (Source: Sports Illustrated 1998 Sports Almanac.)

11. In the 1994 Olympics, Vladimir Smimov of Kazakhstan won the 50-kilometer nordic skiing event in 2.07 hours. What was his average rate?

12. Bonnie Blair won the women's 500-meter speed skating event in the 1994 Olympics. Her average rate was 12.74 meters per second. What was her time?

13. The winner of the 1997 Indianapolis 500 (mile) race was Arie Luyendyk, with an average rate of 145.827 miles per hour. What was his time in hours?

14. In 1997, Jeff Gordon drove his Chevrolet to victory in the Daytona 500 (mile) stock car race. His average rate was 148.295 miles per hour. What was his time in hours?

15. The NCAA Division I women's champion for the 400-meter dash in 1997 was LaTarsha Stroman of Louisiana State University. Her winning time was 50.60 seconds. What was her rate in meters per second?

16. The NCAA Division I women's champion for the 200-yard freestyle event in 1996–1997 was Martina Moracova of Southern Methodist University. Her time was 48.18 seconds. What was her rate in yards per second?

Set up the equation you would use to solve each problem. Do not actually solve. See Example 3.

17. Julio flew his airplane 500 miles against the wind in the same time it took him to fly it 600 miles with the wind. If the speed of the wind was 10 miles per hour, what was the average speed of his plane? (Let $x =$ speed of the plane in still air.)

	d	r	t
Against the wind	500	$x - 10$	
With the wind	600	$x + 10$	

18. Luvenia can row 4 miles per hour in still water. It takes as long to row 8 miles upstream as 24 miles downstream. How fast is the current? (Let $x =$ speed of the current.)

	d	r	t
Upstream	8	$4 - x$	
Downstream	24	$4 + x$	

Solve each problem. See Example 3.

19. Suppose Stephanie walks D miles at R miles per hour in the same time that Wally walks d miles at r miles per hour. Give an equation relating D, R, d, and r.

20. If a migrating hawk travels m miles per hour in still air, what is its rate when it flies into a steady headwind of 5 miles per hour? What is its rate with a tailwind of 5 miles per hour?

21. A boat can go 20 miles against a current in the same time that it can go 60 miles with the current. The current is 4 miles per hour. Find the speed of the boat in still water.

22. A plane flies 350 miles with the wind in the same time that it can fly 310 miles against the wind. The plane has a still-air speed of 165 miles per hour. Find the speed of the wind.

23. The distance from Seattle, Washington, to Victoria, British Columbia, is about 148 miles by ferry. It takes about 4 hours less to travel by the same ferry from Victoria to Vancouver, British Columbia, a distance of about 74 miles. What is the average speed of the ferry?

24. Sandi Goldstein flew from Dallas to Indianapolis at 180 miles per hour and then flew back at 150 miles per hour. The trip at the slower speed took 1 hour longer than the trip at the higher speed. Find the distance between the two cities.

25. Cosmas N'Deti of Kenya won the 1995 Boston Marathon (a 26-mile race) in (about) .8 of an hour less time than the winner of the first Boston Marathon in 1897, John H. McDermott of New York. Find each runner's speed, if McDermott's speed was (about) .73 times N'Deti's speed. (*Hint:* Don't round until the end of the calculations.)

26. Women first ran in the Boston Marathon in 1972, when Nina Kuscsik of New York won the race. In 1995, the winner was Uta Pippig of Germany, whose time was .8 of an hour less than Kuscsik's in 1972. If Pippig ran $\frac{4}{3}$ as fast as Kuscsik, find each runner's speed. (See Exercise 25 for distance.)

27. If it takes Elayn 10 hours to do a job, what is her rate?

28. If it takes Clay 12 hours to do a job, how much of the job does he do in 8 hours?

In Exercises 29 and 30, set up the equation you would use to solve each problem. Do not actually solve. See Example 4.

29. Working alone, Jorge can paint a room in 8 hours. Caterina can paint the same room working alone in 6 hours. How long will it take them if they work together? (Let x represent the time working together.)

30. Edwin Bedford can tune up his Chevy in 2 hours working alone. Beau can do the job in 3 hours working alone. How long would it take them if they worked together? (Let t represent the time working together.)

Solve each problem. See Example 4.

31. Geraldo and Luisa Hernandez operate a small laundry. Luisa, working alone, can clean a day's laundry in 9 hours. Geraldo can clean a day's laundry in 8 hours. How long would it take them if they work together?

32. Lea can groom the horses in her boarding stable in 5 hours, while Tran needs 4 hours. How long will it take them to groom the horses if they work together?

33. A pump can pump the water out of a flooded basement in 10 hours. A smaller pump takes 12 hours. How long would it take to pump the water from the basement using both pumps?

34. Doug Todd's copier can do a printing job in 7 hours. Scott's copier can do the same job in 12 hours. How long would it take to do the job using both copiers?

35. An experienced employee can enter tax data into a computer twice as fast as a new employee. Working together, it takes the employees 2 hours. How long would it take the experienced employee working alone?

36. One roofer can put a new roof on a house three times faster than another. Working together they can roof a house in 4 days. How long would it take the faster roofer working alone?

37. One pipe can fill a swimming pool in 6 hours, and another pipe can do it in 9 hours. How long will it take the two pipes working together to fill the pool $\frac{3}{4}$ full?

38. An inlet pipe can fill a swimming pool in 9 hours, and an outlet pipe can empty the pool in 12 hours. Through an error, both pipes are left open. How long will it take to fill the pool?

39. A cold water faucet can fill a sink in 12 minutes, and a hot water faucet can fill it in 15. The drain can empty the sink in 25 minutes. If both faucets are on and the drain is open, how long will it take to fill the sink?

40. Refer to Exercise 38. Assume the error was discovered after both pipes had been running for 3 hours, and the outlet pipe was then closed. How much more time would then be required to fill the pool? (*Hint:* How much of the job had been done when the error was discovered?)

*Use personal experience or intuition to determine whether the variation between the indicated quantities is direct or inverse.**

41. The number of different lottery tickets you buy and your probability of winning that lottery

42. The rate and the distance traveled by a pickup truck in 3 hours

43. The amount of pressure put on the accelerator of a car and the speed of the car

44. The number of days from now until December 25 and the magnitude of the frenzy of Christmas shopping

45. The surface area of a balloon and its diameter

46. Your age and the probability that you believe in Santa Claus

47. The number of days until the end of the baseball season and the number of home runs that Sammy Sosa has

48. The amount of gasoline you pump and the amount you will pay

Solve each problem involving variation. See Example 6 in Section 2.6 and Examples 5 and 6 in this section.

49. If x varies directly as y, and $x = 27$ when $y = 6$, find x when $y = 2$.

50. If z varies directly as x, and $z = 30$ when $x = 8$, find z when $x = 4$.

51. If m varies inversely as p^2, and $m = 20$ when $p = 2$, find m when p is 5.

52. If a varies inversely as b^2, and $a = 48$ when $b = 4$, find a when $b = 7$.

53. If p varies inversely as q^2, and $p = 4$ when $q = \frac{1}{2}$, find p when $q = \frac{3}{2}$.

54. If z varies inversely as x^2, and $z = 9$ when $x = \frac{2}{3}$, find z when $x = \frac{5}{4}$.

55. If the constant of variation is positive and y varies directly as x, then as x increases, y ______________________.
(increases/decreases)

56. If the constant of variation is positive and y varies inversely as x, then as x increases, y ______________________.
(increases/decreases)

*The authors thank Linda Kodama of Kapiolani Community College for suggesting the inclusion of exercises of this type.

57. Over a specified distance, speed varies inversely with time. If a Dodge Viper on a test track goes a certain distance in one-half minute at 160 miles per hour, what speed is needed to go the same distance in three-fourths of a minute?

58. The pressure exerted by water at a given point varies directly with the depth of the point beneath the surface of the water. Water exerts 4.34 pounds per square inch for every 10 feet traveled below the water's surface. What is the pressure exerted on a scuba diver at 20 feet?

59. In the inversion of raw sugar, the rate of change of the amount of raw sugar varies directly as the amount of raw sugar remaining. The rate is 200 kilograms per hour when there are 800 kilograms left. What is the rate of change per hour when only 100 kilograms are left?

60. The current in a simple electrical circuit varies inversely as the resistance. If the current is 20 amps when the resistance is 5 ohms, find the current when the resistance is 8 ohms.

61. If the temperature is constant, the pressure of a gas in a container varies inversely as the volume of the container. If the pressure is 10 pounds per square foot in a container with 3 cubic feet, what is the pressure in a container with 1.5 cubic feet?

62. If the volume is constant, the pressure of gas in a container varies directly as the temperature. If the pressure is 5 pounds per square inch at a temperature of 200 Kelvin, what is the pressure at a temperature of 300 Kelvin?

63. For a constant area, the length of a rectangle varies inversely as the width. The length of a rectangle is 27 feet when the width is 10 feet. Find the width of a rectangle with the same area if the length is 18 feet.

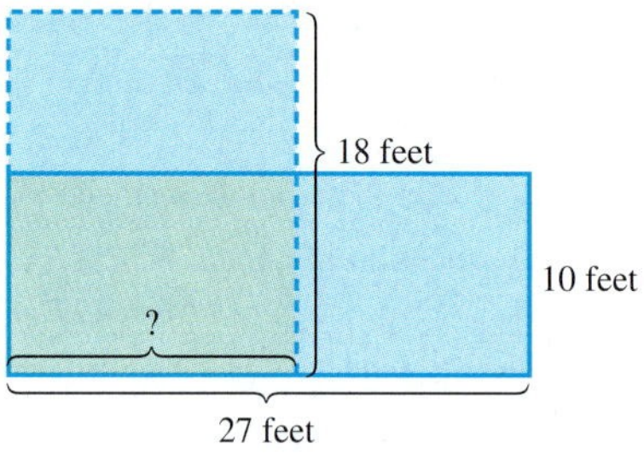

64. Hooke's law for an elastic spring states that the distance a spring stretches varies directly with the force applied. If a force of 75 pounds stretches a certain spring 16 inches, how much will a force of 200 pounds stretch the spring?

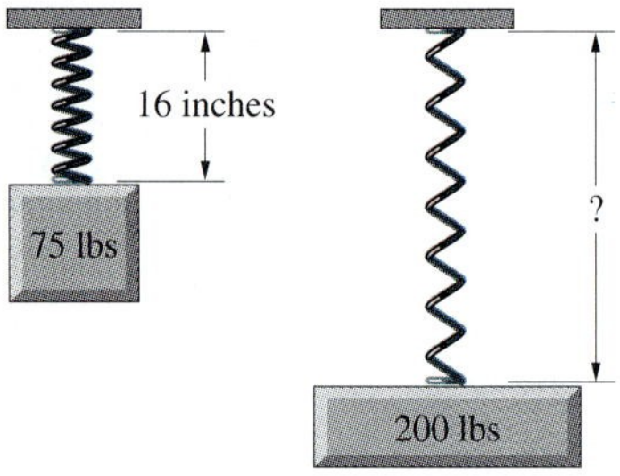

65. For a body falling freely from rest (disregarding air resistance), the distance the body falls varies directly as the square of the time. If an object is dropped from the top of a tower 400 feet high and hits the ground in 5 seconds, how far did it fall in the first 3 seconds?

66. The illumination produced by a light source varies inversely as the square of the distance from the source. If the illumination produced 4 feet from a light source is 75 foot-candles, find the illumination produced 9 feet from the same source.

TECHNOLOGY INSIGHTS (EXERCISES 67–70)

In each table, Y_1 either varies directly or varies inversely as x. Tell which type of variation is illustrated.

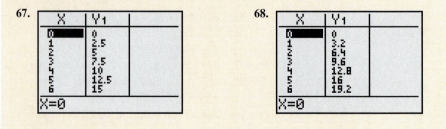

67.

X	Y1	
0	0	
1	2.5	
2	5	
3	7.5	
4	10	
5	12.5	
6	15	
X=0		

68.

X	Y1	
0	0	
1	3.2	
2	6.4	
3	9.6	
4	12.8	
5	16	
6	19.2	
X=0		

TECHNOLOGY INSIGHTS (EXERCISES 67–70) (CONTINUED)

69.

X	Y_1
1	24
2	12
3	8
4	6
5	4.8
6	4
7	3.4286

X=1

70.

X	Y_1
1	48
2	24
3	16
4	12
5	9.6
6	8
7	6.8571

X=1

*Refer to the discussion in the chapter opener pertaining to heart transplants. Use the strategy discussed there to solve each of these problems. (*Sources: *U.S. Department of Health and Human Services, Public Health Service, Division of Organ Transplantation, and United Network for Organ Sharing.)*

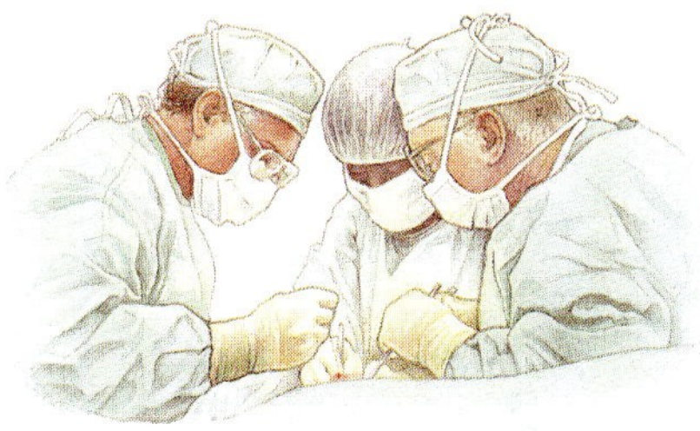

71. In 1995, the number of heart transplants continued to show the increase it had exhibited during the decade preceding it. In the years 1985 and 1995, the total number of heart transplants was 3080, and the ratio of the number in 1995 to the number in 1985 was approximately 33 to 10. How many heart transplants were performed in each of these two years?

72. In the years 1985 and 1995, the total number of liver transplants was 4526. The ratio of the number in 1995 to the number in 1985 was approximately 65 to 10. How many liver transplants were performed in each of these two years?

73. Students often wonder how teachers and textbook authors make up problems for them to solve. One way to do this is to start with the answer and work backwards. For example, suppose that we start with the fraction $\frac{7}{3}$ and add 3 to both the numerator and the denominator. We get $\frac{10}{6}$, which simplifies to $\frac{5}{3}$. Based on this observation, we can write the following problem.

> If a number is added to both the numerator and the denominator of $\frac{7}{3}$, the resulting fraction is equal to $\frac{5}{3}$. What is the number?

Because of how we constructed the problem, the answer must be 3.

Make up your own problem similar to this one, and then solve it using an equation involving rational expressions.

74. Refer to the chart in Exercise 18. Suppose that a student made the error of interchanging the positions of the expressions $4 - x$ and $4 + x$, but used the correct method of setting up the equation, applying the formula $t = \frac{d}{r}$. Solve the equation the student used, and explain how the student should immediately know that there is something wrong in the setup.

CHAPTER 6 GROUP ACTIVITY

How Much Fat Do You Need?

Objective: Use ratios and proportions to calculate total fat grams needed in an individual's daily diet.

Increased risks for some leading causes of death such as heart disease, diabetes, high blood pressure, and some cancers are associated with diets high in fat. Studies show that diets low in fat are linked to the lowest rates of heart disease and some forms of cancer. How do you decide how much fat is necessary for each individual? This activity will show how to calculate total fat grams necessary to maintain or reach ideal body weight.

To calculate daily caloric need, follow these steps.

Step 1 Determine activity level factor.

A sedentary person's activity level factor = 12.

An active person's activity level factor = 15.

A very active person's activity level factor = 18.

Step 2 Determine daily caloric need.

Multiply the activity level from Step 1 by your weight in pounds. (You can use your actual weight or your ideal weight.)

A. As a group, use the above information and the following ratio to write a proportion that can be used to calculate maximum total fat grams needed daily to maintain a person's weight or to reach his/her ideal weight.

$$\frac{400 \text{ calories}}{11 \text{ total fat grams}}$$

B. Use the previous information to complete the chart below. Round answers to the nearest tenth.

Total Fat Grams Needed Daily

Person	*Weight*	*Sedentary*	*Active*	*Very Active*
A	134			
B	200			
C	153			

C. Now complete the chart for the members in your group. Complete the three activity levels. (Actual weight or ideal weight can be used.)

Total Fat Grams Needed Daily

Person	*Weight*	*Sedentary*	*Active*	*Very Active*

CHAPTER 6 SUMMARY

KEY TERMS

6.1 rational expression lowest terms	**6.3** least common denominator (LCD)	**6.5** complex fraction **6.7** constant of variation	inverse variation

TEST YOUR WORD POWER

See how well you have learned the vocabulary in this chapter. Answers, with examples, are given at the bottom of the page.

1. A **rational expression** is

(a) an algebraic expression made up of a term or the sum of a finite number of terms with real coefficients and whole number exponents

(b) a polynomial equation of degree two

(c) an expression with one or more fractions in the numerator, denominator, or both

(d) the quotient of two polynomials with denominator not zero.

2. A **complex fraction** is

(a) an algebraic expression made up of a term or the sum of a finite number of terms with real coefficients and whole number exponents

(b) a polynomial equation of degree two

(c) a rational expression with one or more fractions in the numerator, denominator, or both

(d) the quotient of two polynomials with denominator not zero.

QUICK REVIEW

CONCEPTS	EXAMPLES
6.1 THE FUNDAMENTAL PROPERTY OF RATIONAL EXPRESSIONS	
To find the values for which a rational expression is not defined, set the denominator equal to zero and solve the equation.	Find the values for which the expression $\frac{x-4}{x^2-16}$ is not defined. $x^2 - 16 = 0$ $(x-4)(x+4) = 0$ $x - 4 = 0$ or $x + 4 = 0$ $x = 4$ or $x = -4$ The rational expression is not defined for 4 or -4.
To write a rational expression in lowest terms, (1) factor, and (2) use the fundamental property to remove common factors from the numerator and denominator.	Write $\frac{x^2-1}{(x-1)^2}$ in lowest terms. $\frac{x^2-1}{(x-1)^2} = \frac{(x-1)(x+1)}{(x-1)(x-1)}$ $= \frac{x+1}{x-1}$

(continued)

Answers to Test Your Word Power

1. (d) *Examples:* $-\frac{3}{4y}$, $\frac{5x^3}{x+2}$, $\frac{a+3}{a^2-4a-5}$ **2.** (c) *Examples:* $\frac{\frac{2}{3}}{\frac{4}{7}}$, $\frac{x-\frac{1}{y}}{x+\frac{1}{y}}$, $\frac{\frac{2}{a+1}}{a^2-1}$

CONCEPTS	EXAMPLES

There are often several different equivalent forms of a rational expression.

Give two equivalent forms of $-\dfrac{x-1}{x+2}$.

Distribute the $-$ sign in the numerator to get $\dfrac{-x+1}{x+2}$; do so in the denominator to get $\dfrac{x-1}{-x-2}$. (There are other forms as well.)

6.2 MULTIPLICATION AND DIVISION OF RATIONAL EXPRESSIONS

Multiplication

1. Factor.

2. Multiply numerators and multiply denominators.

3. Write in lowest terms.

Multiply. $\dfrac{3x+9}{x-5} \cdot \dfrac{x^2-3x-10}{x^2-9}$

$$= \dfrac{3(x+3)}{x-5} \cdot \dfrac{(x-5)(x+2)}{(x+3)(x-3)}$$

$$= \dfrac{3(x+3)(x-5)(x+2)}{(x-5)(x+3)(x-3)}$$

$$= \dfrac{3(x+2)}{x-3}$$

Division

1. Factor.

2. Multiply the first rational expression by the reciprocal of the second.

3. Write in lowest terms.

Divide. $\dfrac{2x+1}{x+5} \div \dfrac{6x^2-x-2}{x^2-25}$

$$= \dfrac{2x+1}{x+5} \div \dfrac{(2x+1)(3x-2)}{(x+5)(x-5)}$$

$$= \dfrac{2x+1}{x+5} \cdot \dfrac{(x+5)(x-5)}{(2x+1)(3x-2)}$$

$$= \dfrac{x-5}{3x-2}$$

6.3 THE LEAST COMMON DENOMINATOR

Finding the LCD

1. Factor each denominator into prime factors.

2. List each different factor the greatest number of times it appears.

3. Multiply the factors from Step 2 to get the LCD.

Find the LCD for $\dfrac{3}{k^2-8k+16}$ and $\dfrac{1}{4k^2-16k}$.

$$k^2 - 8k + 16 = (k-4)^2$$
$$4k^2 - 16k = 4k(k-4)$$
$$\text{LCD} = (k-4)^2 \cdot 4 \cdot k$$
$$= 4k(k-4)^2$$

Writing a Rational Expression with the LCD as Denominator

1. Factor both denominators.

2. Decide what factors the denominator must be multiplied by to equal the LCD.

3. Multiply the rational expression by that factor divided by itself (multiply by 1).

Find the numerator: $\dfrac{5}{2z^2-6z} = \dfrac{}{4z^3-12z^2}$.

$$\dfrac{5}{2z(z-3)} = \dfrac{1}{4z^2(z-3)}$$

$2z(z-3)$ must be multiplied by $2z$.

$$\dfrac{5}{2z(z-3)} \cdot \dfrac{2z}{2z} = \dfrac{10z}{4z^2(z-3)} = \dfrac{10z}{4z^3-12z^2}$$

CONCEPTS	EXAMPLES
6.4 ADDITION AND SUBTRACTION OF RATIONAL EXPRESSIONS	
Adding Rational Expressions	Add. $\frac{2}{3m+6} + \frac{m}{m^2-4}$
1. Find the LCD.	$3m + 6 = 3(m+2)$ $m^2 - 4 = (m+2)(m-2)$ The LCD is $3(m+2)(m-2)$.
2. Rewrite each rational expression with the LCD as denominator.	$= \frac{2(m-2)}{3(m+2)(m-2)} + \frac{3m}{3(m+2)(m-2)}$
3. Add the numerators to get the numerator of the sum. The common denominator is the denominator of the sum.	$= \frac{2m - 4 + 3m}{3(m+2)(m-2)}$
4. Write in lowest terms.	$= \frac{5m-4}{3(m+2)(m-2)}$
Subtracting Rational Expressions Follow the same steps as for addition, but subtract in Step 3.	Subtract. $\frac{6}{k+4} - \frac{2}{k}$ The LCD is $k(k+4)$. $\frac{6k}{(k+4)k} - \frac{2(k+4)}{k(k+4)} = \frac{6k - 2(k+4)}{k(k+4)}$ $= \frac{6k - 2k - 8}{k(k+4)} = \frac{4k-8}{k(k+4)}$
6.5 COMPLEX FRACTIONS	
Simplifying Complex Fractions *Method 1* Simplify the numerator and denominator separately. Then divide the simplified numerator by the simplified denominator.	Simplify. **(1)** $\frac{\frac{1}{a} - a}{1-a} = \frac{\frac{1}{a} - \frac{a^2}{a}}{1-a} = \frac{\frac{1-a^2}{a}}{1-a}$ $= \frac{1-a^2}{a} \cdot \frac{1}{1-a}$ $= \frac{(1-a)(1+a)}{a(1-a)} = \frac{1+a}{a}$
Method 2 Multiply numerator and denominator of the complex fraction by the LCD of all the denominators in the complex fraction. Write in lowest terms.	**(2)** $\frac{\frac{1}{a} - a}{1-a} = \frac{\frac{1}{a} - a}{1-a} \cdot \frac{a}{a} = \frac{\frac{a}{a} - a^2}{(1-a)a}$ $= \frac{1-a^2}{(1-a)a} = \frac{(1+a)(1-a)}{(1-a)a}$ $= \frac{1+a}{a}$
6.6 SOLVING EQUATIONS INVOLVING RATIONAL EXPRESSIONS	
Solving Equations with Rational Expressions	Solve $\frac{2}{x-1} + \frac{3}{4} = \frac{5}{x-1}$.
1. Find the LCD of all denominators in the equation.	The LCD is $4(x-1)$. Note that 1 cannot be a solution.
2. Multiply each side of the equation by the LCD.	$4(x-1)\left(\frac{2}{x-1} + \frac{3}{4}\right) = 4(x-1)\left(\frac{5}{x-1}\right)$ $4(x-1)\left(\frac{2}{x-1}\right) + 4(x-1)\left(\frac{3}{4}\right) = 4(x-1)\left(\frac{5}{x-1}\right)$

(continued)

CONCEPTS	EXAMPLES
3. Solve the resulting equation, which should have no fractions.	$8 + 3(x - 1) = 20$ $8 + 3x - 3 = 20$ $3x = 15$ $x = 5$
4. Check each proposed solution.	The proposed solution, 5, checks. The solution set is $\{5\}$.

6.7 APPLICATIONS OF RATIONAL EXPRESSIONS

Solving Problems about Distance

Use the six-step method.

Step 1 State what the variable represents.

On a trip from Sacramento to Monterey, Marge traveled at an average speed of 60 miles per hour. The return trip, at an average speed of 64 miles per hour, took $\frac{1}{4}$ hour less. How far did she travel between the two cities?

Let $x =$ the unknown distance.

Step 2 Use a chart to identify distance, rate, and time. Solve $d = rt$ for the unknown quantity in the chart.

	d	r	$t = \frac{d}{r}$
Going	x	60	$\frac{x}{60}$
Returning	x	64	$\frac{x}{64}$

Step 3 From the wording in the problem, decide the relationship between the quantities. Use those expressions to write an equation.

Since the time for the return trip was $\frac{1}{4}$ hour less, the time going equals the time returning plus $\frac{1}{4}$.

$$\frac{x}{60} = \frac{x}{64} + \frac{1}{4}$$

Step 4 Solve the equation.

$$16x = 15x + 240 \quad \text{Multiply by 960.}$$

$$x = 240 \quad \text{Subtract } 15x.$$

Steps 5 and 6 Answer the question; check the solution.

She traveled 240 miles.

The trip there took $\frac{240}{60} = 4$ hours while the return trip took $\frac{240}{64} = 3.75$ hours, which is $\frac{1}{4}$ hour less time. The solution checks.

Solving Problems about Work

Use the six-step method.

Step 1 State what the variable represents.

It takes the regular mail carrier 6 hours to cover her route. A substitute takes 8 hours to cover the same route. How long would it take them to cover the route together?

Let $x =$ the number of hours it would take them working together.

Step 2 Put the information from the problem in a chart. If a job is done in t units of time, the rate is $\frac{1}{t}$.

The rate of the regular carrier is $\frac{1}{6}$ job per hour; the rate of the substitute is $\frac{1}{8}$ job per hour. Multiply rate times time to get the fractional part of the job done.

	Rate	Time	Part of the Job Done
Regular	$\frac{1}{6}$	x	$\frac{1}{6}x$
Substitute	$\frac{1}{8}$	x	$\frac{1}{8}x$

CONCEPTS	EXAMPLES
Step 3 Write the equation. The sum of the fractional parts should equal 1 (whole job).	The equation is $\frac{1}{6}x + \frac{1}{8}x = 1.$
Step 4 Solve the equation.	The solution of the equation is $\frac{24}{7}$.
Steps 5 and 6 Answer the question; check the solution.	It would take them $\frac{24}{7}$ or $3\frac{3}{7}$ hours to cover the route together.
Solving Inverse Variation Problems 1. Write the variation equation $y = \frac{k}{x}$.	If a varies inversely as b, and $a = 4$ when $b = 4$, find a when $b = 6$. The equation for inverse variation is $a = \frac{k}{b}$.
2. Find k by substituting the given values of x and y into the equation.	Substitute $a = 4$ and $b = 4$. $4 = \frac{k}{4}$ $k = 16$
3. Write the equation with the value of k from Step 2 and the given value of x or y. Solve for the remaining variable.	Let $k = 16$ and $b = 6$ in the variation equation. $a = \frac{16}{6} = \frac{8}{3}$

CHAPTER 6 REVIEW EXERCISES

[6.1] *Find any values of the variable for which the rational expression is undefined.*

1. $\frac{4}{x-3}$
2. $\frac{y+3}{2y}$
3. $\frac{2k+1}{3k^2+17k+10}$
4. How would you determine the values of the variable for which a rational expression is undefined?

Find the numerical value of each rational expression when **(a)** $x = -2$ *and* **(b)** $x = 4$.

5. $\frac{4x-3}{5x+2}$
6. $\frac{3x}{x^2-4}$

Write each rational expression in lowest terms.

7. $\frac{5a^3b^3}{15a^4b^2}$
8. $\frac{m-4}{4-m}$
9. $\frac{4x^2-9}{6-4x}$
10. $\frac{4p^2+8pq-5q^2}{10p^2-3pq-q^2}$

Write four equivalent expressions for each of the following.

11. $-\frac{4x-9}{2x+3}$
12. $\frac{8-3x}{3+6x}$

[6.2] *Find each product or quotient and write the answer in lowest terms.*

13. $\frac{18p^3}{6} \cdot \frac{24}{p^4}$
14. $\frac{8x^2}{12x^5} \cdot \frac{6x^4}{2x}$
15. $\frac{x-3}{4} \cdot \frac{5}{2x-6}$
16. $\frac{2r+3}{r-4} \cdot \frac{r^2-16}{6r+9}$
17. $\frac{6a^2+7a-3}{2a^2-a-6} \div \frac{a+5}{a-2}$
18. $\frac{y^2-6y+8}{y^2+3y-18} \div \frac{y-4}{y+6}$

19. $\dfrac{2p^2 + 13p + 20}{p^2 + p - 12} \cdot \dfrac{p^2 + 2p - 15}{2p^2 + 7p + 5}$

20. $\dfrac{3z^2 + 5z - 2}{9z^2 - 1} \cdot \dfrac{9z^2 + 6z + 1}{z^2 + 5z + 6}$

[6.3] *Find the least common denominator for each list of fractions.*

21. $\dfrac{4}{9y}, \dfrac{7}{12y^2}, \dfrac{5}{27y^4}$

22. $\dfrac{3}{x^2 + 4x + 3}, \dfrac{5}{x^2 + 5x + 4}$

Rewrite each rational expression with the given denominator.

23. $\dfrac{3}{2a^3} = \dfrac{}{10a^4}$

24. $\dfrac{9}{x - 3} = \dfrac{}{18 - 6x}$

25. $\dfrac{-3y}{2y - 10} = \dfrac{}{50 - 10y}$

26. $\dfrac{4b}{b^2 + 2b - 3} = \dfrac{}{(b + 3)(b - 1)(b + 2)}$

[6.4] *Add or subtract and write each answer in lowest terms.*

27. $\dfrac{10}{x} + \dfrac{5}{x}$

28. $\dfrac{6}{3p} - \dfrac{12}{3p}$

29. $\dfrac{9}{k} - \dfrac{5}{k - 5}$

30. $\dfrac{4}{y} + \dfrac{7}{7 + y}$

31. $\dfrac{m}{3} - \dfrac{2 + 5m}{6}$

32. $\dfrac{12}{x^2} - \dfrac{3}{4x}$

33. $\dfrac{5}{a - 2b} + \dfrac{2}{a + 2b}$

34. $\dfrac{4}{k^2 - 9} - \dfrac{k + 3}{3k - 9}$

35. $\dfrac{8}{z^2 + 6z} - \dfrac{3}{z^2 + 4z - 12}$

36. $\dfrac{11}{2p - p^2} - \dfrac{2}{p^2 - 5p + 6}$

[6.5]

37. Simplify the complex fraction $\dfrac{\frac{a^4}{b^2}}{\frac{a^3}{b}}$ by

(a) Method 1 as described in Section 6.5.
(b) Method 2 as described in Section 6.5.
(c) Explain which method you prefer, and why.

Simplify each complex fraction.

38. $\dfrac{\frac{2}{3} - \frac{1}{6}}{\frac{1}{4} + \frac{2}{5}}$

39. $\dfrac{\frac{y - 3}{y}}{\frac{y + 3}{4y}}$

40. $\dfrac{\frac{1}{p} - \frac{1}{q}}{\frac{1}{q - p}}$

41. $\dfrac{x + \frac{1}{w}}{x - \frac{1}{w}}$

42. $\dfrac{\frac{1}{r + t} - 1}{\frac{1}{r + t} + 1}$

[6.6]

43. Before even beginning the solution process, how do you know that 2 cannot be a solution to the equation found in Exercise 46 below?

Solve each equation and check your solutions.

44. $\dfrac{4 - z}{z} + \dfrac{3}{2} = \dfrac{-4}{z}$

45. $\dfrac{3y - 1}{y - 2} = \dfrac{5}{y - 2} + 1$

46. $\dfrac{3}{m - 2} + \dfrac{1}{m - 1} = \dfrac{7}{m^2 - 3m + 2}$

Solve each formula for the specified variable.

47. $m = \dfrac{Ry}{t}$ for t

48. $x = \dfrac{3y - 5}{4}$ for y

49. $p^2 = \dfrac{4}{3m - q}$ for m

[6.7] *Solve each problem. Use the six-step method.*

50. In a certain fraction, the denominator is 4 less than the numerator. If 3 is added to both the numerator and the denominator, the resulting fraction is equal to $\dfrac{3}{2}$. Find the original fraction.

51. The denominator of a certain fraction is 3 times the numerator. If 2 is added to the numerator and subtracted from the denominator, the resulting fraction is equal to 1. Find the original fraction.

52. On August 18, 1996, Scott Sharp won the True Value 200-mile Indy race driving a Ford with an average speed of 130.934 miles per hour. What was his time? (*Source: Sports Illustrated 1998 Sports Almanac.*)

53. A man can plant his garden in 5 hours, working alone. His daughter can do the same job in 8 hours. How long would it take them if they worked together?

54. The head gardener can mow the lawns in the city park twice as fast as his assistant. Working together, they can complete the job in $1\dfrac{1}{3}$ hours. How long would it take the head gardener working alone?

55. The longer the term of your subscription to *Monitoring Times,* the less you will have to pay per year. Is this an example of direct or inverse variation?

56. In 1994, 38,505 American deaths were due to firearms. Of this total, the approximate ratio of male deaths to female deaths was 63 to 10. How many males and how many females died as a result of using firearms? (*Source:* National Center for Health Statistics, Monthly Vital Statistics Reports.)

57. If a parallelogram has a fixed area, the height varies inversely as the base. A parallelogram has a height of 8 centimeters and a base of 12 centimeters. Find the height if the base is changed to 24 centimeters.

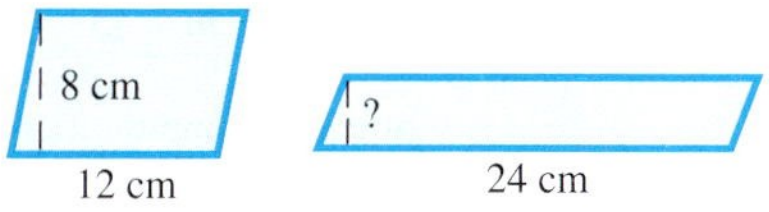

58. At a given hour, two steamboats leave a city in the same direction on a straight canal. One travels at 18 miles per hour, and the other travels at 25 miles per hour. In how many hours will the boats be 35 miles apart?

RELATING CONCEPTS (EXERCISES 59–68)

In these exercises, we summarize the various concepts involving rational expressions we have covered.

Work Exercises 59–68 in order.

Let P, Q, and R be rational expressions defined as follows.

$$P = \frac{6}{x + 3} \qquad Q = \frac{5}{x + 1} \qquad R = \frac{4x}{x^2 + 4x + 3}$$

59. Find the value or values for which the expression is undefined.

(a) P **(b)** Q **(c)** R

23. A man can paint a room in his house, working alone, in 5 hours. His wife can do the job in 4 hours. How long will it take them to paint the room if they work together?

24. Under certain conditions, the length of time that it takes for fruit to ripen during the growing season varies inversely as the average maximum temperature during the season. If it takes 25 days for fruit to ripen with an average maximum temperature of 80°, find the number of days it would take at 75°. Round your answer to the nearest whole number.

CUMULATIVE REVIEW EXERCISES CHAPTERS 1–6

1. Use the order of operations to evaluate $3 + 4\left(\frac{1}{2} - \frac{3}{4}\right)$.

Solve.

2. $3(2y - 5) = 2 + 5y$

3. $A = \frac{1}{2}bh$ for b

4. $\frac{2 + m}{2 - m} = \frac{3}{4}$

5. $5y \leq 6y + 8$

6. $5m - 9 > 2m + 3$

7. For the graph of $4x + 3y = -12$,
(a) what is the x-intercept and **(b)** what is the y-intercept?

Sketch each graph.

8. $y = -3x + 2$

9. $y = -x^2 + 1$

Simplify each expression. Write with only positive exponents.

10. $\frac{(2x^3)^{-1} \cdot x}{2^3x^5}$

11. $\frac{(m^{-2})^3m}{m^5m^{-4}}$

12. $\frac{2p^3q^4}{8p^5q^3}$

Perform the indicated operations.

13. $(2k^2 + 3k) - (k^2 + k - 1)$

14. $8x^2y^2(9x^4y^5)$

15. $(2a - b)^2$

16. $(y^2 + 3y + 5)(3y - 1)$

17. $\frac{12p^3 + 2p^2 - 12p + 4}{2p - 2}$

18. A computer can do one operation in 1.4×10^{-7} seconds. How long would it take for the computer to do one trillion (10^{12}) operations?

Factor completely.

19. $8t^2 + 10tv + 3v^2$

20. $8r^2 - 9rs + 12s^2$

21. $16x^4 - 1$

Solve each equation.

22. $r^2 = 2r + 15$

23. $(r - 5)(2r + 1)(3r - 2) = 0$

Solve each problem.

24. One number is 4 more than another. The product of the numbers is 2 less than the smaller number. Find the smaller number.

25. The length of a rectangle is 2 meters less than twice the width. The area is 60 square meters. Find the width of the rectangle.

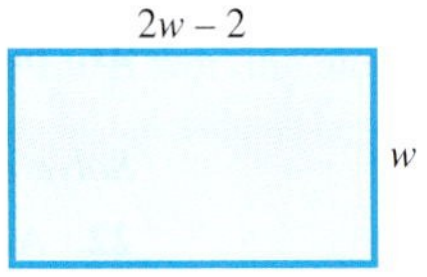

26. In 1996, the U.S. Department of Health and Human Services and the Department of Agriculture set guidelines for adult diets. A maximum percent of one's diet was suggested for calories from fat. It is a number such that when it is divided by 60, the quotient is the same as when it is divided into 15. Find this positive number to discover the maximum suggested percent from fat.

27. One of the following is equal to 1 for *all* real numbers. Which one is it?

(a) $\dfrac{k^2+2}{k^2+2}$ **(b)** $\dfrac{4-m}{4-m}$ **(c)** $\dfrac{2x+9}{2x+9}$ **(d)** $\dfrac{x^2-1}{x^2-1}$

28. Which one of the following rational expressions is *not* equivalent to $\dfrac{4-3x}{7}$?

(a) $-\dfrac{-4+3x}{7}$ **(b)** $-\dfrac{4-3x}{-7}$ **(c)** $\dfrac{-4+3x}{-7}$ **(d)** $\dfrac{-(3x+4)}{7}$

Perform each operation and write the answer in lowest terms.

29. $\dfrac{5}{q}-\dfrac{1}{q}$

30. $\dfrac{3}{7}+\dfrac{4}{r}$

31. $\dfrac{4}{5q-20}-\dfrac{1}{3q-12}$

32. $\dfrac{2}{k^2+k}-\dfrac{3}{k^2-k}$

33. $\dfrac{7z^2+49z+70}{16z^2+72z-40}\div\dfrac{3z+6}{4z^2-1}$

34. Simplify the complex fraction $\dfrac{\dfrac{4}{a}+\dfrac{5}{2a}}{\dfrac{7}{6a}-\dfrac{1}{5a}}$.

35. What values of x cannot possibly be solutions of the equation $\dfrac{1}{x-4}=\dfrac{3}{2x}$?

Solve each equation. Check your solutions.

36. $\dfrac{r+2}{5}=\dfrac{r-3}{3}$

37. $\dfrac{1}{x}=\dfrac{1}{x+1}+\dfrac{1}{2}$

Solve each problem.

38. On a business trip, Arlene traveled to her destination at an average speed of 60 miles per hour. Coming home, her average speed was 50 miles per hour, and the trip took $\dfrac{1}{2}$ hour longer. How far did she travel each way?

39. Juanita can weed the yard in 3 hours. Benito can weed the yard in 2 hours. How long would it take them if they worked together?

40. The force required to compress a spring varies directly as the change in the length of the spring. If a force of 12 pounds is required to compress a certain spring 3 inches, how much force is required to compress the spring 5 inches?

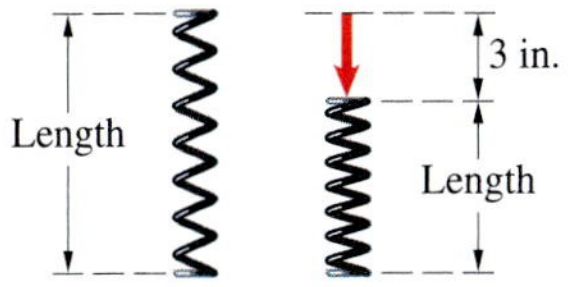

7 Equations of Lines, Inequalities, and Functions

Government

We began the study of linear equations and functions in Chapter 3. Now we continue these topics in this chapter. The Chapter 3 theme was business. One big business is government, our theme for this chapter. The federal government alone employed 2.9 million civilians in 1996 in addition to the armed forces. State and local government employees numbered almost 4 million in 1995, according to the U.S. Justice Department. Government branches collect and disperse vast amounts of mathematical data, much of it in the form of graphs. One example is shown here. From the graph, does it appear that the number of prisoners (in thousands) in state and federal institutions increased linearly over the years from 1988 to 1996? We will return to this example later, in Section 7.3.

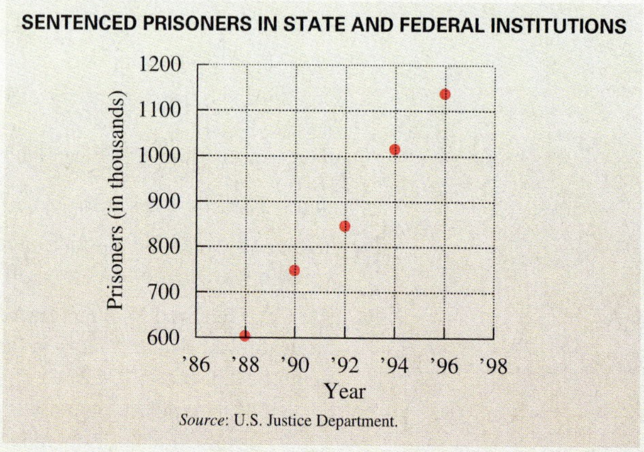

Throughout this chapter we provide many government-related applications of linear functions and their graphs. Other examples include changes in public school expenditures, storage of nuclear waste, and growth in the U.S. foreign-born population.

Visit our Web site at www.LialAlgebra.com

7.1 Equations of a Line

OBJECTIVES

1. Write an equation of a line given its slope and y-intercept.
2. Graph a line given its slope and a point on the line.
3. Write an equation of a line given its slope and any point on the line.
4. Write an equation of a line given two points on the line.
5. Find the equation of a line that fits a data set.

FOR EXTRA HELP

SSG Sec. 7.1
SSM Sec. 7.1
Pass the Test Software
InterAct Math Tutorial Software
Video 12

Section 3.3 showed how to find the slope (steepness) of a line from the equation of the line by solving the equation for y. In that form, the slope is the coefficient of x. For example, the slope of the line with equation $y = 2x + 3$ is 2, the coefficient of x. What does the number 3 represent? If $x = 0$, the equation becomes

$$y = 2(0) + 3 = 0 + 3 = 3.$$

Since $y = 3$ corresponds to $x = 0$, $(0, 3)$ is the y-intercept of the graph of $y = 2x + 3$. An equation like $y = 2x + 3$ that is solved for y is said to be in **slope-intercept form** because both the slope and the y-intercept of the line can be read directly from the equation.

Slope-Intercept Form

The slope-intercept form of the equation of a line with slope m and y-intercept $(0, b)$ is

$$y = mx + b.$$

The slope-intercept form is the most useful form for a linear equation, because it is the one that describes a linear function.

OBJECTIVE 1 Write an equation of a line given its slope and y-intercept. Given the slope and y-intercept of a line, we can use the slope-intercept form to find an equation of the line.

EXAMPLE 1 Finding an Equation of a Line

Find an equation of the line with slope $\frac{2}{3}$ and y-intercept $(0, -1)$.

Here $m = \frac{2}{3}$ and $b = -1$, so the equation is

$$y = mx + b$$

$$y = \frac{2}{3}x - 1.$$

CONNECTIONS

Businesses must consider the amount of value lost, called **depreciation,** during each year of a machine's useful life. The simplest way to calculate depreciation is to assume that an item with a useful life of n years loses $\frac{1}{n}$ of its value each year. Historically, if the equipment had salvage value, the depreciation was calculated on the **net cost,** the difference between the purchase price and the salvage value. If P represents the purchase price and S the salvage value of an item with a useful life of n years, then the annual depreciation would be

$$D = \frac{1}{n}(P - S).$$

Because this is a linear equation, this is called **straight-line depreciation.**

However, from a practical viewpoint, it is often difficult to determine the salvage value when the equipment is new. Also, for some equipment, such as computers, there is no residual dollar value at the end of the useful life due to

CONNECTIONS (CONTINUED)

obsolescence. In actual practice now, it is customary to find straight-line depreciation by the simpler linear equation

$$D = \frac{1}{n}P,$$

which assumes no salvage value.

FOR DISCUSSION OR WRITING

1. Find the depreciation using both methods for a \$50,000 (new) asset with a useful life of 10 years and a salvage value of \$15,000. How much is "written off" in each case over the ten-year period?

2. The depreciation equation is given in slope-intercept form. What does the slope represent here? What does the y-value of the y-intercept represent?

Source: Joel E. Halle, CPA.

OBJECTIVE 2 Graph a line given its slope and a point on the line. We can use the slope and y-intercept to graph a line. For example, to graph $y = \frac{2}{3}x - 1$, first locate the y-intercept, $(0, -1)$, on the y-axis. From the definition of slope and the fact that the slope of this line is $\frac{2}{3}$,

$$m = \frac{\textbf{difference in } y\textbf{-values}}{\textbf{difference in } x\textbf{-values}} = \frac{2}{3}.$$

Another point P on the graph of the line can be found by counting from the y-intercept 2 units up and then counting 3 units to the right. We then draw the line through point P and the y-intercept, as shown in Figure 1. This method can be extended to graph a line given its slope and any point on the line, not just the y-intercept.

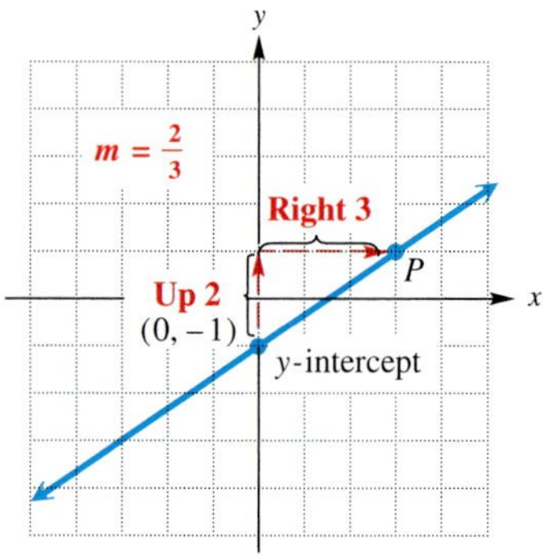

Figure 1

EXAMPLE 2 Graphing a Line Given a Point and the Slope

Graph the line through $(-2, 3)$ with slope -4.

First, locate the point $(-2, 3)$. Write the slope as

$$m = \frac{\text{difference in } y\text{-values}}{\text{difference in } x\text{-values}} = -4 = \frac{-4}{1}.$$

We locate another point on the line by counting 4 units down (because of the negative sign) and then 1 unit to the right. Finally, we draw the line through this new point P and the given point $(-2, 3)$. See Figure 2.

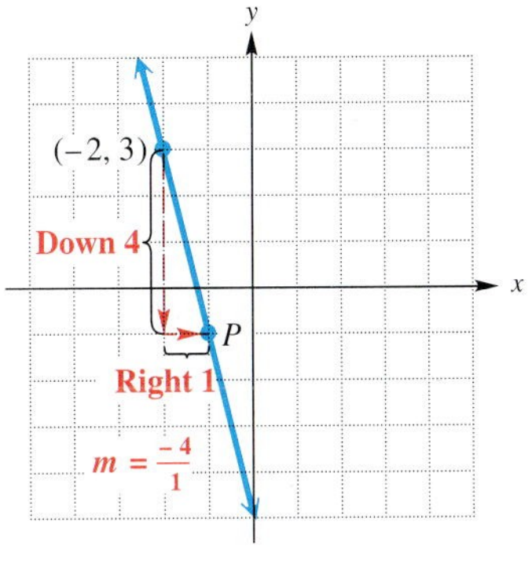

Figure 2

NOTE In Example 2, we could have written the slope as $\frac{4}{-1}$ instead. In this case, we would move 4 units up from $(-2, 3)$ and then 1 unit to the left (because of the negative sign). Verify that this produces the same line.

OBJECTIVE 3 **Write an equation of a line given its slope and any point on the line.** Let m represent the slope of the line and let (x_1, y_1) represent the given point on the line. Let (x, y) represent any other point on the line. Then by the definition of slope,

$$\frac{y - y_1}{x - x_1} = m$$

or

$$y - y_1 = m(x - x_1).$$

This result is the **point-slope form** of the equation of a line.

Point-Slope Form

The point-slope form of the equation of a line with slope m going through (x_1, y_1) is

$$y - y_1 = m(x - x_1)$$

This very important result should be memorized.

EXAMPLE 3 **Using the Point-Slope Form to Write an Equation**

Find an equation of each of the following lines. Write the equation in slope-intercept form.

(a) Through $(-2, 4)$, with slope -3

The given point is $(-2, 4)$ so $x_1 = -2$ and $y_1 = 4$. Also, $m = -3$. Substitute these values into the point-slope form.

$$y - y_1 = m(x - x_1)$$
$$y - 4 = -3[x - (-2)]$$
$$y - 4 = -3(x + 2)$$
$$y - 4 = -3x - 6 \quad \text{Distributive property}$$
$$y = -3x - 2 \quad \text{Add 4.}$$

The last equation is in slope-intercept form.

(b) Through $(4, 2)$, with slope $\frac{3}{5}$

Use $x_1 = 4$, $y_1 = 2$, and $m = \frac{3}{5}$ in the point-slope form.

$$y - y_1 = m(x - x_1)$$
$$y - 2 = \frac{3}{5}(x - 4)$$
$$y - 2 = \frac{3}{5}x - \frac{12}{5} \quad \text{Distributive property}$$
$$y = \frac{3}{5}x - \frac{12}{5} + \frac{10}{5} \quad \text{Add } 2 = \tfrac{10}{5} \text{ to both sides.}$$
$$y = \frac{3}{5}x - \frac{2}{5} \quad \text{Combine terms.}$$

NOTE: We did not clear fractions after the substitution step in Example 3(b) because we want the equation in slope-intercept form—that is, solved for y. Graphing calculators require equations to be in this form, so it has become the most useful.

OBJECTIVE 4 **Write an equation of a line given two points on the line.** We can also use the point-slope form to find an equation of a line when two points on the line are known.

EXAMPLE 4 Finding the Equation of a Line Given Two Points

Find an equation of the line through the points $(-2, 5)$ and $(3, 4)$. Write the equation in slope-intercept form.

First, find the slope of the line, using the definition of slope.

$$\text{slope} = \frac{y_2 - y_1}{x_2 - x_1} = \frac{5 - 4}{-2 - 3} = \frac{1}{-5} = -\frac{1}{5}$$

Now we use either $(-2, 5)$ or $(3, 4)$ and the point-slope form. We choose $(3, 4)$.

$$y - y_1 = m(x - x_1)$$
$$y - 4 = -\frac{1}{5}(x - 3) \quad \text{Let } y_1 = 4,\ m = -\tfrac{1}{5},\ x_1 = 3.$$
$$y - 4 = -\frac{1}{5}x + \frac{3}{5} \quad \text{Distributive property}$$

$$y = -\frac{1}{5}x + \frac{3}{5} + \frac{20}{5} \qquad \text{Add } 4 = \tfrac{20}{5} \text{ to both sides.}$$

$$y = -\frac{1}{5}x + \frac{23}{5} \qquad \text{Combine terms.}$$

The same result would be found by using $(-2, 5)$ for (x_1, y_1).

In Chapter 3, most linear equations were given in the form $Ax + By = C$. A linear equation can be written in this form in many different, equally correct, ways. To avoid confusion, we define the following *standard form of a linear equation.*

Standard Form

A linear equation is in standard form if it is written as

$$Ax + By = C,$$

where A, B, and C are integers and $A > 0$, $B \neq 0$.

CAUTION

The above definition of "standard form" is not the same in all texts. Also, for example, $3x + 4y = 12$, $6x + 8y = 24$, and $9x + 12y = 36$ all represent the same set of ordered pairs. Let us agree that $3x + 4y = 12$ is preferable to the other forms because the greatest common factor of 3, 4, and 12 is 1.

A summary of the types of linear equations is given here.

Linear Equations

$x = k$	**Vertical line** Slope is undefined. x-intercept is $(k, 0)$.
$y = k$	**Horizontal line** Slope is 0. y-intercept is $(0, k)$.
$y = mx + b$	**Slope-intercept form** Slope is m. y-intercept is $(0, b)$.
$y - y_1 = m(x - x_1)$	**Point-slope form** Slope is m. Line goes through (x_1, y_1).
$Ax + By = C$	**Standard form** Slope is $-\frac{A}{B}$. x-intercept is $\left(\frac{C}{A}, 0\right)$. y-intercept is $\left(0, \frac{C}{B}\right)$.

OBJECTIVE 5 **Find the equation of a line that fits a data set.** Given a set of data or a graph, we often want to find an equation that can be used to estimate or predict output values that are reasonable for additional input values. Example 5 discusses a procedure to accomplish this for data that fit a linear pattern—that is, whose graph consists of points lying close to a line.

EXAMPLE 5 Finding an Equation of a Line That Describes Data

The total expenditures (in billions of dollars) for public elementary and secondary education for each year in the first half of the nineties are given below. Plot the data and find an equation that approximates it.

School Year	x	*Expenditures*
1990	0	229
1991	1	241
1992	2	253
1993	3	265
1994	4	279
1995*	5	291

*Estimated
Source: U.S. Department of Education.

Letting y represent the total expenditures in year x, where $x = 0$ represents 1990, $x = 1$ represents 1991, and so on, we plot the data as shown in Figure 3.

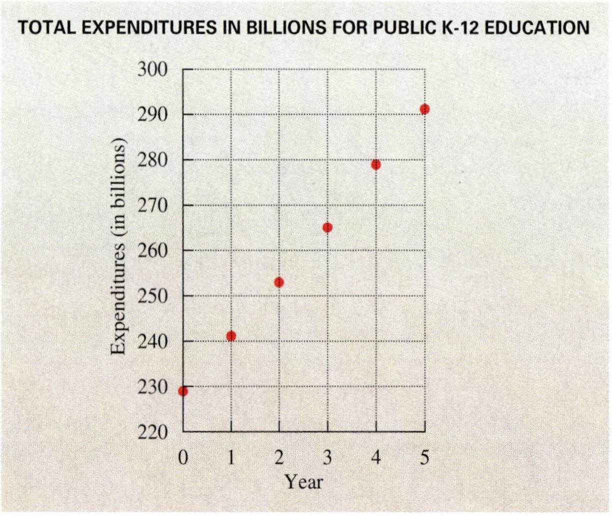

Figure 3

The points appear to lie approximately in a straight line. We can use two of the data pairs and the point-slope form of the equation of a line to get an equation that describes the relationship between the year and the expenditures. Suppose we choose from the table the ordered pairs (0, 229) and (4, 279). First, we need to find the slope of the line through these points.

$$m = \frac{y_2 - y_1}{x_2 - x_1} \qquad \text{Slope formula}$$

$$m = \frac{279 - 229}{4 - 0} \qquad \text{Let } (4, 279) = (x_2, y_2) \text{ and } (0, 229) = (x_1, y_1).$$

$$m = 12.5$$

Now, use the point-slope form to find the equation of the line.

$$y - y_1 = m(x - x_1) \quad \text{Point-slope form}$$
$$y - \mathbf{229} = \mathbf{12.5}(x - \mathbf{0}) \quad \text{Substitute for } y_1, m, \text{ and } x_1.$$
$$y - 229 = 12.5x \quad \text{Distributive property}$$
$$y = 12.5x + 229 \quad \text{Add 229 to both sides.}$$

To see how well the equation $y = 12.5x + 229$ approximates the ordered pairs (x, y) in the table of data, let $x = 2$ (for 1992) and find y.

$$y = 12.5x + 229 \quad \text{Equation of the line}$$
$$y = 12.5(\mathbf{2}) + 229 \quad \text{Substitute 2 for } x.$$
$$y = 254$$

The corresponding value in the table for 1992 is 253, so the equation appears to approximate the data quite well. Thus, the equation could be used with some caution to predict values for years that are not included in the table.

NOTE In Example 5, if we had chosen two different data points, we would get a slightly different equation.

Here is a summary of what is needed to find the equation of a line.

Finding the Equation of a Line

To find the equation of a line, you need

1. a point on the line, and
2. the slope of the line.

If two points are known, first find the slope, and then use the point-slope form.

7.1 EXERCISES

Match the correct equation in Column II with the description given in Column I.

I	II
1. slope $= -2$, through the point $(4, 1)$	**A.** $y = 4x$
2. slope $= -2$, y-intercept $(0, 1)$	**B.** $y = \frac{1}{4}x$
3. passing through the points $(0, 0)$ and $(4, 1)$	**C.** $y = -2x + 1$
4. passing through the points $(0, 0)$ and $(1, 4)$	**D.** $y - 1 = -2(x - 4)$

5. Explain why the equation of a vertical line cannot be written in the form $y = mx + b$.

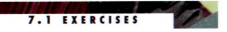

6. Match each equation with the graph that would most closely resemble its graph.

(a) $y = x + 3$
(b) $y = -x + 3$
(c) $y = x - 3$
(d) $y = -x - 3$

A.

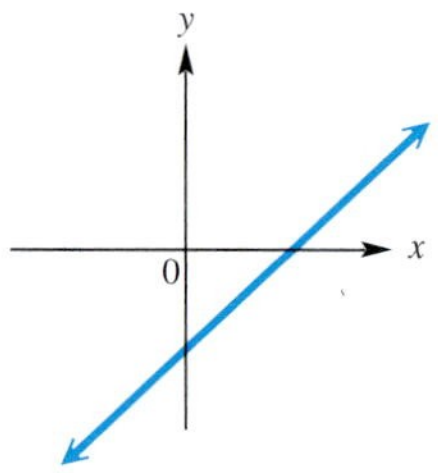

B.

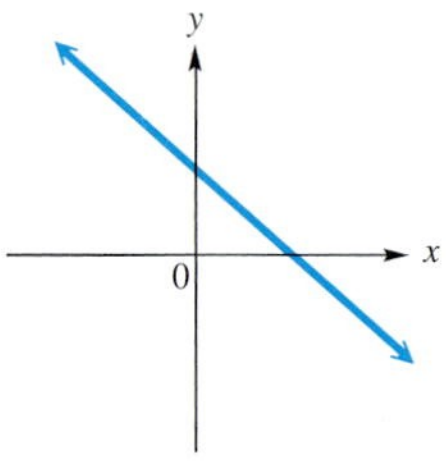

C.

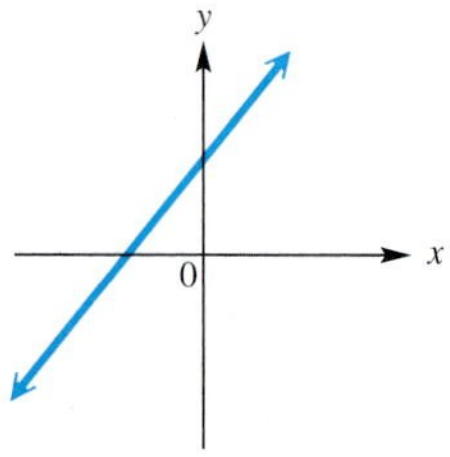

D.

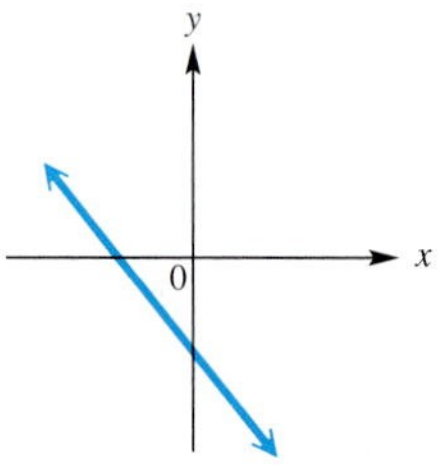

*Use the geometric interpretation of slope (*rise *divided by* run, *from Section 3.3) to find the slope of each line. Then, by identifying the y-intercept from the graph, write the slope-intercept form of the equation of the line.*

7.

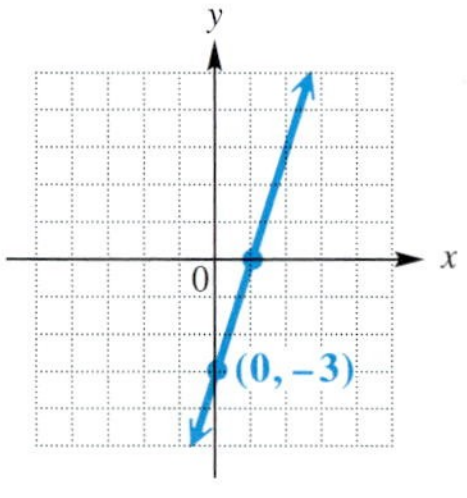

8.

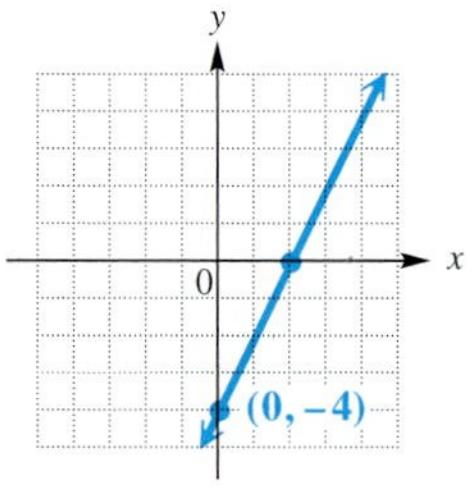

9.

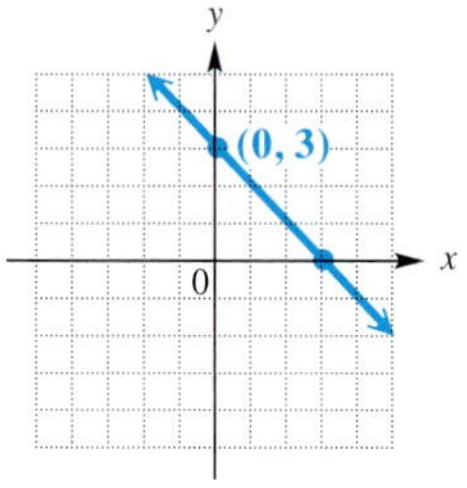

10.

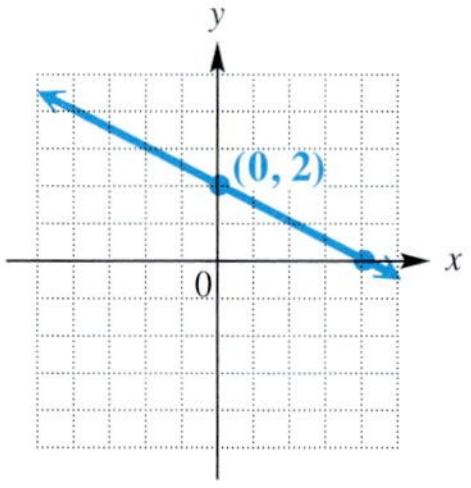

Write the equation of each line with the given slope and y-intercept. See Example 1.

11. $m = 4$, $(0, -3)$

12. $m = -5$, $(0, 6)$

13. $m = 0$, $(0, 3)$

14. $m = 3$, $(0, 0)$

Graph each line going through the given point and having the given slope. (In Exercises 19–22, recall the types of lines having slope 0 and undefined slope.) See Example 2.

15. $(0, 2)$, $m = 3$

16 $(0, -5)$, $m = -2$

17. $(1, -5)$, $m = -\frac{2}{5}$

18. $(2, -1)$, $m = -\frac{1}{3}$

19. $(3, 2)$, $m = 0$

20. $(-2, 3)$, $m = 0$

21. $(3, -2)$, undefined slope

22. $(2, 4)$, undefined slope

23. What is the common name given to a vertical line whose x-intercept is the origin?

24. What is the common name given to a line with slope 0 whose y-intercept is the origin?

Write an equation for each line passing through the given point and having the given slope. Write the equation in slope-intercept form. See Example 3.

25. $(4, 1)$, $m = 2$

26. $(2, 7)$, $m = 3$

27. $(-2, 5)$, $m = \frac{2}{3}$

28. $(-4, 1)$, $m = \frac{3}{4}$

29. $(6, -3)$, $m = -\frac{4}{5}$

30. $(7, -2)$, $m = -\frac{7}{2}$

RELATING CONCEPTS (EXERCISES 31–34)

In the examples of this section, we used the point-slope form $y - y_1 = m(x - x_1)$ to find the equation of a line given a point on the line and the slope of the line. Another method uses the slope-intercept (or function) form, $y = mx + b$. To develop this method, we will use the same problem as in Example 3(a): Find an equation of the line through $(-2, 4)$ with slope -3.

Work Exercises 31–34 in order.

31. Substitute the given information for m, x, and y into the equation $y = mx + b$.

32. Use your answer from Exercise 31 to find the value of b.

33. Using the given slope and the value of b from Exercise 32, what is the equation of the line in the form $y = mx + b$?

34. Compare your answer from Exercise 33 with the equation found in Example 3(a). What do you notice?

Did you make the connection between using the two different forms of linear equations to write the equation of a line?

35. If a line passes through the origin and a second point whose x- and y-coordinates are equal, what is an equation of the line?

36. Describe in your own words the slope-intercept and point-slope forms of the equation of a line. Tell what information must be given to use each form to write an equation. Include examples.

Write an equation for the line passing through the given pair of points. Write each equation in slope-intercept form. See Example 4.

37. $(8, 5)$ and $(9, 6)$

38. $(4, 10)$ and $(6, 12)$

39. $(-1, -7)$ and $(-8, -2)$

40. $(-2, -1)$ and $(3, -4)$

41. $(0, -2)$ and $(-3, 0)$

42. $(-4, 0)$ and $(0, 2)$

43. $\left(\frac{1}{2}, \frac{3}{2}\right)$ and $\left(-\frac{1}{4}, \frac{5}{4}\right)$

44. $\left(-\frac{2}{3}, \frac{8}{3}\right)$ and $\left(\frac{1}{3}, \frac{7}{3}\right)$

45. The table on the next page gives the heavy metal nuclear waste (in thousands of metric tons) from spent reactor fuel now stored temporarily at reactor sites, awaiting permanent storage. The figures for the year 2000 and beyond are estimates of the U.S. Department of Energy. (*Source:* "Burial of Radioactive Nuclear Waste under the Seabed," *Scientific American,* January 1998, p. 62.)

Heavy Metal Nuclear Waste

Year x	Waste y
1995	32
2000	42
2010	61
2020	76

Let $x = 0$ represent 1995, $x = 5$ represent 2000 (since $2000 - 1995 = 5$), and so on. Refer to Example 5 and do the following.

(a) For 1995, the ordered pair is (0, 32). Write ordered pairs for the data for the other years given in the table.

(b) Plot the ordered pairs (x, y). Do the points lie approximately in a line?

(c) Use the ordered pairs (0, 32) and (25, 76) to find the equation of a line that approximates the other ordered pairs. Use the form $y = mx + b$.

(d) Use the equation from part (c) to estimate the amount of nuclear waste in 2005. (*Hint:* What is the value of x for 2005?)

46. The Waste Management and Recycling Division of Sacramento County is responsible for managing the disposal of solid waste, including the operation of a landfill. The graph shows the remaining capacity (in millions of tons) at the county landfill at several times during the past few years. These points appear to lie close to a line.

(a) Write an equation in slope-intercept form of the line shown through the points (0, 5.6) and (2.5, 3.2).

(b) Use the equation from part (a) to estimate the remaining capacity of the landfill in 2001.

(c) During what year will the capacity of the landfill be used up? (*Hint:* Find the point where $y = 0$—that is, the x-intercept.) Estimate the month when it will occur.

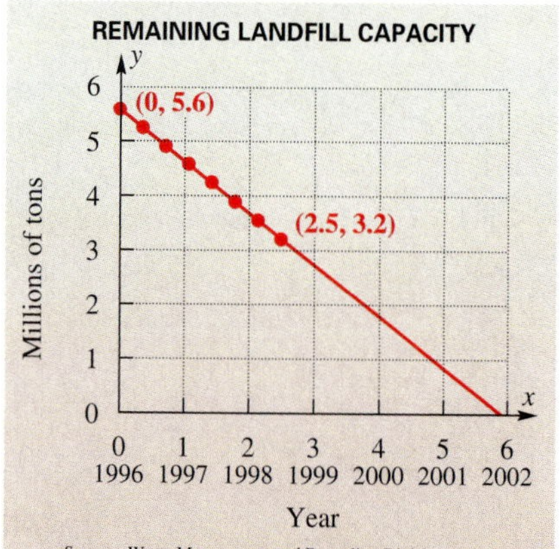

Source: Waste Management and Recycling Division, Sacramento County Public Works Agency, 1998 Report.

The cost to produce x items, in some cases, is expressed as $y = mx + b$. The number b gives the fixed cost *(that is, the cost that is the same no matter how many items are produced), and the number m is the* variable cost *(the cost to produce an additional item). Write the cost equation for each of the following, and answer the questions.*

47. It costs \$400 to start up a business selling ice cream cones. Each cone costs \$.25 to produce.

(a) What is the fixed cost?

(b) What is the variable cost?

(c) Write the cost equation.

(d) What will be the cost to produce 100 cones, based on the cost equation?

(e) How many cones will be produced if total cost is \$775?

48. It costs $2000 to purchase a copier and each copy costs $.02 to make.
- **(a)** What is the fixed cost?
- **(b)** What is the variable cost?
- **(c)** Write the cost equation.
- **(d)** What will be the cost to produce 10,000 copies, based on the cost equation?
- **(e)** How many copies will be produced if total cost is $2600?

The sales of a company for a given year can be written as an ordered pair in which the first number, x, gives the year and the second number, y, gives the sales for that year. If the sales increase at the same rate each year, a linear equation for sales (in millions of dollars) can be found. Merck & Co. sales for 1996 (year 1) and 1997 (year 2) are given for two different products.

Antibiotics

Year x	*Sales* y
1	822.3
2	774.9

Vaccines/Biologicals

Year x	*Sales* y
1	586.8
2	733.6

Source: Merck & Co., Inc. 1997 Annual Report.

49. **(a)** Write two ordered pairs in the form (year, sales) for antibiotics.
- **(b)** Write the sales equation in the form $y = mx + b$.
- **(c)** What does the slope, m, represent in this situation?

50. **(a)** Write two ordered pairs in the form (year, sales) for vaccines/biologicals.
- **(b)** Write the sales equation in the form $y = mx + b$.
- **(c)** What does the slope, m, represent in this situation?

TECHNOLOGY INSIGHTS (EXERCISES 51–54)

Two views of the same line are shown on a calculator screen.

Use the displays at the bottom of the screen to find an equation of the form $y = mx + b$ for each line.

51.

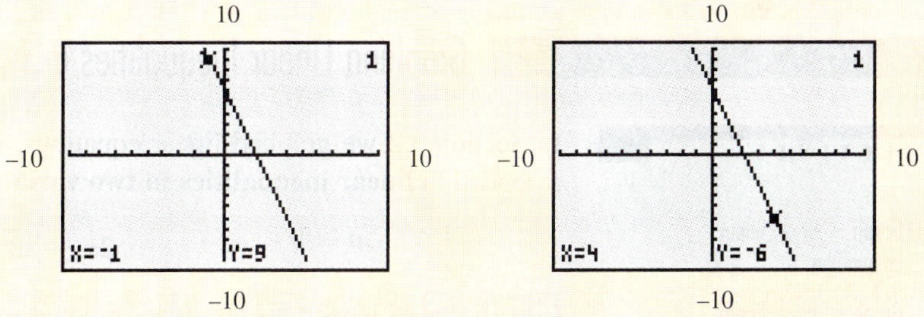

52.

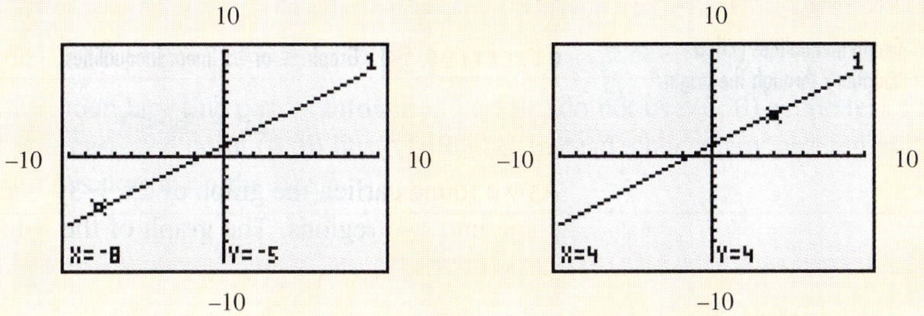

(continued)

EXAMPLE 1 Graphing a Linear Inequality

Graph the inequality $x - y > 5$.

This inequality does not include the equals sign. Therefore, the points on the line $x - y = 5$ do *not* belong to the graph. However, the line still serves as a boundary for two regions, one of which satisfies the inequality. To graph the inequality, first graph the equation $x - y = 5$. We use a dashed line to show that the points on the line are *not* solutions of the inequality $x - y > 5$. Now we choose a test point to see which side of the line satisfies the inequality. We choose $(1, -2)$ this time.

$$x - y > 5 \qquad \text{Original inequality}$$
$$1 - (-2) > 5 \quad ? \qquad \text{Let } x = 1 \text{ and } y = -2.$$
$$3 > 5 \qquad \text{False}$$

Since $3 > 5$ is false, the graph of the inequality is *not* the region that contains $(1, -2)$. We shade the other region, as shown in Figure 5. This shaded region is the required graph. To check that the correct region is shaded, select a test point in the shaded region and substitute for x and y in the inequality $x - y > 5$. For example, we use $(4, -3)$ from the shaded region as follows.

$$x - y > 5$$
$$4 - (-3) > 5 \quad ?$$
$$7 > 5 \qquad \text{True}$$

This verifies that the correct region was shaded in Figure 5.

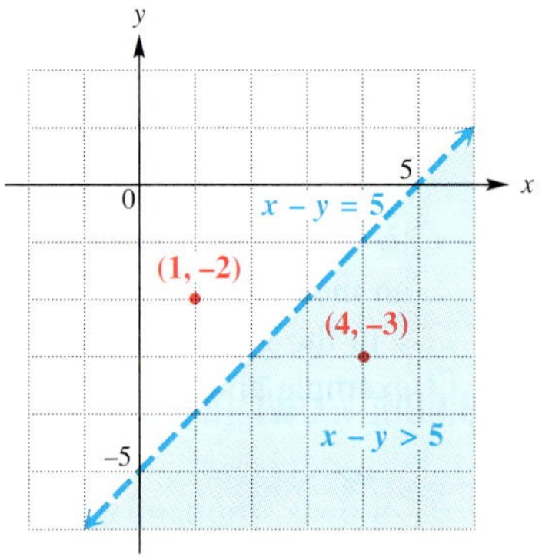

Figure 5

A summary of the steps used to graph a linear inequality in two variables follows.

Graphing a Linear Inequality

Step 1 **Graph the boundary.** Graph the line that is the boundary of the region. Use the methods of Section 3.2. Draw a solid line if the inequality has $\leq$ or $\geq$; draw a dashed line if the inequality has $<$ or $>$.

Step 2 **Shade the appropriate side.** Use any point not on the line as a test point. Substitute for x and y in the *inequality.* If a true statement results, shade the side containing the test point. If a false statement results, shade the other side.

EXAMPLE 2 Graphing a Linear Inequality with a Vertical Boundary Line

Graph the inequality $x \leq 3$.

First, we graph $x = 3$, a vertical line going through the point (3, 0). We use a solid line (why?) and choose (0, 0) as a test point.

$x \leq 3$		Original inequality
$0 \leq 3$	?	Let $x = 0$.
$0 \leq 3$		True

Since $0 \leq 3$ is true, we shade the region containing (0, 0), as in Figure 6.

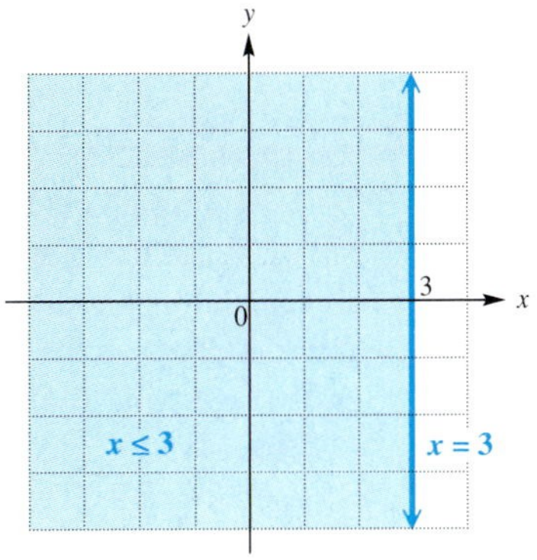

Figure 6

OBJECTIVE 3 **Graph inequalities with a boundary through the origin.** Remember that when we graph an inequality having a boundary line that goes through the origin, (0, 0) cannot be used as a test point.

EXAMPLE 3 Graphing a Linear Inequality with a Boundary Line through the Origin

Graph the inequality $x \leq 2y$.

Begin by graphing $x = 2y$. Some ordered pairs that can be used to graph this line are (0, 0), (6, 3), and (4, 2). Draw a solid line. Since (0, 0) is on the line $x = 2y$, it cannot be used as a test point. We must choose a test point off the line. For example, choose (1, 3), which is not on the line.

$x \leq 2y$		Original inequality
$1 \leq 2(3)$	?	Let $x = 1$ and $y = 3$.
$1 \leq 6$		True

Since $1 \leq 6$ is true, we shade the side of the graph containing the test point (1, 3). (See Figure 7 on the next page.)

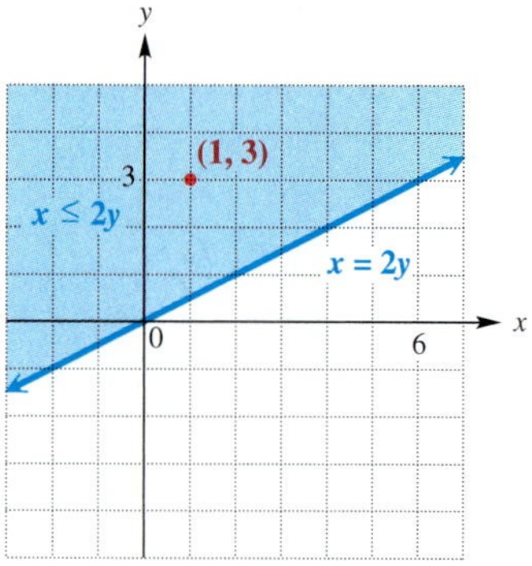

Figure 7

CONNECTIONS

Graphing calculators have a feature that allows us to shade regions in the plane, so they can be used to graph a linear inequality in two variables. Since all calculators are different, consult the manual for directions for your calculator. The calculator will not draw the graph as a dashed line, so it is still necessary to understand what is and is not included in the solution set.

To solve the inequalities in one variable, $-2x + 4 > 0$ and $-2x + 4 < 0$, we use the graph of $y = -2x + 4$, shown below. For $y = -2x + 4 > 0$, we want the values of x such that $y > 0$, so that the line is *above* the x-axis. From the graph, we see that this is the case for $x < 2$. Thus, the solution set is $(-\infty, 2)$. Similarly, the solution set of $y = -2x + 4 < 0$ is $(2, \infty)$, because the line is *below* the x-axis when $x > 2$.

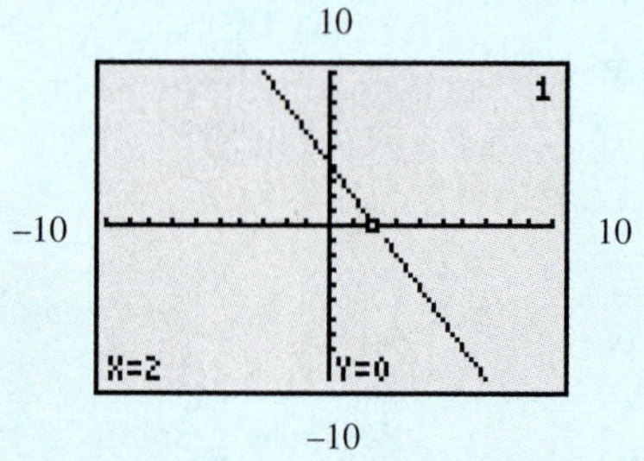

FOR DISCUSSION OR WRITING

1. Discuss the pros and cons of using a calculator to solve a linear inequality in one variable.
2. Use a graphing calculator to solve the following inequalities in one variable from Section 2.7.
   (a) $3x + 2 - 5 > -x + 7 + 2x$ (Example 4)
   (b) $3x + 2 - 5 < -x + 7 + 2x$ (Use the result from part (a).)
   (c) $4 \le 3x - 5 < 6$ (Example 7)

7.2 EXERCISES

The following statements were taken from recent articles in newspapers or magazines. Each includes a phrase that can be symbolized with one of the inequality symbols $<$, $\leq$, $>$, or $\geq$. In Exercises 1–6, complete each sentence with the appropriate inequality symbol.

1. The landslides of November 14 and January 19 caused more than \$1 million in damage to the condominium complex. (*Source:* Bunky Bakutis, "Owners Thank Makaha Flood Heroes," *The Honolulu Advertiser,* January 28, 1997, B3.)
Damages from landslides were _____ \$1 million.

2. In 1989, the U.S. Labor Department's Wage and Hour Division had 1000 compliance officers. Now it has fewer than 800. (*Source:* Michael Doyle, "3-Nation Meeting to Assess Battle against Child Labor," *The Sacramento Bee,* February 24, 1997.)
The department has _____ 800 officers.

3. More than \$3 billion in student loans were at least 60 days past due in 1994. (*Source:* Gilbert Chan, "Lessons on Loans," *The Sacramento Bee,* February 13, 1997.)
Student loans were _____ 60 days past due.

4. The committees supervise housing projects with up to 100 families. (*Source:* "Doug Is It," *Fortune,* May 25, 1998, p. 82.)
The housing projects have _____ 100 families.

5. Internal combustion engines can reach fuel efficiencies as high as 40 percent at a constant speed. (*Source:* "The Case for Electric Vehicles," *Scientific American,* November 1996.)
Fuel efficiencies are _____ 40 percent.

6. A car must travel through the intersection at no less than 15 miles an hour before the driver receives a ticket. (*Source:* "The Case for Electric Vehicles," *Scientific American,* November 1996.)
The speed must be _____ 15 miles per hour.

Answer true or false to each of the following.

7. The point (4, 0) lies on the graph of $3x - 4y < 12$.

8. The point (4, 0) lies on the graph of $3x - 4y \leq 12$.

9. Both points (4, 1) and (0, 0) lie on the graph of $3x - 2y \geq 0$.

10. The graph of $y > x$ does not contain points in quadrant IV.

In Exercises 11–16, the straight line boundary has been drawn. Complete the graph by shading the correct region. See Examples 1–3.

11. $x + 2y \geq 7$

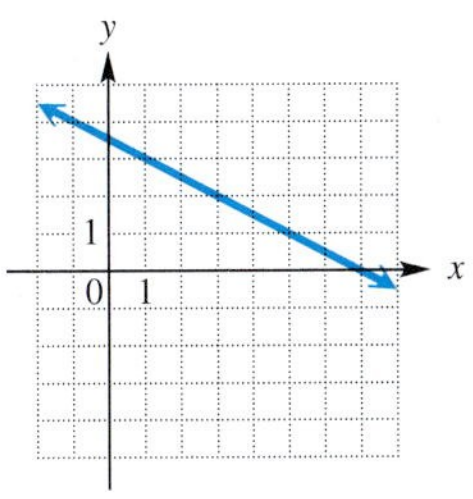

12. $2x + y \geq 5$

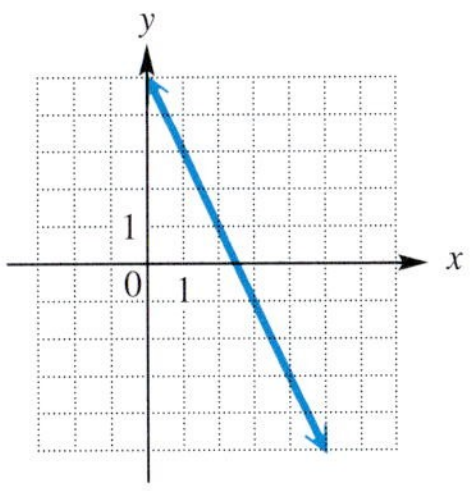

13. $-3x + 4y > 12$

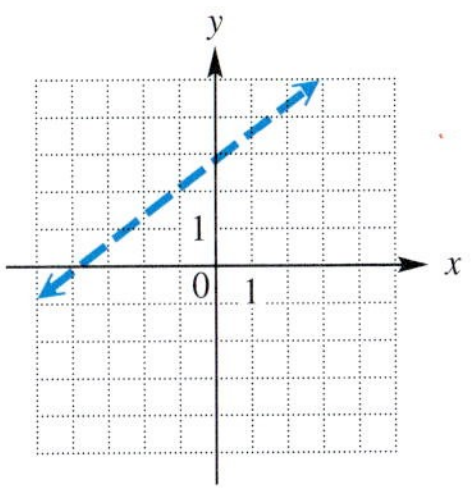

14. $x > 4$

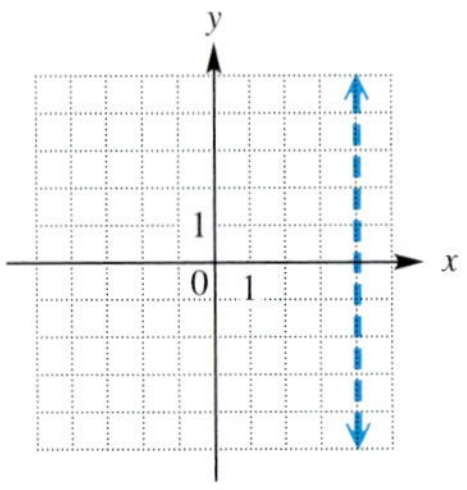

15. $y < -1$

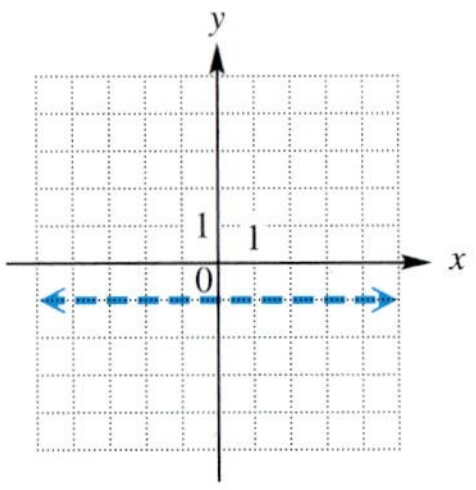

16. $x \leq 3y$

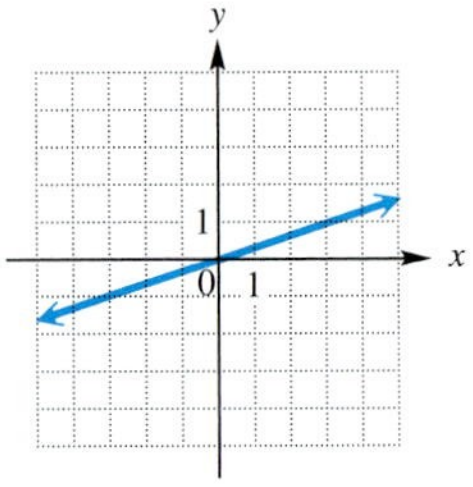

17. Explain in your own words how to graph a linear inequality in two variables.

18. Explain why the point (0, 0) is not an appropriate choice for a test point when graphing an inequality whose boundary goes through the origin.

Graph each linear inequality. See Examples 1–3.

19. $x + y \leq 5$ **20.** $x + y \geq 3$ **21.** $2x + 3y > -6$ **22.** $3x + 4y < 12$

23. $y \geq 2x + 1$ **24.** $y < -3x + 1$ **25.** $x \leq -2$ **26.** $x \geq 1$

27. $y < 5$ **28.** $y < -3$ **29.** $y \geq 4x$ **30.** $y \leq 2x$

31. Explain why the graph of $y > x$ cannot lie in quadrant IV.

32. Explain why the graph of $y < x$ cannot lie in quadrant II.

TECHNOLOGY INSIGHTS (EXERCISES 33-36)

A calculator was used to generate the shaded graphs in choices A–D.

A.

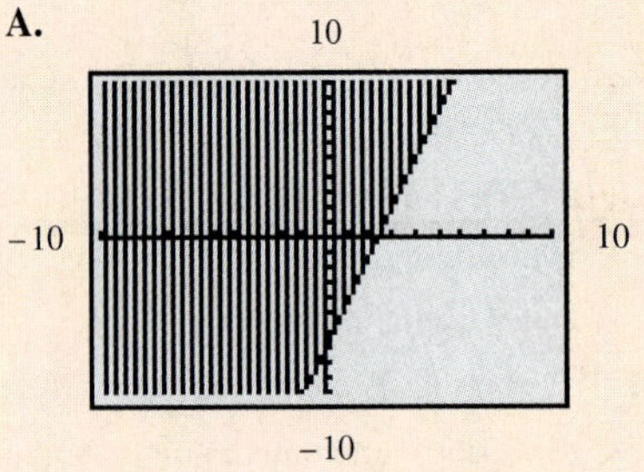

B.

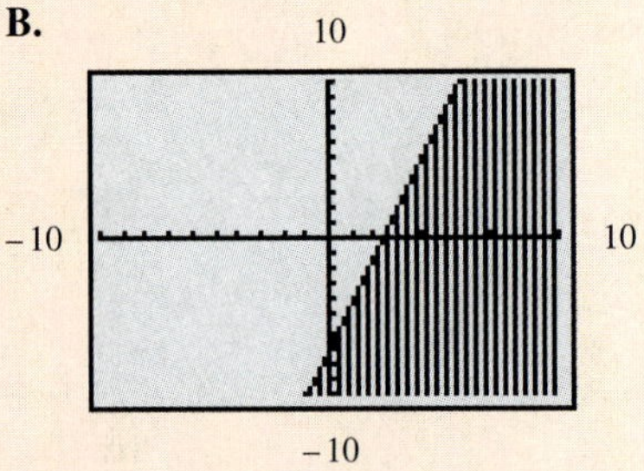

C.

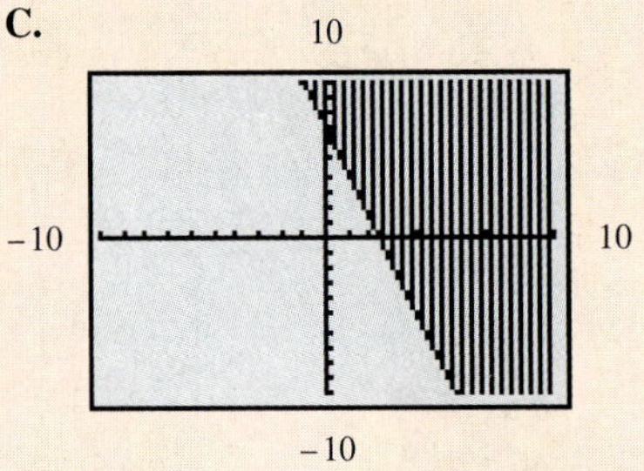

D.

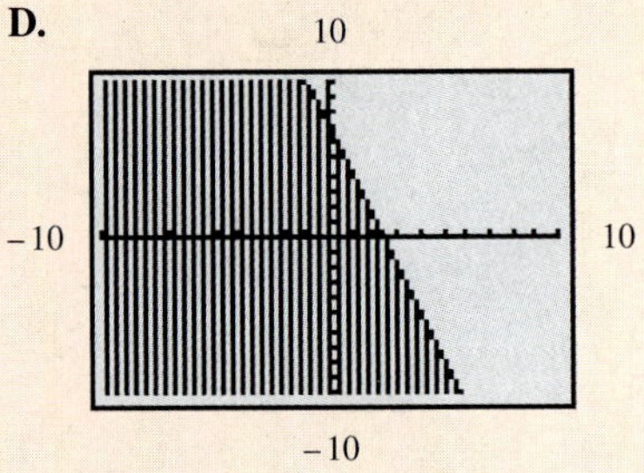

Match each inequality with the appropriate choice.

33. $y \geq 3x - 7$ **34.** $y \leq 3x - 7$ **35.** $y \geq -3x + 7$ **36.** $y \leq -3x + 7$

*For the given information **(a)** graph the inequality. Here $x \geq 0$ and $y \geq 0$, so graph only the part of the inequality in quadrant I. **(b)** Give some ordered pairs that satisfy the inequality.*

37. A company will ship x units of merchandise to outlet I and y units of merchandise to outlet II. The company must ship a total of at least 500 units to these two outlets. This can be expressed by writing

$$x + y \geq 500.$$

38. A toy manufacturer makes stuffed bears and geese. It takes 20 minutes to sew a bear and 30 minutes to sew a goose. There is a total of 480 minutes of sewing time available to make x bears and y geese. These restrictions lead to the inequality

$$20x + 30y \leq 480.$$

39. The number of civilian federal employees (in thousands) is approximated by $y = -51.1x + 3153$, where x represents the year. Here, $x = 1$ corresponds to 1991, $x = 2$ corresponds to 1992, and so on. Thus, the total number of federal employees (including noncivilians) in each year is described by the inequality

$$y \geq -51.1x + 3153.$$

(*Hint:* Use tick marks of 1 to 10 on the x-axis and 2500 to 3500, at intervals of 200, on the y-axis.) (*Source:* U.S. Bureau of the Census.)

40. The equation $y = 57.1x + 1798$ approximates the number of free lunches served in public schools each year in millions. The year 1992 is represented by $x = 2$, 1993 is represented by $x = 3$, and so on. Free school breakfasts are also served in some (but not all) of the same schools. The number of children receiving both meals is described by the inequality

$$y \leq 57.1x + 1798.$$

(*Hint:* Use tick marks of 1 to 10 on the x-axis and 1800, 1900, 2000, 2100, 2200, 2300, and 2400 on the y-axis.) (*Source:* U.S. Department of Agriculture, Food and Consumer Service.)

TECHNOLOGY INSIGHTS (EXERCISES 41–44)

The calculator graph at the top of the next page is that of $y = -2(x + 1) + 4(x - 1)$. If we solve the equation

$$-2(x + 1) + 4(x - 1) = 0,$$

we obtain the solution 3. Because the graph of the line lies *above* the x-axis for values of x greater than 3, the solution set of

"y is greater than 0" means *above* the x-axis
↓
$$-2(x + 1) + 4(x - 1) > 0$$

is the set of numbers greater than 3. On the other hand, because the graph of the line lies *below* the x-axis for values of x less than 3, the solution set of

"y is less than 0" means *below* the x-axis
↓
$$-2(x + 1) + 4(x - 1) < 0$$

(continued)

CHAPTER 7 GROUP ACTIVITY

Taxes as Functions

Objective: Use data to determine if a relation is a function, determine domain and range, and graph the function.

Different forms of taxation fund governments. There are various types of taxes, including income, property, estate, and sales taxes. Taxes are determined in different ways. This activity looks at some of these taxes and asks you to determine if they can be modeled by linear functions. Take turns in your group graphing and answering the questions below.

A. The following table is for sales tax in Sierra County, NM.

x = Cost of Item	*y = Sales Tax*
$ 4.00	$ .23
$ 8.00	$ .46
$24.00	$1.38

1. Plot the points on a graph. (Determine the scale for the domain and range of your graph.)

2. Using the criteria for functions, determine if this is a function.

3. Is this a linear function?

4. What is the domain? What is the range?

5. Write the equation of the line that fits the data. Use $f(x)$ notation if applicable. (*Hint:* Start by finding the slope of the line.)

B. The table below is for income tax for a family of four in the U.S. in 1994.

x = Income	*y = Tax*
$10,000	−$2527
$20,000	−$358
$30,000	$2078
$40,000	$3578
$50,000	$5078

Source: 1995 Information Please Almanac.

Complete parts (1)–(4) above for this tax.

(continued)

C. In the 1996 presidential election, one candidate proposed a flat income tax. For this activity, consider a flat tax rate of 10% of a given taxpayer's income.

1. Write the function that would represent this tax.
2. Using the incomes given in the table in B, determine the flat tax for each income. Make your own table.
3. Plot these points.
4. How does the flat tax compare to the 1994 taxes?
5. Who might oppose/support a flat tax?

CHAPTER 7 SUMMARY

KEY TERMS

7.2 **linear inequality in two variables**
boundary line

7.3 **components**
relation
domain
range
function

NEW SYMBOLS

$f(x)$ function f of x

TEST YOUR WORD POWER

See how well you have learned the vocabulary in this chapter. Answers, with examples, are given at the bottom of the page.

1. A **relation** is
(a) any set of ordered pairs
(b) a set of ordered pairs in which each first component corresponds to exactly one second component
(c) two sets of ordered pairs that are related
(d) a graph of ordered pairs.

2. The **domain** of a relation is
(a) the set of all x- and y-values in the ordered pairs of the relation
(b) the difference between the components in an ordered pair of the relation
(c) the set of all first components in the ordered pairs of the relation
(d) the set of all second components in the ordered pairs of the relation.

3. The **range** of a relation is
(a) the set of all x- and y-values in the ordered pairs of the relation
(b) the difference between the components in an ordered pair of the relation
(c) the set of all first components in the ordered pairs of the relation
(d) the set of all second components in the ordered pairs of the relation.

4. A **function** is
(a) any set of ordered pairs
(b) a set of ordered pairs in which each first component corresponds to exactly one second component
(c) two sets of ordered pairs that are related
(d) a graph of ordered pairs.

Answers to Test Your Word Power

1. (a) *Example:* $\{(0, 2), (2, 4), (3, 6), (-1, 3)\}$ **2.** (c) *Example:* The domain in the relation given above is the set of x-values, that is, $\{0, 2, 3, -1\}$. **3.** (d) *Example:* The range of the relation given above is the set of y-values, that is, $\{2, 4, 6, 3\}$. **4.** (b) *Example:* The relation given above is a function since each x-value corresponds to exactly one y-value.

To solve a system by substitution, follow these steps.

Solving Linear Systems by Substitution

Step 1 **Solve for a variable.** Solve one of the equations for either variable. (If one of the variables has coefficient 1 or -1, choose it, since the substitution method is usually easier this way.)

Step 2 **Substitute.** Substitute for that variable in the other equation. The result should be an equation with just one variable.

Step 3 **Solve.** Solve the equation from Step 2.

Step 4 **Substitute.** Substitute the result from Step 3 into the equation from Step 1 to find the value of the other variable.

Step 5 **Find the solution set.** Check the solution in both of the given equations. Then write the solution set.

EXAMPLE 2 Solving a System by Substitution

Use substitution to solve the system

$$2x = 8 - 3y$$
$$4x + 3y = 4.$$

Step 1 The substitution method requires that an equation be solved for one of the variables. Choose the first equation of the system and solve it for x.

$$2x = 8 - 3y$$
$$x = \frac{8 - 3y}{2} \qquad \text{Divide both sides by 2.}$$

Step 2 Now we substitute this value for x in the second equation of the system.

$$4x + 3y = 4$$
$$4\left(\frac{8 - 3y}{2}\right) + 3y = 4 \qquad \text{Let } x = \frac{8 - 3y}{2}.$$

Step 3 Solve this equation.

$$2(8 - 3y) + 3y = 4 \qquad \text{Divide 4 by 2.}$$
$$16 - 6y + 3y = 4 \qquad \text{Distributive property}$$
$$-3y = -12 \qquad \text{Combine terms; subtract 16.}$$
$$y = 4 \qquad \text{Divide by } -3.$$

Step 4 Find x by letting $y = 4$ in $x = \dfrac{8 - 3y}{2}$.

$$x = \frac{8 - 3 \cdot 4}{2} = \frac{8 - 12}{2} = \frac{-4}{2} = -2$$

Step 5 The solution set of the given system is $\{(-2, 4)\}$. Check the solution in both equations. ■

OBJECTIVE 2 Solve special systems. In the previous section we solved inconsistent systems with graphs that are parallel lines and systems of dependent equations with graphs that are the same line. We can also solve these special systems with the substitution method.

EXAMPLE 3 Solving an Inconsistent System by Substitution

Use substitution to solve the system

$$x = 5 - 2y$$
$$2x + 4y = 6.$$

Substitute $5 - 2y$ for x in the second equation.

$$2x + 4y = 6$$
$$2(5 - 2y) + 4y = 6 \quad \text{Let } x = 5 - 2y.$$
$$10 - 4y + 4y = 6 \quad \text{Distributive property}$$
$$10 = 6 \quad \text{False}$$

This false result means that the system is inconsistent and its solution set is $\emptyset$. The equations of the system have graphs that are parallel lines. See Figure 5.

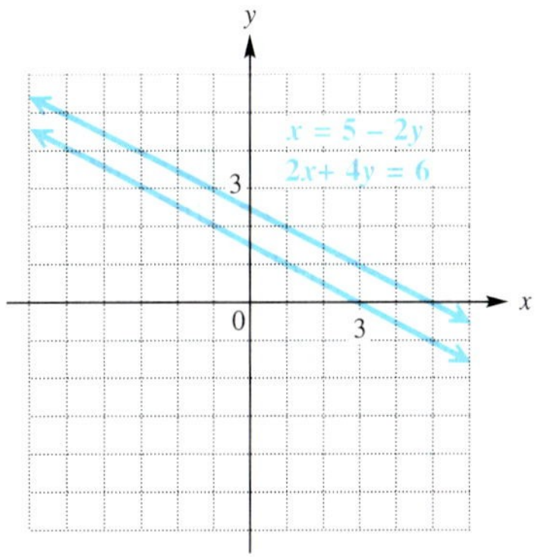

Figure 5

CAUTION It is a common error to give "false" as the answer to an inconsistent system. The correct response is $\emptyset$.

EXAMPLE 4 Solving a System with Dependent Equations by Substitution

Solve the following system by the substitution method.

$$3x - y = 4 \quad (1)$$
$$-9x + 3y = -12 \quad (2)$$

Begin by solving the first equation for y to get $y = 3x - 4$. Substitute $3x - 4$ for y in equation (2) and solve the resulting equation.

$$-9x + 3(3x - 4) = -12$$
$$-9x + 9x - 12 = -12 \quad \text{Distributive property}$$
$$0 = 0 \quad \text{Add 12; combine terms.}$$

TECHNOLOGY INSIGHTS (EXERCISES 29–32)

Match each system of inequalities with its calculator-generated solution set. See Example 4.

A.

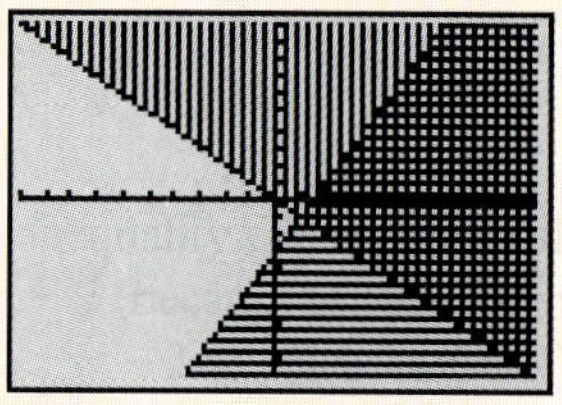

B.

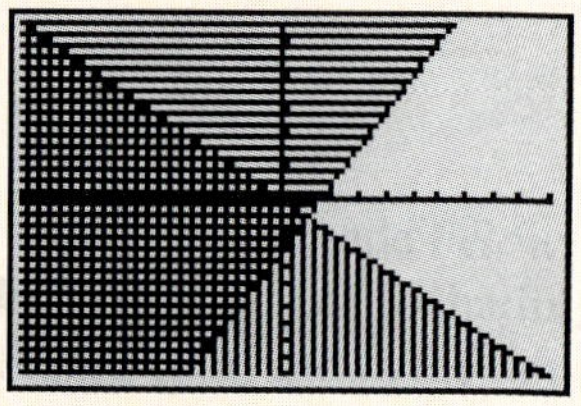

C.

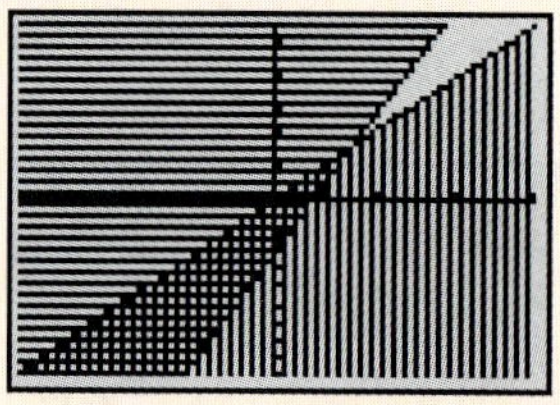

D.

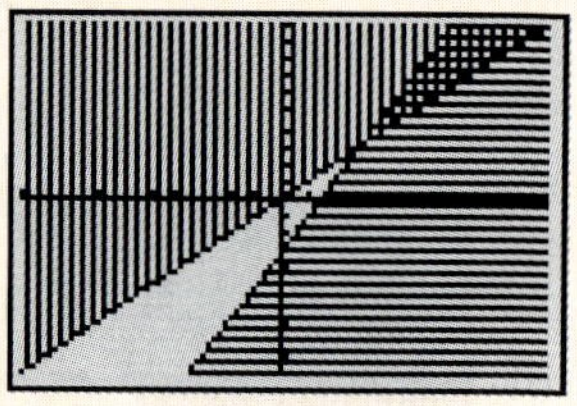

29. $y \geq x$
$y \leq 2x - 3$

30. $y \leq x$
$y \geq 2x - 3$

31. $y \geq -x$
$y \leq 2x - 3$

32. $y \leq -x$
$y \geq 2x - 3$

CHAPTER 8 GROUP ACTIVITY

Top Concert Tours

Objective: Use systems of equations to analyze data.

In addition to movies and television as forms of entertainment, Americans continue to attend concerts. This activity will use systems of equations to analyze the top grossing concert tours of 1994 and 1997.

- In 1994 the top grossing North American concert tours were The Rolling Stones and Pink Floyd. Together they grossed \$224.7 million. The Rolling Stones grossed \$17.7 million more than Pink Floyd. They performed a total of 119 shows, and Pink Floyd performed one less show than The Rolling Stones. (*Source:* Pollstar.)
- In 1997 the top grossing North American concert tours were The Rolling Stones and U2. Together they grossed \$169.2 million. The Rolling Stones grossed \$9.4 million more than U2. They performed a total of 79 shows, and U2 performed 13 more shows than The Rolling Stones. (*Source:* Pollstar.)

Using the information given above, complete the chart. One student should complete the information for 1994 and the other student for 1997.

A. Write a system of equations to find the total gross for each group.

B. Write a system of equations to find the number of shows per year for each group.

C. Use the information from Exercises A and B to find the gross per show.

Year	Group	Total Gross	Shows per Year	Gross per Show
1994	Rolling Stones			
1994	Pink Floyd			
1997	Rolling Stones			
1997	U2			

D. Once the chart is completed, compare the results. Answer these questions.

1. What differences do you notice between 1994 and 1997?
2. Which group had the highest total gross?
3. Which group had the highest gross per show?

CHAPTER 8 SUMMARY

KEY TERMS

8.1
- **system of linear equations**
- **solution of a system**
- **solution set of a system**
- **inconsistent system**
- **independent equations**
- **dependent equations**
- **consistent system**

8.5
- **system of linear inequalities**
- **solution set of a system of linear inequalities**

TEST YOUR WORD POWER

See how well you have learned the vocabulary in this chapter. Answers, with examples, are given at the bottom of the page.

1. A **system of linear equations** consists of
(a) at least two linear equations with different variables
(b) two or more linear equations that have an infinite number of solutions
(c) two or more linear equations with the same variables
(d) two or more linear inequalities.

2. A **solution of a system** of linear equations is
(a) an ordered pair that makes one equation of the system true
(b) an ordered pair that makes all the equations of the system true at the same time
(c) any ordered pair that makes one or the other or both equations of the system true
(d) the set of values that make all the equations of the system false.

3. A **consistent system** is a system of equations
(a) with one solution
(b) with no solutions
(c) with an infinite number of solutions
(d) that have the same graph.

4. An **inconsistent system** is a system of equations
(a) with one solution
(b) with no solutions
(c) with an infinite number of solutions
(d) that have the same graph.

5. **Dependent equations**
(a) have different graphs
(b) have no solutions
(c) have one solution
(d) are different forms of the same equation.

Answers to Test Your Word Power

1. (c) *Example:* $2x + y = 7,\quad 3x - y = 3$ **2.** (b) *Example:* The ordered pair (2, 3) satisfies both equations of the above system, so it is a solution of the system. **3.** (a) *Example:* The above system is consistent. The graphs of the equations intersect at exactly one point, in this case the solution (2, 3). **4.** (b) *Example:* The equations of two parallel lines make up an inconsistent system; their graphs never intersect, so there is no solution to the system. **5.** (d) *Example:* The equations $4x - y = 8$ and $8x - 2y = 16$ are dependent because their graphs are the same line.

QUICK REVIEW

CONCEPTS | **EXAMPLES**

8.1 SOLVING SYSTEMS OF LINEAR EQUATIONS BY GRAPHING

An ordered pair is a solution of a system if it makes all equations of the system true at the same time.

Is $(4, -1)$ a solution of the following system?

$$x + y = 3$$
$$2x - y = 9$$

Yes, because $4 + (-1) = 3$, and $2(4) - (-1) = 9$ are both true.

If the graphs of the equations of a system are both sketched on the same axes, the points of intersection, if any, form the solution set of the system.

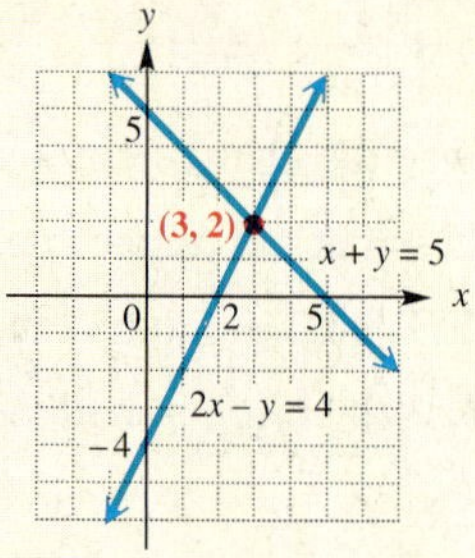

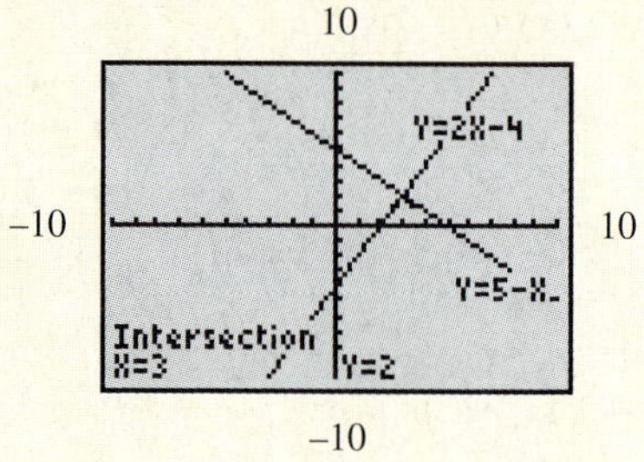

A graphing calculator can find the solution of a system by locating the point of intersection of the graphs.

$\{(3, 2)\}$ is the solution set of the system

$$x + y = 5$$
$$2x - y = 4.$$

8.2 SOLVING SYSTEMS OF LINEAR EQUATIONS BY SUBSTITUTION

Step 1 Solve one equation for one variable.

Solve by substitution.

$$x + 2y = -5 \quad (1)$$
$$y = -2x - 1 \quad (2)$$

Equation (2) is already solved for y.

Step 2 Substitute for that variable in the other equation to get an equation in one variable.

Substitute $-2x - 1$ for y in equation (1).

$$x + 2(-2x - 1) = -5$$

Step 3 Solve the equation from Step 2.

Solve to get $x = 1$.

Step 4 Substitute the result into the equation from Step 1 to get the value of the other variable.

To find y, let $x = 1$ in equation (2).

$$y = -2(1) - 1 = -3$$

Step 5 Check. Write the solution set.

The solution, $(1, -3)$, checks so $\{(1, -3)\}$ is the solution set.

8.3 SOLVING SYSTEMS OF LINEAR EQUATIONS BY ELIMINATION

Step 1 Write both equations in the form

$$Ax + By = C.$$

Solve by elimination.

$$x + 3y = 7$$
$$3x - y = 1$$

Step 2 Multiply one or both equations by appropriate numbers (if necessary) so that the coefficients of x (or y) are negatives of each other.

Multiply the first equation by -3 to eliminate the x-terms by addition.

CONCEPTS	EXAMPLES
Step 3 Add the equations to get an equation with only one variable.	$\begin{aligned} -3x - 9y &= -21 \\ 3x - y &= 1 \\ \hline -10y &= -20 \end{aligned}$ Add.
Step 4 Solve the equation from Step 3.	$y = 2$ Divide by -10.
Step 5 Substitute the solution from Step 4 into either of the original equations to find the value of the remaining variable.	Substitute to get the value of x. $x + 3y = 7$; $x + 3(2) = 7$ Let $y = 2$.; $x + 6 = 7$ Multiply.; $x = 1$ Subtract 6.
Step 6 Write the solution set containing an ordered pair, and check the answer.	The solution, (1, 2), checks. The solution set is $\{(1, 2)\}$.
If the result of the addition step is a false statement, such as $0 = 4$, the graphs are parallel lines and *the solution set is* $\emptyset$.	$\begin{aligned} x - 2y &= 6 \\ -x + 2y &= -2 \\ \hline 0 &= 4 \end{aligned}$ Solution set: $\emptyset$
If the result is a true statement, such as $0 = 0$, the graphs are the same line, and an *infinite number of ordered pairs are solutions.*	$\begin{aligned} x - 2y &= 6 \\ -x + 2y &= -6 \\ \hline 0 &= 0 \end{aligned}$ Infinite number of solutions
8.4 APPLICATIONS OF LINEAR SYSTEMS	
Use the six-step method.	The sum of two numbers is 30. Their difference is 6. Find the numbers.
Step 1 Choose a variable to represent each unknown value.	Let x represent one number. Let y represent the other number.
Step 2 Draw a figure or a diagram if it will help.	
Step 3 Translate the problem into a system of two equations using both variables.	$\begin{aligned} x + y &= 30 \\ x - y &= 6 \\ \hline 2x \quad &= 36 \end{aligned}$ Add.
Step 4 Solve the system.	$x = 18$ Divide by 2. Let $x = 18$ in the top equation: $18 + y = 30$. Solve to get $y = 12$.
Step 5 Answer the question or questions asked.	The two numbers are 18 and 12.
Step 6 Check the solution in the words of the problem.	$18 + 12 = 30$ and $18 - 12 = 6$, so the solution checks.
8.5 SOLVING SYSTEMS OF LINEAR INEQUALITIES	
To solve a system of linear inequalities, graph each inequality on the same axes. (This was explained in Section 7.2.) The solution set of the system is formed by the overlap of the regions of the two graphs.	The shaded region shows the solution set of the system $2x + 4y \geq 5$, $x \geq 1$. 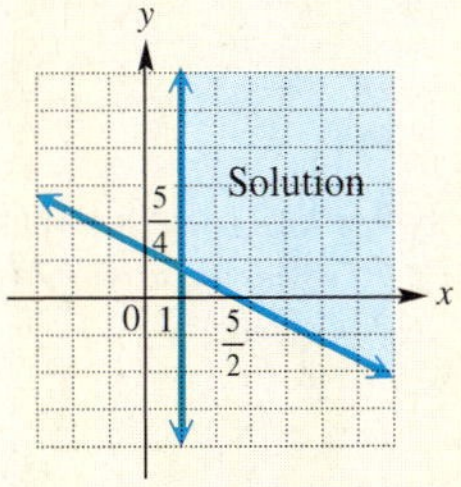

(continued)

CONCEPTS	EXAMPLES
Graphing calculators also can represent solutions of systems of inequalities.	The cross-hatched region shows the solution set of the system $y > 2x + 1$, $y < 3$. The boundary lines are not included.

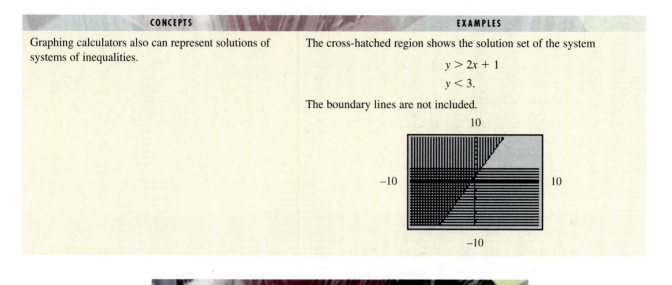

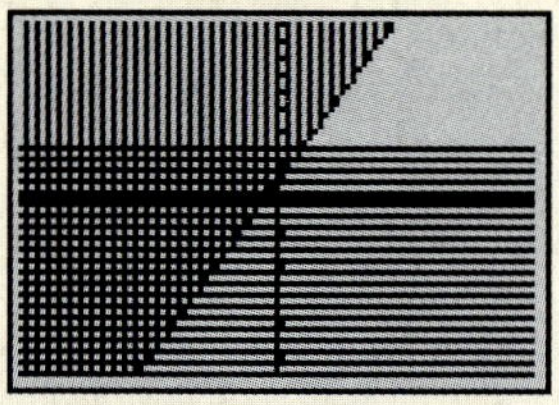

CHAPTER 8 REVIEW EXERCISES

[8.1] *Decide whether the given ordered pair is a solution of the given system.*

1. (3, 4)
$$\begin{aligned} 4x - 2y &= 4 \\ 5x + y &= 19 \end{aligned}$$

2. (−5, 2)
$$\begin{aligned} x - 4y &= -13 \\ 2x + 3y &= 4 \end{aligned}$$

Solve each system by graphing.

3. $$\begin{aligned} x + y &= 4 \\ 2x - y &= 5 \end{aligned}$$

4. $$\begin{aligned} 2x + 4 &= 2y \\ y - x &= -3 \end{aligned}$$

[8.2]

5. A student solves the system $\begin{aligned} 2x + y &= 6 \\ -2x - y &= 4 \end{aligned}$ and gets the equation $0 = 10$. The student gives the solution set as $\{(0, 10)\}$. Is this correct? Explain.

6. Can a system of two linear equations in two unknowns have exactly three solutions? Explain.

Solve each system by the substitution method.

7. $$\begin{aligned} 3x + y &= 7 \\ x &= 2y \end{aligned}$$

8. $$\begin{aligned} 2x - 5y &= -19 \\ y &= x + 2 \end{aligned}$$

9. $$\begin{aligned} 5x + 15y &= 30 \\ x + 3y &= 6 \end{aligned}$$

10. $$\begin{aligned} \frac{1}{3}x + \frac{1}{7}y &= \frac{52}{21} \\ \frac{1}{2}x - \frac{1}{3}y &= \frac{19}{6} \end{aligned}$$

11. After solving a system of linear equations by the substitution method, a student obtained the equation "$0 = 0$." He gave the solution set of the system as $\{(0, 0)\}$. Was his answer correct? Why or why not?

12. Suppose that you were asked to solve the system $\begin{aligned} 5x - 3y &= 7 \\ -x + 2y &= 4 \end{aligned}$ by substitution. Which variable in which equation would be easiest to solve for in your first step? Why?

[8.3]

13. Only one of the following systems does not require that we multiply one or both equations by a constant in order to solve the system by the elimination method. Which one is it?

(a) $\begin{aligned} -4x + 3y &= 7 \\ 3x - 4y &= 4 \end{aligned}$ **(b)** $\begin{aligned} 5x + 8y &= 13 \\ 12x + 24y &= 36 \end{aligned}$ **(c)** $\begin{aligned} 2x + 3y &= 5 \\ x - 3y &= 12 \end{aligned}$ **(d)** $\begin{aligned} x + 2y &= 9 \\ 3x - y &= 6 \end{aligned}$

14. For the system

$$\begin{aligned} 2x + 12y &= 7 \\ 3x + 4y &= 1, \end{aligned}$$

if we were to multiply the first equation by -3, by what number would we have to multiply the second equation in order to

(a) eliminate the x-terms when solving by the elimination method?

(b) eliminate the y-terms when solving by the elimination method?

Solve each system by the elimination method.

15. $2x - y = 13$
$x + y = 8$

16. $-4x + 3y = 25$
$6x - 5y = -39$

17. $3x - 4y = 9$
$6x - 8y = 18$

18. $2x + y = 3$
$-4x - 2y = 6$

[8.1–8.3] *Solve each system by any method.*

19. $2x + 3y = -5$
$3x + 4y = -8$

20. $6x - 9y = 0$
$2x - 3y = 0$

21. $2x + y - x = 3y + 5$
$y + 2 = x - 5$

22. $\frac{x}{2} + \frac{y}{3} = 7$
$\frac{x}{4} + \frac{2y}{3} = 8$

23. What are the three methods of solving systems discussed in this chapter? Choose one and discuss its drawbacks and advantages.

24. Why would it be easier to solve system B by the substitution method than system A?

A: $-5x + 6y = 7$
$2x + 5y = -5$

B: $2x + 9y = 13$
$y = 3x - 2$

[8.4] *Solve each problem by using a system of equations. Use the six-step method.*

25. A popular leisure activity of Americans is reading. In 1996, two popular fiction titles were *How Stella Got Her Groove Back* by Terry McMillan and *The Deep End of the Ocean* by Jacquelyn Mitchard. Together, these two titles sold 1,622,962 copies. The Mitchard book sold 57,564 more than the McMillan book. How many copies of each title were sold? (*Source: Publishers Weekly.*)

26. When people are not reading fiction during their leisure time, they are often reading magazines. Two of the most popular magazines in the United States are *Modern Maturity* and *Reader's Digest.* Together, the average total circulation for these two magazines during July–December 1996 was 35.6 million. *Reader's Digest* circulation was 5.4 million less than that of *Modern Maturity.* What were the circulation figures for each magazine? (*Source:* Audit Bureau of Circulations and Magazine Publishers of America.)

27. The two tallest buildings in Houston, Texas, are the Texas Commerce Tower and the First Interstate Plaza. The first of these is 4 stories taller than the second. Together the buildings have 146 stories. How many stories tall are the individual buildings? (*Source: The World Almanac and Book of Facts,* 1998.)

28. In the 1996 presidential election, Bill Clinton received 220 more electoral votes than Bob Dole. Together, the two candidates received a total of 538 votes. How many votes did each receive? (*Source: The World Almanac and Book of Facts,* 1998.)

29. The perimeter of a rectangle is 90 meters. Its length is $1\frac{1}{2}$ times its width. Find the length and width of the rectangle.

30. A cashier has 20 bills, all of which are \$10 or \$20 bills. The total value of the money is \$330. How many of each type does the cashier have?

31. Candy that sells for \$1.30 per pound is to be mixed with candy selling for \$.90 per pound to get 100 pounds of a mix that will sell for \$1 per pound. How much of each type should be used?

9 Roots and Radicals

Electronics

The electronics industry encompasses producing, selling, and servicing the computers, TVs, VCRs, CD players, and cell phones that pervade our lives. The industry boomed for many years, but recently industry sales have leveled off. The graph, which gives the percent change in orders from year to year, shows the turbulence in the electronics industry from 1983 to 1998.*

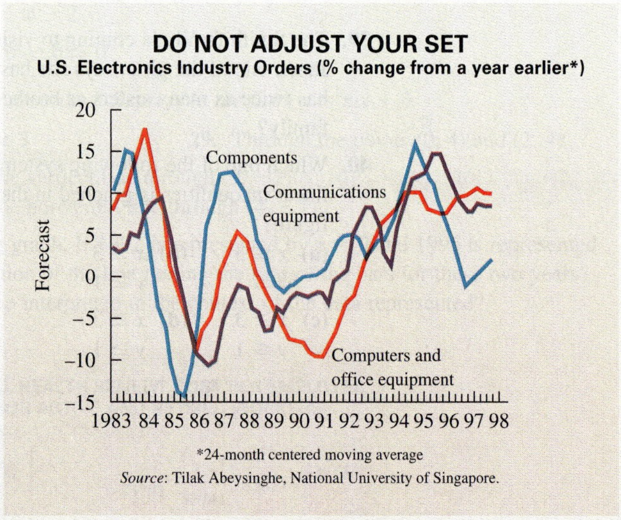

A major reason for the downturn in electronics business is the increase in the number of companies, particularly Asian companies, that have entered the industry. This large increase in supply has caused a glut of chips, with resulting price declines. This, in turn, has contributed to the collapse of export growth throughout Eastern Asia. In the exercises for Section 9.3, we examine data on U.S. exports and imports of electronics.

*"The silicon tigers' electric shocker," *The Economist,* November 9, 1996.

Visit our Web site at www.LialAlgebra.com

9.1 Evaluating Roots

OBJECTIVES

1. Find square roots.
2. Decide whether a given root is rational, irrational, or not a real number.
3. Find decimal approximations for irrational square roots.
4. Use the Pythagorean formula.
5. Use the distance formula.
6. Find higher roots.

FOR EXTRA HELP

SSG Sec. 9.1
SSM Sec. 9.1
Pass the Test Software
InterAct Math Tutorial Software
Video 14

In Section 1.2 we discussed the idea of the *square* of a number. Recall that squaring a number means multiplying the number by itself.

$$\text{If } a = 7, \text{ then } a^2 = 7 \cdot 7 = 49.$$
$$\text{If } a = -5, \text{ then } a^2 = (-5) \cdot (-5) = 25.$$
$$\text{If } a = -\frac{1}{2}, \text{ then } a^2 = \left(-\frac{1}{2}\right) \cdot \left(-\frac{1}{2}\right) = \frac{1}{4}.$$

In this chapter the opposite problem is considered.

$$\text{If } a^2 = 49, \text{ then } a = ?$$
$$\text{If } a^2 = 25, \text{ then } a = ?$$
$$\text{If } a^2 = \frac{1}{4}, \text{ then } a = ?$$

OBJECTIVE 1 **Find square roots.** To find a in the three statements above, we must find a number that when multiplied by itself results in the given number. The number a is called a **square root** of the number a^2.

EXAMPLE 1 Finding All Square Roots of a Number

Find all square roots of 49.

To find a square root of 49, think of a number that when multiplied by itself gives 49. One square root is 7, since $7 \cdot 7 = 49$. Another square root of 49 is -7, since $(-7)(-7) = 49$. The number 49 has two square roots, 7 and -7; one is positive and one is negative.

The positive square root of a number is written with the symbol $\sqrt{}$. For example, the positive square root of 49 is 7, written

$$\sqrt{49} = 7.$$

The symbol $-\sqrt{}$ is used for the negative square root of a number. For example, the negative square root of 49 is -7, written

$$-\sqrt{49} = -7.$$

Most calculators have a square root key, usually labeled $\boxed{\sqrt{x}}$, that allows us to find the square root of a number. For example, if we enter 49 and use the square root key, the display will show 7.

The symbol $\sqrt{}$ is called a **radical sign** and always represents the nonnegative square root. The number inside the radical sign is called the **radicand** and the entire expression, radical sign and radicand, is called a **radical.** An algebraic expression containing a radical is called a **radical expression.**

OBJECTIVE 4 Use the Pythagorean formula. One application of square roots uses the Pythagorean formula. Recall from Section 5.6 that by this formula, if c is the length of the hypotenuse of a right triangle, and a and b are the lengths of the two legs, then

$$c^2 = a^2 + b^2.$$

See Figure 1.

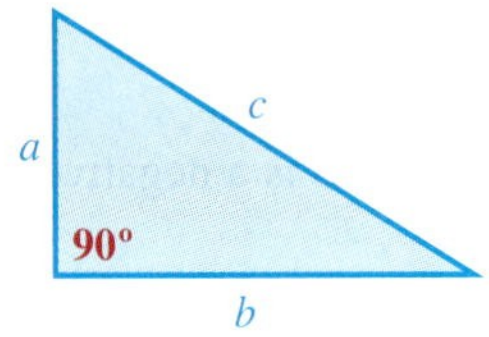

Figure 1

EXAMPLE 6 Using the Pythagorean Formula

Find the unknown length of the third side of each right triangle with sides a, b, and c, where c is the hypotenuse.

(a) $a = 3, b = 4$

Use the formula to find c^2 first.

$$\begin{aligned} c^2 &= a^2 + b^2 \\ &= 3^2 + 4^2 && \text{Let } a = 3 \text{ and } b = 4. \\ &= 9 + 16 = 25 && \text{Square and add.} \end{aligned}$$

Now find the positive square root of 25 to get c.

$$c = \sqrt{25} = 5$$

(Although -5 is also a square root of 25, the length of a side of a triangle must be a positive number.)

(b) $c = 9, b = 5$

Substitute the given values in the formula $c^2 = a^2 + b^2$. Then solve for a^2.

$$\begin{aligned} 9^2 &= a^2 + 5^2 && \text{Let } c = 9 \text{ and } b = 5. \\ 81 &= a^2 + 25 && \text{Square.} \\ 56 &= a^2 && \text{Subtract 25.} \end{aligned}$$

Again, we want only the positive root $a = \sqrt{56} \approx 7.483$.

CAUTION

Be careful not to make the common mistake of thinking that $\sqrt{a^2 + b^2}$ equals $a + b$. As Example 6(a) shows,

$$\sqrt{9 + 16} = \sqrt{25} = 5 \neq \sqrt{9} + \sqrt{16} = 3 + 4,$$

so that, in general,

$$\sqrt{a^2 + b^2} \neq a + b.$$

CONNECTIONS

Pythagoras did not actually discover the Pythagorean formula. While he may have written the first proof, there is evidence that the Babylonians knew the concept quite well. The figure on the left illustrates the formula by using a tile pattern. In the figure, the side of the square along the hypotenuse measures 5 units, while the sides along the legs measure 3 and 4 units. If we let $a = 3$, $b = 4$, and $c = 5$, the equation of the Pythagorean formula is satisfied.

$$a^2 + b^2 = c^2$$

$$3^2 + 4^2 = 5^2 \quad ?$$

$$25 = 25 \quad \text{True}$$

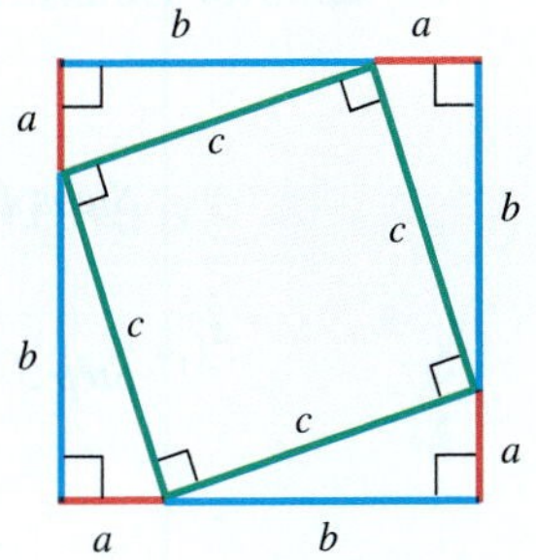

FOR DISCUSSION OR WRITING

The diagram on the right can be used to verify the Pythagorean formula. To do so, express the area of the figure in two ways: first, as the area of the large square, and then as the sum of the areas of the smaller square and the four right triangles. Finally, set the areas equal and simplify the equation.

PROBLEM SOLVING

The Pythagorean formula can be used to solve applied problems that involve right triangles. A good way to begin the solution is to sketch the triangle and label the three sides appropriately, using a variable as needed. Then use the Pythagorean formula to write an equation. This procedure is simply Steps 1–3 of the six-step problem-solving method given in Chapter 2, and used throughout the book. In Steps 4–6, we solve the equation, answer the question(s), and check the solution(s).

EXAMPLE 7 Solving an Application

A ladder 10 feet long leans against a wall. The foot of the ladder is 6 feet from the base of the wall. How high up the wall does the top of the ladder rest?

EXAMPLE 9 Finding Cube Roots

Find each cube root.

(a) $\sqrt[3]{8}$

Look for a number that can be cubed to give 8. Since $2^3 = 8$, then $\sqrt[3]{8} = 2$.

(b) $\sqrt[3]{-8}$

$\sqrt[3]{-8} = -2$ because $(-2)^3 = -8$.

As Example 9 suggests, the cube root of a positive number is positive, and the cube root of a negative number is negative. *There is only one real number cube root for each real number.*

When the index of the radical is even (square root, fourth root, and so on), the radicand must be nonnegative to get a real number root. Also, for even indexes the symbols $\sqrt{\ }$, $\sqrt[4]{\ }$, $\sqrt[6]{\ }$, and so on are used for the *nonnegative* roots, which are called **principal roots.** The symbols $-\sqrt{\ }$, $-\sqrt[4]{\ }$, $-\sqrt[6]{\ }$, and so on are used for the negative roots.

EXAMPLE 10 Finding Higher Roots

Find each root.

(a) $\sqrt[4]{16}$

$\sqrt[4]{16} = 2$ because 2 is positive and $2^4 = 16$.

(b) $-\sqrt[4]{16}$

From part (a), $\sqrt[4]{16} = 2$, so the negative root $-\sqrt[4]{16} = -2$.

(c) $\sqrt[4]{-16}$

To find the fourth root, the radicand must be nonnegative. There is no real number that equals $\sqrt[4]{-16}$.

(d) $\sqrt[3]{64} = 4$ since $4^3 = 64$.

(e) $-\sqrt[5]{32}$

First find $\sqrt[5]{32}$. The prime factorization of 32 as 2^5 shows that $\sqrt[5]{32} = 2$. If $\sqrt[5]{32} = 2$, then $-\sqrt[5]{32} = -2$.

9.1 EXERCISES

Decide whether each statement is true or false. If false, tell why.

1. Every nonnegative number has two square roots.
2. A negative number has negative square roots.
3. Every positive number has two real square roots.
4. Every positive number has three real cube roots.
5. The cube root of every real number has the same sign as the number itself.
6. The positive square root of a positive number is its principal square root.

Find all square roots of each number. See Example 1.

7. 16 **8.** 9 **9.** 144 **10.** 225

11. $\frac{25}{196}$ **12.** $\frac{81}{400}$ **13.** 900 **14.** 1600

Find each square root that is a real number. See Examples 2 and 4(e).

15. $\sqrt{49}$ **16.** $\sqrt{81}$ **17.** $-\sqrt{121}$ **18.** $\sqrt{196}$

19. $-\sqrt{\frac{144}{121}}$ **20.** $-\sqrt{\frac{49}{36}}$ **21.** $\sqrt{-121}$ **22.** $\sqrt{-49}$

Find the square of each radical expression. See Example 3.

23. $\sqrt{100}$ **24.** $\sqrt{36}$ **25.** $-\sqrt{19}$

26. $-\sqrt{99}$ **27.** $\sqrt{3x^2 + 4}$ **28.** $\sqrt{9y^2 + 3}$

What must be true about the variable a for each statement to be true?

29. $\sqrt{a}$ represents a positive number. **30.** $-\sqrt{a}$ represents a negative number.

31. $\sqrt{a}$ is not a real number. **32.** $-\sqrt{a}$ is not a real number.

Write rational, irrational, *or* not a real number *for each number. If the number is rational, give its exact value. If the number is irrational, give a decimal approximation to the nearest thousandth. Use a calculator as necessary. See Examples 4 and 5.*

33. $\sqrt{25}$ **34.** $\sqrt{169}$ **35.** $\sqrt{29}$ **36.** $\sqrt{33}$

37. $-\sqrt{64}$ **38.** $-\sqrt{500}$ **39.** $\sqrt{-29}$ **40.** $\sqrt{-47}$

41. Explain why the answers to Exercises 17 and 21 are different.

42. Explain why $\sqrt[3]{-8}$ and $-\sqrt[3]{8}$ represent the same number.

Use a calculator with a square root key to find each root. Round to the nearest thousandth. See Example 5.

43. $\sqrt{571}$ **44.** $\sqrt{693}$ **45.** $\sqrt{798}$

46. $\sqrt{453}$ **47.** $\sqrt{3.94}$ **48.** $\sqrt{.00895}$

Find each square root. Use a calculator and round to the nearest thousandth, if necessary. (Hint: *First simplify the radicand to a single number.*)

49. $\sqrt{3^2 + 4^2}$ **50.** $\sqrt{6^2 + 8^2}$ **51.** $\sqrt{8^2 + 15^2}$

52. $\sqrt{5^2 + 12^2}$ **53.** $\sqrt{2^2 + 3^2}$ **54.** $\sqrt{(-1)^2 + 5^2}$

Use a calculator with a cube root key to find each root. Round to the nearest thousandth. (In Exercises 59 and 60, you may have to use the fact that if $a > 0$, $\sqrt[3]{-a} = -\sqrt[3]{a}$.)

55. $\sqrt[3]{12}$ **56.** $\sqrt[3]{74}$ **57.** $\sqrt[3]{130.6}$

58. $\sqrt[3]{251.8}$ **59.** $\sqrt[3]{-87}$ **60.** $\sqrt[3]{-95}$

Find the length of the unknown side of each right triangle with legs a and b and hypotenuse c. In Exercises 65 and 66, use a calculator and round to the nearest thousandth. See Example 6.

61. $a = 8, b = 15$ **62.** $a = 24, b = 10$ **63.** $a = 6, c = 10$

64. $b = 12, c = 13$ **65.** $a = 11, b = 4$ **66.** $a = 13, b = 9$

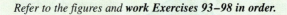

RELATING CONCEPTS (EXERCISES 93–98)

One of the many proofs of the Pythagorean formula was given in the Connections box in this section. Here is another one, attributed to the Hindu mathematician Bhāskara.

Refer to the figures and ***work Exercises 93–98 in order.***

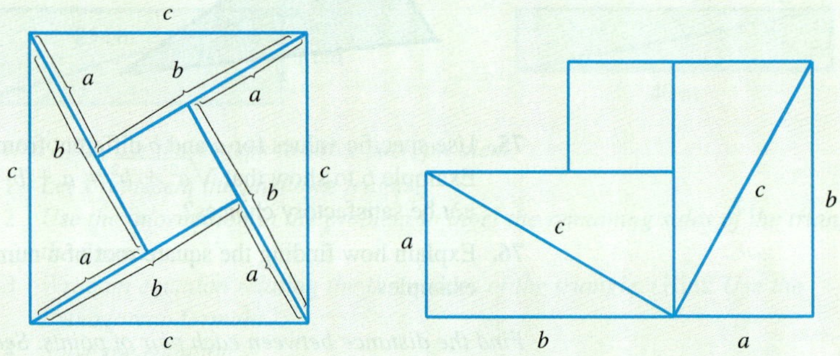

93. What is the area of the square on the left in terms of c?

94. What is the area of the small square in the middle of the figure on the left, in terms of $(b - a)$?

95. What is the sum of the areas of the two rectangles made up of triangles in the figure on the right?

96. What is the area of the small square in the figure on the right in terms of a and b?

97. The figure on the left is made up of the same square and triangles as the figure on the right. Write an equation setting the answer to Exercise 93 equal to the sum of the answers in Exercises 95 and 96.

98. Simplify the expressions you obtained in Exercise 97. What is your final result?

Did you make the connection that geometric figures can be used to prove algebraic formulas?

9.2 Multiplication and Division of Radicals

OBJECTIVES

1. Multiply radicals.
2. Simplify radicals using the product rule.
3. Simplify radical quotients using the quotient rule.
4. Use the product and quotient rules to simplify higher roots.

FOR EXTRA HELP

SSG Sec. 9.2
SSM Sec. 9.2
Pass the Test Software
InterAct Math Tutorial Software
Video 14

CONNECTIONS

The sixteenth century German radical symbol $\sqrt{\ }$ we use today is probably derived from the letter *R*. The radical symbol on the left below comes from the Latin word for root, *radix*. It was first used by Leonardo da Pisa (Fibonnaci) in 1220.

The cube root symbol shown on the right above was used by the German mathematician Christoff Rudolff in 1525. The symbol used today originated in the seventeenth century in France.

CONNECTIONS (CONTINUED)

FOR DISCUSSION OR WRITING

1. In the radical sign shown on the left, the *R* referred to above is clear. What other letter do you think is part of the symbol? What would the equivalent be in our modern notation?
2. How is the cube root symbol on the right related to our modern radical sign?

OBJECTIVE 1 Multiply radicals. Several useful rules for finding products and quotients of radicals are developed in this section. To illustrate the rule for products, notice that

$$\sqrt{4} \cdot \sqrt{9} = 2 \cdot 3 = 6 \quad \text{and} \quad \sqrt{4 \cdot 9} = \sqrt{36} = 6,$$

showing that

$$\sqrt{4} \cdot \sqrt{9} = \sqrt{4 \cdot 9}.$$

This result is a particular case of the more general product rule for radicals.

Product Rule for Radicals

For nonnegative real numbers x and y,

$$\sqrt{x} \cdot \sqrt{y} = \sqrt{x \cdot y} \quad \text{and} \quad \sqrt{x \cdot y} = \sqrt{x} \cdot \sqrt{y}.$$

That is, the product of two radicals is the radical of the product.

CAUTION

In general, $\sqrt{x + y} \neq \sqrt{x} + \sqrt{y}$. To see why this is so, let $x = 16$ and $y = 9$.

$$\sqrt{16 + 9} = \sqrt{25} = 5$$

but

$$\sqrt{16} + \sqrt{9} = 4 + 3 = 7.$$

EXAMPLE 1 Using the Product Rule to Multiply Radicals

Use the product rule for radicals to find each product.

(a) $\sqrt{2} \cdot \sqrt{3} = \sqrt{2 \cdot 3} = \sqrt{6}$

(b) $\sqrt{7} \cdot \sqrt{5} = \sqrt{35}$

(c) $\sqrt{11} \cdot \sqrt{a} = \sqrt{11a}$ Assume $a \geq 0$.

OBJECTIVE 2 Simplify radicals using the product rule. A square root radical is **simplified** when no perfect square factor remains under the radical sign. This is accomplished by using the product rule as shown in Example 2.

Some problems require both the product and quotient rules, as Example 6 shows.

EXAMPLE 6 Using Both the Product and Quotient Rules

Simplify $\sqrt{\frac{3}{5}} \cdot \sqrt{\frac{4}{5}}$.

$$\sqrt{\frac{3}{5}} \cdot \sqrt{\frac{4}{5}} = \frac{\sqrt{3}}{\sqrt{5}} \cdot \frac{\sqrt{4}}{\sqrt{5}} \qquad \text{Quotient rule}$$

$$= \frac{\sqrt{3} \cdot \sqrt{4}}{\sqrt{5} \cdot \sqrt{5}} \qquad \text{Multiply fractions.}$$

$$= \frac{\sqrt{3} \cdot 2}{\sqrt{25}} \qquad \text{Product rule; } \sqrt{4} = 2$$

$$= \frac{2\sqrt{3}}{5} \qquad \sqrt{25} = 5$$

The product and quotient rules also apply when variables appear under the radical sign, as long as all the variables represent only nonnegative numbers. For example, $\sqrt{5^2} = 5$, but $\sqrt{(-5)^2} \neq -5$.

For a real number a, $\sqrt{a^2} = a$ only if a is nonnegative.

EXAMPLE 7 Simplifying Radicals Involving Variables

Simplify each radical. Assume all variables represent positive real numbers.

(a) $\sqrt{25m^4} = \sqrt{25} \cdot \sqrt{m^4}$ Product rule

$= 5m^2$

(b) $\sqrt{64p^{10}} = 8p^5$ Product rule

(c) $\sqrt{r^9} = \sqrt{r^8 \cdot r}$

$= \sqrt{r^8} \cdot \sqrt{r} = r^4\sqrt{r}$ Product rule

(d) $\sqrt{\frac{5}{x^2}} = \frac{\sqrt{5}}{\sqrt{x^2}} = \frac{\sqrt{5}}{x}$ Quotient rule

OBJECTIVE 4 Use the product and quotient rules to simplify higher roots. The product and quotient rules for radicals also work for other roots, as shown in Example 8. To simplify cube roots, look for factors that are *perfect cubes.* A **perfect cube** is a number with a rational cube root. For example, $\sqrt[3]{64} = 4$, and since 4 is a rational number, 64 is a perfect cube. Higher roots are handled in a similar manner.

Properties of Radicals

For all real numbers where the indicated roots exist,

$$\sqrt[n]{x} \cdot \sqrt[n]{y} = \sqrt[n]{xy} \quad \text{and} \quad \frac{\sqrt[n]{x}}{\sqrt[n]{y}} = \sqrt[n]{\frac{x}{y}}, y \neq 0.$$

EXAMPLE 8 Simplifying Higher Roots

Simplify each radical.

(a) $\sqrt[3]{32} = \sqrt[3]{8 \cdot 4}$ 8 is a perfect cube.

$= \sqrt[3]{8} \cdot \sqrt[3]{4} = 2\sqrt[3]{4}$

(b) $\sqrt[3]{108} = \sqrt[3]{27 \cdot 4}$ 27 is a perfect cube.

$= \sqrt[3]{27} \cdot \sqrt[3]{4} = 3\sqrt[3]{4}$

(c) $\sqrt[4]{32} = \sqrt[4]{16} \cdot \sqrt[4]{2} = 2\sqrt[4]{2}$ 16 is a perfect fourth power.

(d) $\sqrt[3]{\frac{8}{125}} = \frac{\sqrt[3]{8}}{\sqrt[3]{125}} = \frac{2}{5}$

(e) $\sqrt[4]{\frac{16}{625}} = \frac{\sqrt[4]{16}}{\sqrt[4]{625}} = \frac{2}{5}$

(f) $\sqrt[3]{7} \cdot \sqrt[3]{49} = \sqrt[3]{7 \cdot 49} = \sqrt[3]{7 \cdot 7^2} = \sqrt[3]{7^3} = 7$

9.2 EXERCISES

Decide whether each statement is true or false for real numbers. If false, tell why.

1. $\sqrt{4} = \pm 2$
2. $\sqrt{(-6)^2} = -6$
3. $\sqrt{-6} \cdot \sqrt{6} = -6$
4. $\sqrt[3]{(-6)^3} = -6$
5. $\sqrt[3]{3} \cdot \sqrt[3]{2} = \sqrt[3]{6}$
6. $\sqrt[3]{4} \cdot \sqrt[3]{4} = 4$
7. $\sqrt{4} \cdot \sqrt{9} = \sqrt{4 \cdot 9}$
8. $\sqrt{4} + \sqrt{9} = \sqrt{4 + 9}$

Use the product rule for radicals to find each product. See Example 1.

9. $\sqrt{3} \cdot \sqrt{27}$
10. $\sqrt{2} \cdot \sqrt{8}$
11. $\sqrt{6} \cdot \sqrt{15}$
12. $\sqrt{10} \cdot \sqrt{15}$
13. $\sqrt{13} \cdot \sqrt{13}$
14. $\sqrt{17} \cdot \sqrt{17}$
15. $\sqrt{13} \cdot \sqrt{r}, r \geq 0$
16. $\sqrt{19} \cdot \sqrt{k}, k \geq 0$

17. Which one of the following radicals is simplified according to the guidelines of Objective 2?

(a) $\sqrt{47}$ **(b)** $\sqrt{45}$ **(c)** $\sqrt{48}$ **(d)** $\sqrt{44}$

18. If p is a prime number, is $\sqrt{p}$ in simplified form? Explain your answer.

Simplify each radical according to the method described in Objective 2. See Example 2.

19. $\sqrt{45}$
20. $\sqrt{56}$
21. $\sqrt{75}$
22. $\sqrt{18}$
23. $\sqrt{125}$
24. $\sqrt{80}$
25. $-\sqrt{700}$
26. $-\sqrt{600}$
27. $3\sqrt{27}$
28. $9\sqrt{8}$

Find each product and simplify. See Example 3.

29. $\sqrt{3} \cdot \sqrt{18}$
30. $\sqrt{3} \cdot \sqrt{21}$
31. $\sqrt{12} \cdot \sqrt{48}$
32. $\sqrt{50} \cdot \sqrt{72}$
33. $\sqrt{12} \cdot \sqrt{30}$
34. $\sqrt{30} \cdot \sqrt{24}$

35. In your own words, describe the product and quotient rules.

36. Simplify the radical $\sqrt{288}$ in two ways. First, factor 288 as $144 \cdot 2$ and then simplify completely. Second, factor 288 as $48 \cdot 6$ and then simplify completely. How do the answers compare? Make a conjecture concerning the quickest way to simplify such a radical.

Use the quotient rule and the product rule, as necessary, to simplify each radical expression. See Examples 4–6.

37. $\sqrt{\frac{16}{225}}$
38. $\sqrt{\frac{9}{100}}$
39. $\sqrt{\frac{7}{16}}$
40. $\sqrt{\frac{13}{25}}$

41. $\sqrt{\frac{5}{7}} \cdot \sqrt{35}$
42. $\sqrt{\frac{10}{13}} \cdot \sqrt{130}$
43. $\sqrt{\frac{5}{2}} \cdot \sqrt{\frac{125}{8}}$
44. $\sqrt{\frac{8}{3}} \cdot \sqrt{\frac{512}{27}}$
45. $\frac{30\sqrt{10}}{5\sqrt{2}}$
46. $\frac{50\sqrt{20}}{2\sqrt{10}}$

Simplify each radical. Assume that all variables represent nonnegative real numbers. See Example 7.

47. $\sqrt{m^2}$
48. $\sqrt{k^2}$
49. $\sqrt{y^4}$
50. $\sqrt{s^4}$
51. $\sqrt{36z^2}$
52. $\sqrt{49n^2}$
53. $\sqrt{400x^6}$
54. $\sqrt{900y^8}$
55. $\sqrt{z^5}$
56. $\sqrt{a^{13}}$
57. $\sqrt{x^6y^{12}}$
58. $\sqrt{a^8b^{10}}$

Simplify each radical. See Example 8.

59. $\sqrt[3]{40}$
60. $\sqrt[3]{48}$
61. $\sqrt[3]{54}$
62. $\sqrt[3]{192}$
63. $\sqrt[4]{80}$
64. $\sqrt[4]{243}$
65. $\sqrt[3]{\frac{8}{27}}$
66. $\sqrt[3]{\frac{64}{125}}$
67. $\sqrt[3]{-\frac{216}{125}}$
68. $\sqrt[3]{-\frac{1}{64}}$
69. $\sqrt[3]{5} \cdot \sqrt[3]{25}$
70. $\sqrt[3]{4} \cdot \sqrt[3]{16}$
71. $\sqrt[4]{4} \cdot \sqrt[4]{3}$
72. $\sqrt[4]{7} \cdot \sqrt[4]{4}$
73. $\sqrt[3]{4x} \cdot \sqrt[3]{8x^2}$
74. $\sqrt[3]{25p} \cdot \sqrt[3]{125p^3}$

75. In Example 2(a) we showed *algebraically* that $\sqrt{20}$ is equal to $2\sqrt{5}$. To give *numerical support* to this result, use a calculator to do the following:
   (a) Find a decimal approximation for $\sqrt{20}$ using your calculator. Record as many digits as the calculator shows.
   (b) Find a decimal approximation for $\sqrt{5}$ using your calculator, and then multiply the result by 2. Record as many digits as the calculator shows.
   (c) Your results in parts (a) and (b) should be the same. A mathematician would not accept this numerical exercise as *proof* that $\sqrt{20}$ is equal to $2\sqrt{5}$. Explain why.

76. On your calculator, multiply the approximations for $\sqrt{3}$ and $\sqrt{5}$. Now, predict what your calculator will show when you find an approximation for $\sqrt{15}$. What rule stated in this section justifies your answer?

The volume of a cube is found with the formula $V = s^3$, where s is the length of an edge of the cube. Use this information in Exercises 77 and 78.

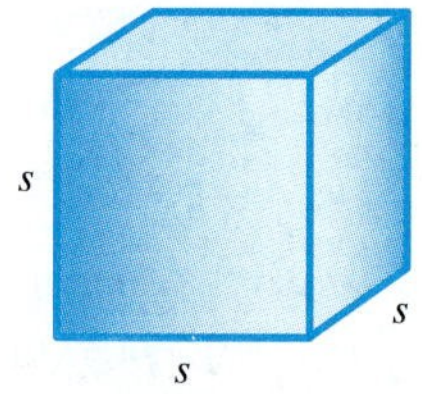

77. A container in the shape of a cube has a volume of 216 cubic centimeters. What is the depth of the container?

78. A cube-shaped box must be constructed to contain 128 cubic feet. What should the dimensions (height, width, and length) of the box be?

The volume of a sphere is found with the formula $V = \frac{4}{3}\pi r^3$, where r is the length of the radius of the sphere. Use this information in Exercises 79 and 80.

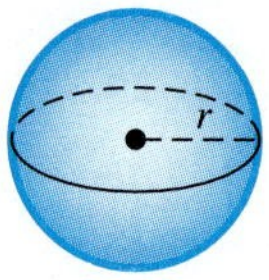

79. A ball in the shape of a sphere has a volume of 288π cubic inches. What is the radius of the ball?

80. Suppose that the volume of the ball described in Exercise 79 is multiplied by 8. How is the radius affected?

81. When we multiply two radicals with variables under the radical sign, such as $\sqrt{a} \cdot \sqrt{b} = \sqrt{ab}$, why is it important to know that both a and b represent nonnegative numbers?

82. Is it necessary to restrict k to a nonnegative number to say that $\sqrt[3]{k} \cdot \sqrt[3]{k} \cdot \sqrt[3]{k} = k$? Why?

RELATING CONCEPTS (EXERCISES 83–86)

An interesting way to represent the lengths corresponding to $\sqrt{2}$, $\sqrt{3}$, $\sqrt{4}$, $\sqrt{5}$, and so on is shown in the figure.

Work the following exercises in order.

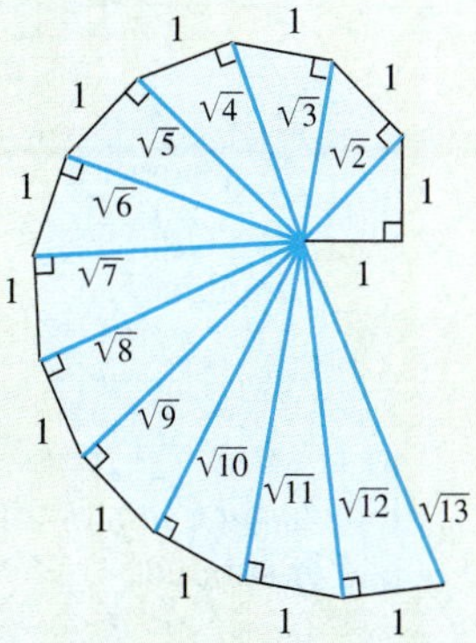

83. Use the Pythagorean formula to verify the lengths in the figure.

84. Which of the lengths indicated as radicals equal whole numbers? If the figure is continued in the same way, what would the next two whole number lengths be?

85. Find the consecutive differences between the radicands in Exercise 84. (*Hint:* The first difference is $9 - 4 = 5$.)

86. Look for a pattern that determines where these whole number lengths occur. Use this pattern to predict where the next whole number length and the one after that will occur.

Did you make the connection between the lengths of the sides of the triangles and the Pythagorean formula?

9.3 Addition and Subtraction of Radicals

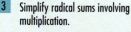

OBJECTIVES

1. Add and subtract radicals.
2. Simplify radical sums and differences.
3. Simplify radical sums involving multiplication.

FOR EXTRA HELP

SSG Sec. 9.3
SSM Sec. 9.3
Pass the Test Software
InterAct Math Tutorial Software
Video 14

OBJECTIVE 1 **Add and subtract radicals.** We add or subtract radicals by using the distributive property. For example,

$$8\sqrt{3} + 6\sqrt{3} = (8 + 6)\sqrt{3} \quad \text{Distributive property}$$
$$= 14\sqrt{3}.$$

Also,

$$2\sqrt{11} - 7\sqrt{11} = -5\sqrt{11}.$$

Like radicals are terms that have multiples of the *same root* of the *same number.* Only like radicals can be combined using the distributive property. In the example above, the like radicals are $2\sqrt{11}$ and $-7\sqrt{11}$. On the other hand, examples of *unlike radicals* are

$$2\sqrt{5} \quad \text{and} \quad 2\sqrt{3}, \quad \text{Different radicands}$$

as well as

$$2\sqrt{3} \quad \text{and} \quad 2\sqrt[3]{3}. \quad \text{Different indexes}$$

EXAMPLE 1 Adding and Subtracting Like Radicals

Add or subtract, as indicated.

(a) $3\sqrt{6} + 5\sqrt{6} = (3 + 5)\sqrt{6} = 8\sqrt{6}$ Distributive property

(b) $5\sqrt{10} - 7\sqrt{10} = (5 - 7)\sqrt{10} = -2\sqrt{10}$

(c) $\sqrt[3]{5} + \sqrt[3]{5} = 1\sqrt[3]{5} + 1\sqrt[3]{5} = (1 + 1)\sqrt[3]{5} = 2\sqrt[3]{5}$

(d) $\sqrt[4]{7} + 2\sqrt[4]{7} = 1\sqrt[4]{7} + 2\sqrt[4]{7} = 3\sqrt[4]{7}$

(e) $\sqrt{3} + \sqrt{13}$ cannot be added using the distributive property.

CAUTION In general, $\sqrt{x} + \sqrt{y} \neq \sqrt{x + y}$. In Example 1(e), it would be **incorrect** to try to simplify $\sqrt{3} + \sqrt{13}$ as $\sqrt{3} + \sqrt{13} = \sqrt{16} = 4$. Only *like radicals* can be combined.

OBJECTIVE 2 Simplify radical sums and differences. Sometimes we must simplify one or more radicals in a sum or difference. Doing this may result in like radicals, which we can then add or subtract.

EXAMPLE 2 Adding and Subtracting Radicals That Require Simplification

Simplify as much as possible.

(a)

$$\begin{aligned} 3\sqrt{2} + \sqrt{8} &= 3\sqrt{2} + \sqrt{4 \cdot 2} && \text{Factor.} \\ &= 3\sqrt{2} + \sqrt{4} \cdot \sqrt{2} && \text{Product rule} \\ &= 3\sqrt{2} + 2\sqrt{2} && \sqrt{4} = 2 \\ &= 5\sqrt{2} && \text{Add like radicals.} \end{aligned}$$

(b)

$$\begin{aligned} \sqrt{18} - \sqrt{27} &= \sqrt{9 \cdot 2} - \sqrt{9 \cdot 3} && \text{Factor.} \\ &= \sqrt{9} \cdot \sqrt{2} - \sqrt{9} \cdot \sqrt{3} && \text{Product rule} \\ &= 3\sqrt{2} - 3\sqrt{3} && \sqrt{9} = 3 \end{aligned}$$

Since $\sqrt{2}$ and $\sqrt{3}$ are unlike radicals, this difference cannot be simplified further.

(c)

$$\begin{aligned} 2\sqrt{12} + 3\sqrt{75} &= 2(\sqrt{4} \cdot \sqrt{3}) + 3(\sqrt{25} \cdot \sqrt{3}) && \text{Product rule} \\ &= 2(2\sqrt{3}) + 3(5\sqrt{3}) && \sqrt{4} = 2 \text{ and } \sqrt{25} = 5 \\ &= 4\sqrt{3} + 15\sqrt{3} && \text{Multiply.} \\ &= 19\sqrt{3} && \text{Add like radicals.} \end{aligned}$$

(d)

$$\begin{aligned} 3\sqrt[3]{16} + 5\sqrt[3]{2} &= 3(\sqrt[3]{8} \cdot \sqrt[3]{2}) + 5\sqrt[3]{2} && \text{Product rule} \\ &= 3(2\sqrt[3]{2}) + 5\sqrt[3]{2} && \sqrt[3]{8} = 2 \\ &= 6\sqrt[3]{2} + 5\sqrt[3]{2} && \text{Multiply.} \\ &= 11\sqrt[3]{2} && \text{Add like radicals.} \end{aligned}$$

OBJECTIVE 3 Simplify radical sums involving multiplication. Some radical expressions require both multiplication and addition (or subtraction). The order of operations presented in Chapter 1 still applies.

EXAMPLE 3 Simplifying Radical Sums Involving Multiplication

Simplify each radical expression. Assume that all variables represent nonnegative real numbers.

(a) $\sqrt{5} \cdot \sqrt{15} + 4\sqrt{3} = \sqrt{5 \cdot 15} + 4\sqrt{3}$ Product rule

$= \sqrt{75} + 4\sqrt{3}$ Multiply.

$= \sqrt{25 \cdot 3} + 4\sqrt{3}$ 25 is a perfect square.

$= \sqrt{25} \cdot \sqrt{3} + 4\sqrt{3}$ Product rule

$= 5\sqrt{3} + 4\sqrt{3}$ $\sqrt{25} = 5$

$= 9\sqrt{3}$ Add like radicals.

(b) $\sqrt{2} \cdot \sqrt{6k} + \sqrt{27k} = \sqrt{12k} + \sqrt{27k}$ Product rule

$= \sqrt{4 \cdot 3k} + \sqrt{9 \cdot 3k}$ Factor.

$= \sqrt{4} \cdot \sqrt{3k} + \sqrt{9} \cdot \sqrt{3k}$ Product rule

$= 2\sqrt{3k} + 3\sqrt{3k}$ $\sqrt{4} = 2$ and $\sqrt{9} = 3$

$= 5\sqrt{3k}$ Add like radicals.

(c) $\sqrt[3]{2} \cdot \sqrt[3]{16m^3} - \sqrt[3]{108m^3} = \sqrt[3]{32m^3} - \sqrt[3]{108m^3}$ Product rule

$= \sqrt[3]{(8m^3)4} - \sqrt[3]{(27m^3)4}$ Factor.

$= 2m\sqrt[3]{4} - 3m\sqrt[3]{4}$ $\sqrt[3]{8m^3} = 2m$ and $\sqrt[3]{27m^3} = 3m$

$= -m\sqrt[3]{4}$ Subtract like radicals.

CAUTION Remember that a sum or difference of radicals can be simplified only if the radicals are *like radicals.* For example, $2\sqrt{3} + 5\sqrt[3]{3}$ cannot be simplified further.

9.3 EXERCISES

Fill in each blank with the correct response.

1. $5\sqrt{2} + 6\sqrt{2} = (5 + 6)\sqrt{2} = 11\sqrt{2}$ is an example of the _____ property.
2. Like radicals have the same _____ of the same _____.
3. $\sqrt{2} + 2\sqrt{3}$ cannot be simplified because the _____ are different.
4. $2\sqrt{3} + 4\sqrt[3]{3}$ cannot be simplified because the _____ are different.

Simplify and add or subtract wherever possible. See Examples 1 and 2.

5. $14\sqrt{7} - 19\sqrt{7}$
6. $16\sqrt{2} - 18\sqrt{2}$
7. $\sqrt{17} + 4\sqrt{17}$
8. $5\sqrt{19} + \sqrt{19}$
9. $6\sqrt{7} - \sqrt{7}$
10. $11\sqrt{14} - \sqrt{14}$
11. $\sqrt{45} + 4\sqrt{20}$
12. $\sqrt{24} + 6\sqrt{54}$
13. $5\sqrt{72} - 3\sqrt{50}$
14. $6\sqrt{18} - 5\sqrt{32}$
15. $-5\sqrt{32} + 2\sqrt{45}$
16. $-4\sqrt{75} + 3\sqrt{24}$

17. $5\sqrt{7} - 3\sqrt{28} + 6\sqrt{63}$

18. $3\sqrt{11} + 5\sqrt{44} - 8\sqrt{99}$

19. $2\sqrt{8} - 5\sqrt{32} - 2\sqrt{48}$

20. $5\sqrt{72} - 3\sqrt{48} + 4\sqrt{128}$

21. $4\sqrt{50} + 3\sqrt{12} - 5\sqrt{45}$

22. $6\sqrt{18} + 2\sqrt{48} + 6\sqrt{28}$

23. $\frac{1}{4}\sqrt{288} + \frac{1}{6}\sqrt{72}$

24. $\frac{2}{3}\sqrt{27} + \frac{3}{4}\sqrt{48}$

25. The distributive property, which says $a(b + c) = ab + ac$ and $ba + ca = (b + c)a$, provides the justification for adding and subtracting like radicals. While we usually skip the step that indicates this property, we could not make the statement $2\sqrt{3} + 4\sqrt{3} = 6\sqrt{3}$ without it. Write an equation showing how the distributive property is actually used in this statement.

26. Refer to Example 1(e), and explain why $\sqrt{3} + \sqrt{7}$ cannot be further simplified. Confirm, by using calculator approximations, that $\sqrt{3} + \sqrt{7}$ is *not* equal to $\sqrt{10}$.

Perform the indicated operations. Assume that all variables represent nonnegative real numbers. See Example 3.

27. $\sqrt{6} \cdot \sqrt{2} + 9\sqrt{3}$

28. $4\sqrt{15} \cdot \sqrt{3} + 4\sqrt{5}$

29. $\sqrt{9x} + \sqrt{49x} - \sqrt{25x}$

30. $\sqrt{4a} - \sqrt{16a} + \sqrt{100a}$

31. $\sqrt{6x^2} + x\sqrt{24}$

32. $\sqrt{75x^2} + x\sqrt{108}$

33. $3\sqrt{8x^2} - 4x\sqrt{2} - x\sqrt{8}$

34. $\sqrt{2b^2} + 3b\sqrt{18} - b\sqrt{200}$

35. $-8\sqrt{32k} + 6\sqrt{8k}$

36. $4\sqrt{12x} + 2\sqrt{27x}$

37. $2\sqrt{125x^2z} + 8x\sqrt{80z}$

38. $\sqrt{48x^2y} + 5x\sqrt{27y}$

39. $4\sqrt[3]{16} - 3\sqrt[3]{54}$

40. $5\sqrt[3]{128} + 3\sqrt[3]{250}$

41. $6\sqrt[3]{8p^2} - 2\sqrt[3]{27p^2}$

42. $8k\sqrt[3]{54k} + 6\sqrt[3]{16k^4}$

43. $5\sqrt[4]{m^3} + 8\sqrt[4]{16m^3}$

44. $5\sqrt[4]{m^5} + 3\sqrt[4]{81m^5}$

45. Describe in your own words how to add and subtract radicals.

46. In the directions for Exercises 27–44, we made the assumption that all variables represent nonnegative real numbers. However, in Exercises 41 and 42, variables actually *may* represent negative numbers. Explain why this is so.

RELATING CONCEPTS (EXERCISES 47–50)

Adding and subtracting like radicals is no different than adding and subtracting other like terms.

Work Exercises 47–50 in order.

47. Combine like terms: $5x^2y + 3x^2y - 14x^2y$.

48. Combine like terms: $5(p - 2q)^2(a + b) + 3(p - 2q)^2(a + b) - 14(p - 2q)^2(a + b)$.

49. Combine like radicals: $5a^2\sqrt{xy} + 3a^2\sqrt{xy} - 14a^2\sqrt{xy}$.

50. Compare your answers in Exercises 47–49. How are they alike? How are they different?

Did you make the connection between adding and subtracting like radicals and adding and subtracting like terms?

Perform the indicated operations following the order of operations we have used throughout the book. Use a calculator and round to the nearest thousandth, if necessary.

51. $\sqrt{(-3 - 6)^2 + (2 - 4)^2}$

52. $\sqrt{(-9 - 3)^2 + (3 - 8)^2}$

53. $\sqrt{(2 - (-2))^2 + (-1 - 2)^2}$

54. $\sqrt{(3 - 1)^2 + (2 - (-1))^2}$

Find the perimeter of each figure.

55.

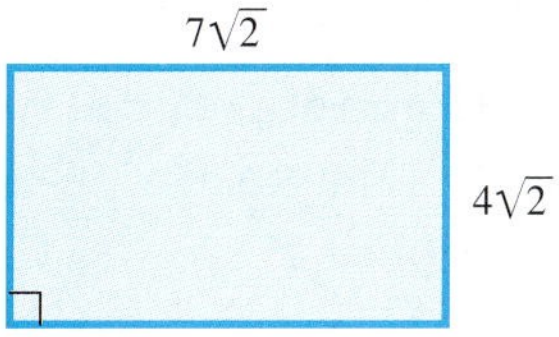

56.

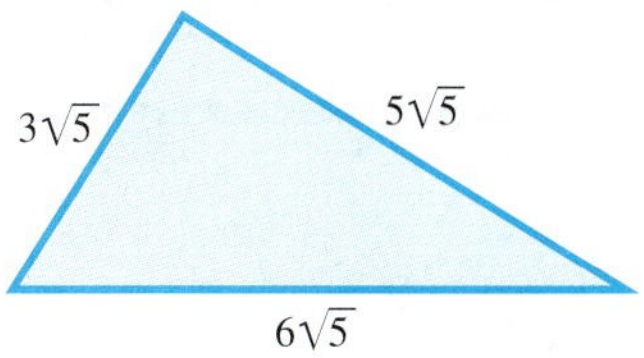

Work each problem.

57. The table shows the year in which U.S. exports of electronics were x billion dollars.

U.S. Exports

Billions of Dollars	*Year*
7.5	1990
19.6	1993
25.8	1994
36.4	1996

Source: U.S. Bureau of the Census; *U.S. Merchandise Trade,* series FT 900, December issue; and unpublished data.

The function defined by $y = 1.4\sqrt{x - 2.5} + 87.5$ gives a good approximation of the year y when exports were x billion dollars. Here $y = 90$ represents 1990, $y = 93$ represents 1993, and so on. Use the equation to find the years in which exports were

(a) \$10 billion

(b) \$50 billion.

58. U.S. imports of electronics for several years are shown in the table.

U.S. Imports

Billions of Dollars	*Year*
11.0	1990
19.4	1993
25.9	1994
36.8	1996

Source: U.S. Bureau of the Census; *U.S. Merchandise Trade,* series FT 900, December issue; and unpublished data.

The year when U.S. imports were x billion dollars is closely approximated by the function $y = 1.6\sqrt{x - 6} + 87$, where $y = 90$ represents 1990, and so on, as in Exercise 57.

(a) Use the equation to find the years when U.S. imports were \$7 billion; \$30 billion.

(b) Using the tables in Exercises 57 and 58, find the differences between U.S. exports and U.S. imports in the years listed in both tables. How do they compare?

9.4 Rationalizing the Denominator

OBJECTIVES

1. Rationalize denominators with square roots.
2. Write radicals in simplified form.
3. Rationalize denominators with cube roots.

FOR EXTRA HELP

SSG Sec. 9.4
SSM Sec. 9.4
Pass the Test Software
InterAct Math Tutorial Software
Video 14

OBJECTIVE 1 Rationalize denominators with square roots. Fractions are simplified by rewriting them without any radical expressions in the denominator. For example, the radical in the denominator of

$$\frac{\sqrt{3}}{\sqrt{2}}$$

can be eliminated by multiplying the numerator and the denominator by $\sqrt{2}$.

$$\frac{\sqrt{3}}{\sqrt{2}} = \frac{\sqrt{3} \cdot \sqrt{2}}{\sqrt{2} \cdot \sqrt{2}} = \frac{\sqrt{6}}{2} \qquad \text{Since } \sqrt{2} \cdot \sqrt{2} = 2$$

This process of changing the denominator from a radical (irrational number) to a rational number is called **rationalizing the denominator.** The value of the number is not changed; only the form of the number is changed, because the expression has been multiplied by 1 in the form $\frac{\sqrt{2}}{\sqrt{2}}$.

EXAMPLE 1 Rationalizing Denominators

Rationalize each denominator.

(a) $\frac{9}{\sqrt{6}}$

Multiply both numerator and denominator by $\sqrt{6}$.

$$\frac{9}{\sqrt{6}} = \frac{9 \cdot \sqrt{6}}{\sqrt{6} \cdot \sqrt{6}}$$

$$= \frac{9\sqrt{6}}{6} \qquad \sqrt{6} \cdot \sqrt{6} = 6$$

$$= \frac{3\sqrt{6}}{2} \qquad \text{Lowest terms}$$

(b) $\frac{12}{\sqrt{8}}$

The denominator here could be rationalized by multiplying by $\sqrt{8}$. However, the result can be found more directly by first simplifying the denominator.

$$\sqrt{8} = \sqrt{4} \cdot \sqrt{2} = 2\sqrt{2}$$

Then multiply numerator and denominator by $\sqrt{2}$.

$$\frac{12}{\sqrt{8}} = \frac{12}{2\sqrt{2}}$$

$$= \frac{12 \cdot \sqrt{2}}{2\sqrt{2} \cdot \sqrt{2}} \qquad \text{Multiply by } \frac{\sqrt{2}}{\sqrt{2}}.$$

$$= \frac{12\sqrt{2}}{2 \cdot 2} \qquad \sqrt{2} \cdot \sqrt{2} = 2$$

$$= \frac{12\sqrt{2}}{4} \qquad \text{Multiply.}$$

$$= 3\sqrt{2} \qquad \text{Lowest terms}$$

OBJECTIVE 2 **Write radicals in simplified form.** A radical is considered to be in simplified form if the following three conditions are met.

Simplified Form of a Radical

1. All nth power factors of the radicand of $\sqrt[n]{\ }$ are removed. (For example, $\sqrt{3^2} = 3$, $\sqrt[3]{5^3} = 5$, and so on.)
2. The radicand has no fractions.
3. No denominator contains a radical.

In the following examples, radicals are simplified according to these conditions.

EXAMPLE 2 Simplifying a Radical with a Fraction

Simplify $\sqrt{\frac{27}{5}}$ by rationalizing the denominator.

First use the quotient rule for radicals.

$$\sqrt{\frac{27}{5}} = \frac{\sqrt{27}}{\sqrt{5}}$$

Now multiply both numerator and denominator by $\sqrt{5}$.

$$\frac{\sqrt{27}}{\sqrt{5}} = \frac{\sqrt{27}\cdot\sqrt{5}}{\sqrt{5}\cdot\sqrt{5}} \quad \text{Rationalize the denominator.}$$

$$= \frac{\sqrt{9\cdot 3}\cdot\sqrt{5}}{5} \quad \sqrt{5}\cdot\sqrt{5} = 5$$

$$= \frac{\sqrt{9}\cdot\sqrt{3}\cdot\sqrt{5}}{5} \quad \text{Product rule}$$

$$= \frac{3\cdot\sqrt{3\cdot 5}}{5} = \frac{3\sqrt{15}}{5} \quad \text{Product rule}$$

EXAMPLE 3 Simplifying a Product of Radicals

Simplify $\sqrt{\frac{5}{8}}\cdot\sqrt{\frac{1}{6}}$.

Use both the product and quotient rules.

$$\sqrt{\frac{5}{8}}\cdot\sqrt{\frac{1}{6}} = \sqrt{\frac{5}{8}\cdot\frac{1}{6}} \quad \text{Product rule}$$

$$= \sqrt{\frac{5}{48}} \quad \text{Multiply.}$$

$$= \frac{\sqrt{5}}{\sqrt{48}} \quad \text{Quotient rule}$$

OBJECTIVE 3 Write radical expressions with quotients in lowest terms. The final example shows this.

EXAMPLE 4 Writing a Radical Quotient in Lowest Terms

Write $\frac{3\sqrt{3} + 15}{12}$ in lowest terms.

Factor the numerator and denominator, and then divide numerator and denominator by any common factors.

$$\frac{3\sqrt{3} + 15}{12} = \frac{3(\sqrt{3} + 5)}{3 \cdot 4} = \frac{\sqrt{3} + 5}{4}$$

This technique is used in Chapter 10.

CAUTION

A common error is to reduce an expression like the one in Example 4 incorrectly to lowest terms before factoring. For example,

$$\frac{4 + 8\sqrt{5}}{4} \neq 1 + 8\sqrt{5}.$$

The correct simplification is $1 + 2\sqrt{5}$. Why?

9.5 EXERCISES

Based on the work so far, many simple operations involving radicals should now be performed mentally. In Exercises 1–8, perform the operations mentally, and write the answer without doing intermediate steps.

1. $\sqrt{49} + \sqrt{36}$
2. $\sqrt{100} - \sqrt{81}$
3. $\sqrt{2} \cdot \sqrt{8}$
4. $\sqrt{8} \cdot \sqrt{8}$
5. $\sqrt{2}(\sqrt{32} - \sqrt{8})$
6. $\sqrt{3}(\sqrt{27} - \sqrt{3})$
7. $\sqrt[3]{8} + \sqrt[3]{27}$
8. $\sqrt{4} - \sqrt[3]{64} + \sqrt[4]{16}$

Simplify each expression. Use the five guidelines given in the text. See Examples 1 and 2.

9. $3\sqrt{5} + 2\sqrt{45}$
10. $2\sqrt{2} + 4\sqrt{18}$
11. $8\sqrt{50} - 4\sqrt{72}$
12. $4\sqrt{80} - 5\sqrt{45}$
13. $\sqrt{5}(\sqrt{3} - \sqrt{7})$
14. $\sqrt{7}(\sqrt{10} + \sqrt{3})$
15. $2\sqrt{5}(\sqrt{2} + 3\sqrt{5})$
16. $3\sqrt{7}(2\sqrt{7} + 4\sqrt{5})$
17. $3\sqrt{14} \cdot \sqrt{2} - \sqrt{28}$
18. $7\sqrt{6} \cdot \sqrt{3} - 2\sqrt{18}$
19. $(2\sqrt{6} + 3)(3\sqrt{6} + 7)$
20. $(4\sqrt{5} - 2)(2\sqrt{5} - 4)$
21. $(5\sqrt{7} - 2\sqrt{3})(3\sqrt{7} + 4\sqrt{3})$
22. $(2\sqrt{10} + 5\sqrt{2})(3\sqrt{10} - 3\sqrt{2})$
23. $(2\sqrt{7} + 3)^2$
24. $(4\sqrt{5} + 5)^2$
25. $(5 - \sqrt{2})(5 + \sqrt{2})$
26. $(3 - \sqrt{5})(3 + \sqrt{5})$
27. $(\sqrt{8} - \sqrt{7})(\sqrt{8} + \sqrt{7})$
28. $(\sqrt{12} - \sqrt{11})(\sqrt{12} + \sqrt{11})$
29. $(\sqrt{2} + \sqrt{3})(\sqrt{6} - \sqrt{2})$
30. $(\sqrt{3} + \sqrt{5})(\sqrt{15} - \sqrt{5})$
31. $(\sqrt{10} - \sqrt{5})(\sqrt{5} + \sqrt{20})$
32. $(\sqrt{6} - \sqrt{3})(\sqrt{3} + \sqrt{18})$
33. $(5\sqrt{7} - 2\sqrt{3})(3\sqrt{7} + 3\sqrt{3})$
34. $(2\sqrt{10} + 5\sqrt{2})(3\sqrt{10} - 4\sqrt{2})$
35. In Example 1(b), the original expression simplifies to $-37 - 2\sqrt{15}$. Students often try to simplify expressions like this by combining the -37 and the -2 to get $-39\sqrt{15}$, which is incorrect. Explain why.

36. If you try to rationalize the denominator of $\frac{2}{4 + \sqrt{3}}$ by multiplying the numerator and denominator by $4 + \sqrt{3}$, what problem arises? What should you multiply by?

Rationalize each denominator. See Example 3.

37. $\frac{1}{3 + \sqrt{2}}$ **38.** $\frac{1}{4 - \sqrt{3}}$ **39.** $\frac{14}{2 - \sqrt{11}}$ **40.** $\frac{19}{5 - \sqrt{6}}$

41. $\frac{\sqrt{2}}{2 - \sqrt{2}}$ **42.** $\frac{\sqrt{7}}{7 - \sqrt{7}}$ **43.** $\frac{\sqrt{5}}{\sqrt{2} + \sqrt{3}}$ **44.** $\frac{\sqrt{3}}{\sqrt{2} + \sqrt{3}}$

45. $\frac{\sqrt{12}}{\sqrt{3} + 1}$ **46.** $\frac{\sqrt{18}}{\sqrt{2} - 1}$ **47.** $\frac{\sqrt{5} + 2}{2 - \sqrt{3}}$ **48.** $\frac{\sqrt{7} + 3}{4 - \sqrt{5}}$

Write each quotient in lowest terms. See Example 4.

49. $\frac{6\sqrt{11} - 12}{6}$ **50.** $\frac{12\sqrt{5} - 24}{12}$ **51.** $\frac{2\sqrt{3} + 10}{16}$

52. $\frac{4\sqrt{6} + 24}{20}$ **53.** $\frac{12 - \sqrt{40}}{4}$ **54.** $\frac{9 - \sqrt{72}}{12}$

Simplify each radical expression. Assume all variables represent nonnegative real numbers.

55. $(\sqrt{5x} + \sqrt{30})(\sqrt{6x} + \sqrt{3})$ **56.** $(\sqrt{10y} - \sqrt{20})(\sqrt{2y} - \sqrt{5})$

57. $(3\sqrt{t} + \sqrt{7})(2\sqrt{t} - \sqrt{14})$ **58.** $(2\sqrt{z} - \sqrt{3})(\sqrt{z} - \sqrt{5})$

59. $(\sqrt{3m} + \sqrt{2n})(\sqrt{5m} - \sqrt{5n})$ **60.** $(\sqrt{4p} - \sqrt{3k})(\sqrt{2p} + \sqrt{9k})$

61. $\sqrt[3]{4}(\sqrt[3]{2} - 3)$ **62.** $\sqrt[3]{5}(4\sqrt[3]{5} - \sqrt[3]{25})$

63. $2\sqrt[4]{2}(3\sqrt[4]{8} + 5\sqrt[4]{4})$ **64.** $6\sqrt[4]{9}(2\sqrt[4]{9} - \sqrt[4]{27})$

65. $(\sqrt[3]{2} - 1)(\sqrt[3]{4} + 3)$ **66.** $(\sqrt[3]{9} + 5)(\sqrt[3]{3} - 4)$

67. $(\sqrt[3]{5} - \sqrt[3]{4})(\sqrt[3]{25} + \sqrt[3]{20} + \sqrt[3]{16})$ **68.** $(\sqrt[3]{4} + \sqrt[3]{2})(\sqrt[3]{16} - \sqrt[3]{8} + \sqrt[3]{4})$

RELATING CONCEPTS (EXERCISES 69–74)

***Work Exercises 69–74 in order.** They are designed to help you see why a common student error is indeed an error.*

69. Use the distributive property to write $6(5 + 3x)$ as a sum.

70. Your answer in Exercise 69 should be $30 + 18x$. Why can we not combine these two terms to get $48x$?

71. Repeat Exercise 22 from earlier in this exercise set.

72. Your answer in Exercise 71 should be $30 + 18\sqrt{5}$. Many students will, in error, try to combine these terms to get $48\sqrt{5}$. Why is this wrong?

73. Write the expression similar to $30 + 18x$ that simplifies to $48x$. Then write the expression similar to $30 + 18\sqrt{5}$ that simplifies to $48\sqrt{5}$.

74. Write a short explanation of the similarities between combining like terms and combining like radicals.

Did you make the connection that the procedure used in combining radical terms is the same as that used in combining variable terms?

RELATING CONCEPTS (EXERCISES 63–68)

The rules for multiplying and dividing radicals presented earlier in this chapter were stated for radicals having the same index. For example, we only multiplied or divided square roots, or multiplied or divided cube roots, and so on. Since we know how to write radicals with fractional exponents and from past work know how to add and subtract fractions with different denominators, we can now multiply and divide radicals having different indexes.

Work Exercises 63–68 in order, *to see how to multiply* $\sqrt{2}$ *by* $\sqrt[3]{2}$.

63. Write $\sqrt{2}$ and $\sqrt[3]{2}$ using fractional exponents.

64. Write the product $\sqrt{2} \cdot \sqrt[3]{2}$ using the expressions you found in Exercise 63.

65. What is the least common denominator of the fractional exponents in Exercise 64?

66. Repeat Exercise 64, but write the fractional exponents with the common denominator from Exercise 65.

67. Use the rule for multiplying exponential expressions with like bases to simplify the product in Exercise 66. (*Hint:* The base remains the same; do not multiply the two bases.)

68. Write the answer you obtained in Exercise 67 as a radical.

Did you make the connection that when radicals are written in exponential form, the rules for exponents may be used to multiply (or divide) unlike radical terms?

Use the exponential key on your calculator to find the following roots. For example, to find $\sqrt[5]{32}$, *enter* 32 *and then raise to the* 1/5 *power. (The exponent* 1/5 *may be entered as* .2 *if you wish.) If the root is irrational, round it to the nearest thousandth.*

69. $\sqrt[6]{64}$ **70.** $\sqrt[5]{243}$ **71.** $\sqrt[7]{84}$ **72.** $\sqrt[9]{16}$

73. $\sqrt[5]{987}$ **74.** $\sqrt[6]{249}$ **75.** $\sqrt[4]{19^3}$ **76.** $\sqrt[5]{27^4}$

TECHNOLOGY INSIGHTS (EXERCISES 77–80)

Each calculator screen gives the simplified real number value (or an approximation) of an exponential expression.

Use a scientific calculator to verify that the radical expression equals the exponential expression.

77.

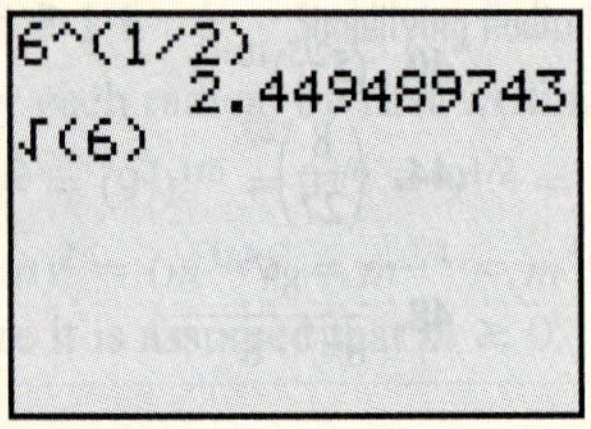

78.

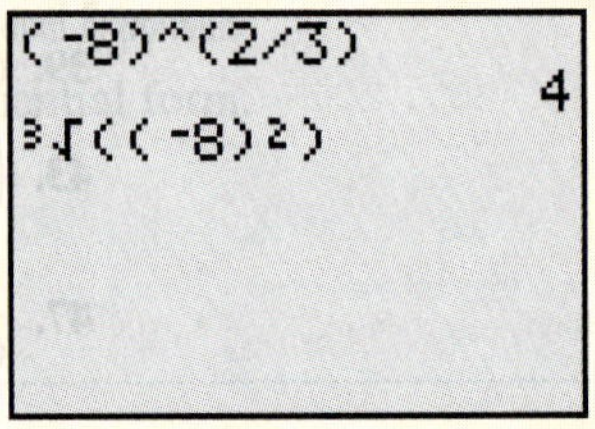

79.

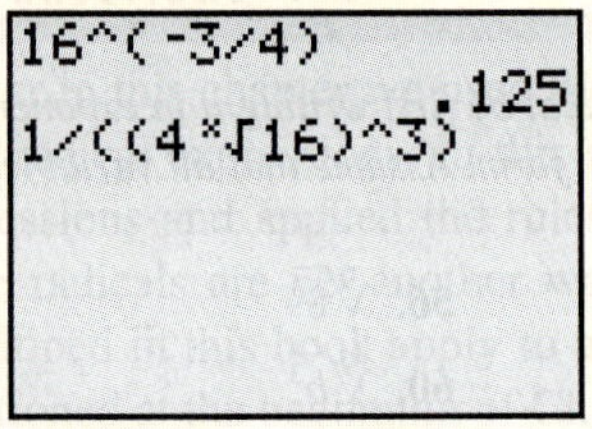

80.

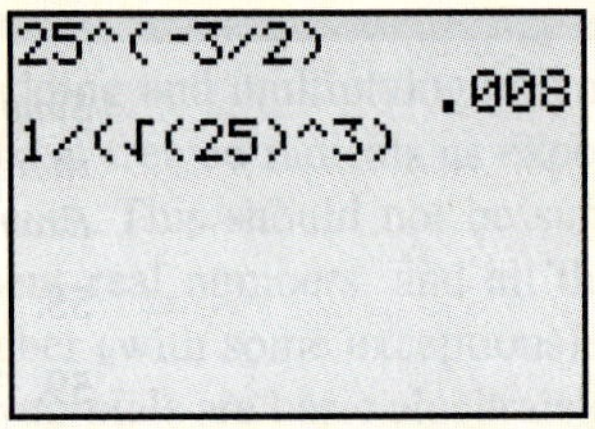

Solve each problem.

81. In Section 9.6, Exercise 50, we gave the formula $d = 1.22\sqrt{x}$ for calculating the distance in miles one can see from an airplane to the horizon on a clear day. Here, x is in feet.
(a) Write the formula using a fractional exponent.
(b) Find d to the nearest hundredth if the altitude x is 30,000 feet.

82. A biologist has shown that the number of different plant species S on a Galápagos Island is related to the area of the island, A (in miles), by

$$S = 28.6A^{1/3}.$$

How many plant species would exist on such an island with the following areas?
(a) 8 square miles **(b)** 27,000 square miles

83. Explain in your own words why $7^{1/2}$ is defined as $\sqrt{7}$.

84. Explain in your own words why $7^{1/3}$ is defined as $\sqrt[3]{7}$.

CHAPTER 9 GROUP ACTIVITY

Comparing Television Sizes

Objective: Use the Pythagorean formula to find the dimensions of different television sets.

Television sets are identified by the diagonal measurement of the viewing screen. For example, a 19-inch TV measures 19 inches from one corner of the viewing screen diagonally to the other corner.

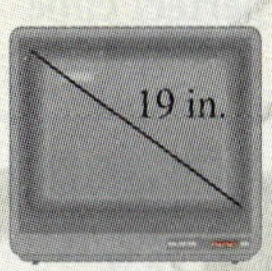

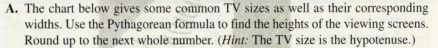

A. The chart below gives some common TV sizes as well as their corresponding widths. Use the Pythagorean formula to find the heights of the viewing screens. Round up to the next whole number. (*Hint:* The TV size is the hypotenuse.)

TV Set	*TV Size*	*Width in Inches*	*Height in Inches*
A	13-inch	10	
B	19-inch	15	
C	25-inch	21	
D	27-inch	22	
E	32-inch	25	
F	36-inch	30	

Source: Data based on information from a Sears catalog.

(continued)

CONCEPTS	EXAMPLES

9.3 ADDITION AND SUBTRACTION OF RADICALS

Add and subtract like radicals by using the distributive property. Only like radicals can be combined in this way.

$$2\sqrt{5} + 4\sqrt{5} = (2 + 4)\sqrt{5} = 6\sqrt{5}$$

$$\sqrt{8} + \sqrt{32} = 2\sqrt{2} + 4\sqrt{2} = 6\sqrt{2}$$

9.4 RATIONALIZING THE DENOMINATOR

The denominator of a radical can be rationalized by multiplying both the numerator and denominator by the same number.

$$\frac{2}{\sqrt{3}} = \frac{2 \cdot \sqrt{3}}{\sqrt{3} \cdot \sqrt{3}} = \frac{2\sqrt{3}}{3}$$

$$\sqrt[3]{\frac{5}{6}} = \frac{\sqrt[3]{5}}{\sqrt[3]{6}} \cdot \frac{\sqrt[3]{6^2}}{\sqrt[3]{6^2}} = \frac{\sqrt[3]{180}}{6}$$

9.5 SIMPLIFYING RADICAL EXPRESSIONS

When appropriate, use the rules for adding and multiplying polynomials to simplify radical expressions.

$$\sqrt{6}(\sqrt{5} - \sqrt{7}) = \sqrt{30} - \sqrt{42}$$

$$(\sqrt{5} - \sqrt{3})(\sqrt{5} + \sqrt{3}) = 5 - 3 = 2$$

Any denominators with radicals should be rationalized.

$$\frac{3}{\sqrt{6}} = \frac{3\sqrt{6}}{6} = \frac{\sqrt{6}}{2}$$

If a radical expression contains two terms in the denominator and at least one of those terms is a radical, multiply both the numerator and the denominator by the conjugate of the denominator.

$$\frac{6}{\sqrt{7} - \sqrt{2}} = \frac{6}{\sqrt{7} - \sqrt{2}} \cdot \frac{\sqrt{7} + \sqrt{2}}{\sqrt{7} + \sqrt{2}}$$

$$= \frac{6(\sqrt{7} + \sqrt{2})}{7 - 2} \quad \text{Multiply fractions.}$$

$$= \frac{6(\sqrt{7} + \sqrt{2})}{5} \quad \text{Simplify.}$$

9.6 SOLVING EQUATIONS WITH RADICALS

Solving an Equation with Radicals

Solve $\sqrt{2x - 3} + x = 3$.

Step 1 Arrange the terms so that a radical is alone on one side of the equation.

$$\sqrt{2x - 3} = 3 - x \quad \text{Isolate radical.}$$

Step 2 Square each side. (By the squaring property of equality, all solutions of the original equation are *among* the solutions of the squared equation.)

$$(\sqrt{2x - 3})^2 = (3 - x)^2 \quad \text{Square.}$$

$$2x - 3 = 9 - 6x + x^2$$

Step 3 Combine like terms.

$$0 = x^2 - 8x + 12 \quad \text{Get one side equal to 0.}$$

Step 4 If there is still a term with a radical, repeat Steps 1–3.

$$0 = (x - 2)(x - 6) \quad \text{Factor.}$$

$$x - 2 = 0 \quad \text{or} \quad x - 6 = 0 \quad \text{Set each factor equal to 0.}$$

Step 5 Solve the equation for potential solutions.

$$x = 2 \quad \text{or} \quad x = 6 \quad \text{Solve.}$$

Step 6 Check all potential solutions from Step 5 in the original equation.

Verify that 2 is the only solution (6 is extraneous).
Solution set: {2}

9.7 FRACTIONAL EXPONENTS

Assume $a \geq 0$, m and n are integers, $n > 0$.

$$a^{1/n} = \sqrt[n]{a}$$

$$a^{m/n} = \sqrt[n]{a^m} = (\sqrt[n]{a})^m$$

$$a^{-m/n} = \frac{1}{a^{m/n}} \quad (a \neq 0)$$

$$8^{1/3} = \sqrt[3]{8} = 2$$

$$(81)^{3/4} = \sqrt[4]{81^3} = (\sqrt[4]{81})^3 = 3^3 = 27$$

$$36^{-3/2} = \frac{1}{36^{3/2}} = \frac{1}{(36^{1/2})^3} = \frac{1}{6^3} = \frac{1}{216}$$

CHAPTER 9 REVIEW EXERCISES

[9.1] *Find all square roots of each number.*

1. 49 2. 81 3. 196 4. 121 5. 225 6. 729

Find each indicated root. If the root is not a real number, say so.

7. $\sqrt{16}$ 8. $-\sqrt{36}$ 9. $\sqrt[3]{1000}$ 10. $\sqrt[4]{81}$

11. $\sqrt{-8100}$ 12. $-\sqrt{4225}$ 13. $\sqrt{\frac{49}{36}}$ 14. $\sqrt{\frac{100}{81}}$

15. If $\sqrt{a}$ is not a real number, then what kind of number must a be?

16. Find the value of x in the figure.

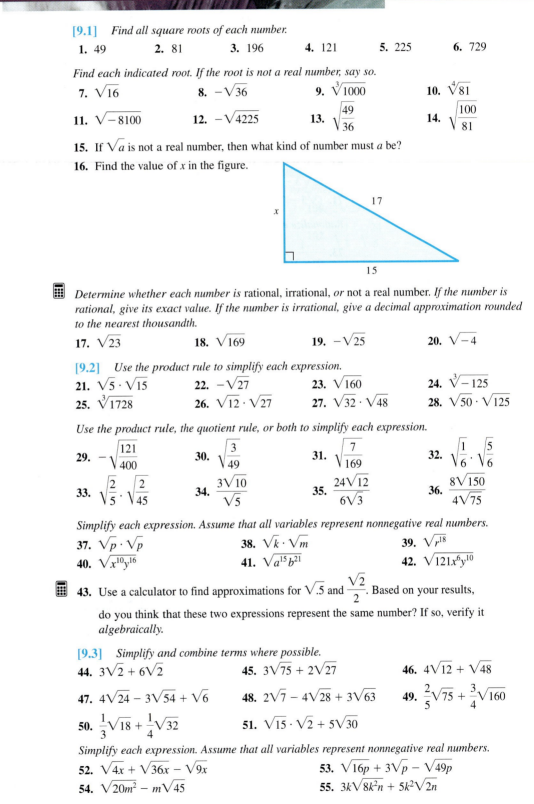

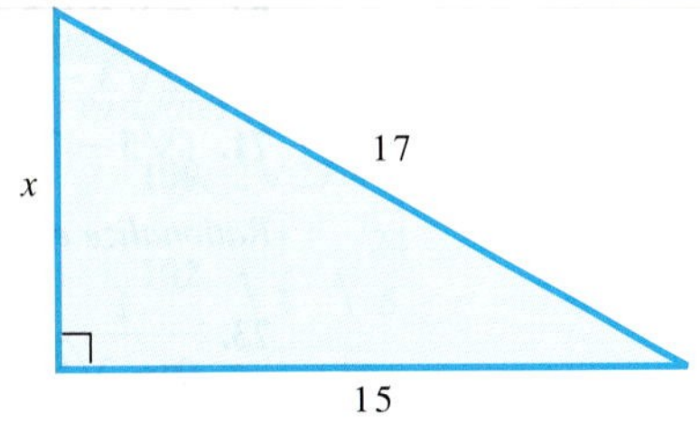

Determine whether each number is rational, irrational, *or* not a real number. *If the number is rational, give its exact value. If the number is irrational, give a decimal approximation rounded to the nearest thousandth.*

17. $\sqrt{23}$ 18. $\sqrt{169}$ 19. $-\sqrt{25}$ 20. $\sqrt{-4}$

[9.2] *Use the product rule to simplify each expression.*

21. $\sqrt{5}\cdot\sqrt{15}$ 22. $-\sqrt{27}$ 23. $\sqrt{160}$ 24. $\sqrt[3]{-125}$

25. $\sqrt[3]{1728}$ 26. $\sqrt{12}\cdot\sqrt{27}$ 27. $\sqrt{32}\cdot\sqrt{48}$ 28. $\sqrt{50}\cdot\sqrt{125}$

Use the product rule, the quotient rule, or both to simplify each expression.

29. $-\sqrt{\frac{121}{400}}$ 30. $\sqrt{\frac{3}{49}}$ 31. $\sqrt{\frac{7}{169}}$ 32. $\sqrt{\frac{1}{6}}\cdot\sqrt{\frac{5}{6}}$

33. $\sqrt{\frac{2}{5}}\cdot\sqrt{\frac{2}{45}}$ 34. $\frac{3\sqrt{10}}{\sqrt{5}}$ 35. $\frac{24\sqrt{12}}{6\sqrt{3}}$ 36. $\frac{8\sqrt{150}}{4\sqrt{75}}$

Simplify each expression. Assume that all variables represent nonnegative real numbers.

37. $\sqrt{p}\cdot\sqrt{p}$ 38. $\sqrt{k}\cdot\sqrt{m}$ 39. $\sqrt{r^{18}}$

40. $\sqrt{x^{10}y^{16}}$ 41. $\sqrt{a^{15}b^{21}}$ 42. $\sqrt{121x^6y^{10}}$

43. Use a calculator to find approximations for $\sqrt{.5}$ and $\frac{\sqrt{2}}{2}$. Based on your results, do you think that these two expressions represent the same number? If so, verify it *algebraically.*

[9.3] *Simplify and combine terms where possible.*

44. $3\sqrt{2}+6\sqrt{2}$ 45. $3\sqrt{75}+2\sqrt{27}$ 46. $4\sqrt{12}+\sqrt{48}$

47. $4\sqrt{24}-3\sqrt{54}+\sqrt{6}$ 48. $2\sqrt{7}-4\sqrt{28}+3\sqrt{63}$ 49. $\frac{2}{5}\sqrt{75}+\frac{3}{4}\sqrt{160}$

50. $\frac{1}{3}\sqrt{18}+\frac{1}{4}\sqrt{32}$ 51. $\sqrt{15}\cdot\sqrt{2}+5\sqrt{30}$

Simplify each expression. Assume that all variables represent nonnegative real numbers.

52. $\sqrt{4x}+\sqrt{36x}-\sqrt{9x}$ 53. $\sqrt{16p}+3\sqrt{p}-\sqrt{49p}$

54. $\sqrt{20m^2}-m\sqrt{45}$ 55. $3k\sqrt{8k^2n}+5k^2\sqrt{2n}$

A complex number written in the form $a + bi$ (or $a + ib$) is in **standard form.** Figure 2 shows the relationships among the various types of numbers discussed in this book. (Compare this figure to Figure 8 in Chapter 1.)

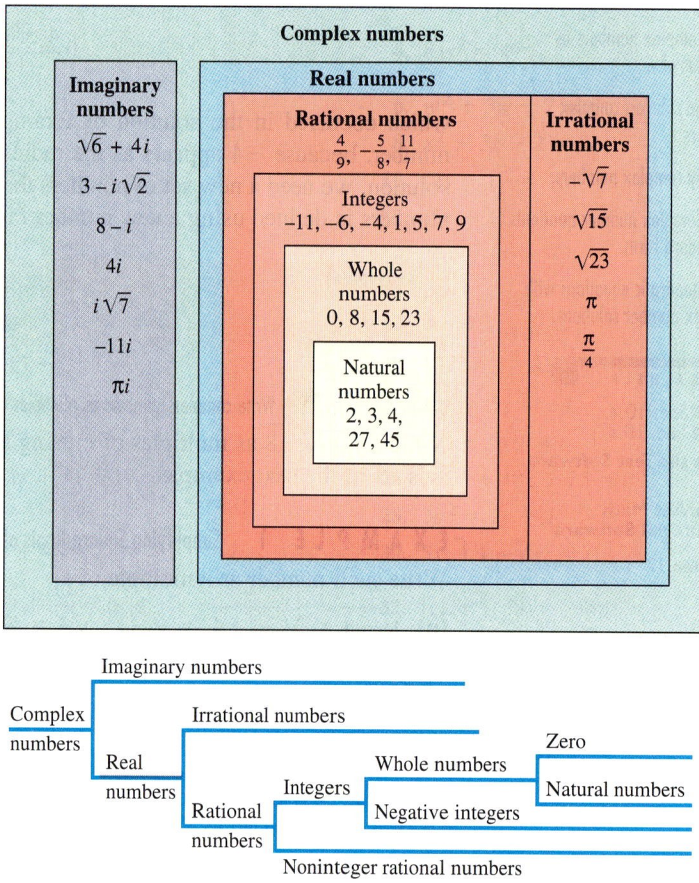

Figure 2

OBJECTIVE 2 Add and subtract complex numbers. Adding and subtracting complex numbers is similar to adding and subtracting binomials.

Addition and Subtraction of Complex Numbers

1. To add complex numbers, add their real parts and add their imaginary parts.
2. To subtract complex numbers, change the number following the subtraction sign to its negative, and then add.

The properties of Section 1.7 (commutative, associative, etc.) also hold for operations with complex numbers.

EXAMPLE 2 Adding and Subtracting Complex Numbers

Add or subtract.

(a) $(2 - 6i) + (7 + 4i) = (2 + 7) + (-6 + 4)i = 9 - 2i$

(b) $3i + (-2 - i) = -2 + (3 - 1)i = -2 + 2i$

(c) $(2 + 6i) - (-4 + i)$
Change $-4 + i$ to its negative, and then add.

$$\begin{aligned}(2 + 6i) - (-4 + i) &= (2 + 6i) + (4 - i) && -(-4 + i) = 4 - i \\ &= (2 + 4) + (6 - 1)i && \text{Commutative, associative, and distributive properties} \\ &= 6 + 5i\end{aligned}$$

(d) $(-1 + 2i) - 4 = (-1 - 4) + 2i = -5 + 2i$

OBJECTIVE 3 **Multiply complex numbers.** We multiply complex numbers as we do polynomials. Since $i^2 = -1$ by definition, whenever i^2 appears, we replace it with -1.

EXAMPLE 3 **Multiplying Complex Numbers**

Find the following products.

(a)
$$\begin{aligned}3i(2 - 5i) &= 6i - 15i^2 && \text{Distributive property} \\ &= 6i - 15(-1) && i^2 = -1 \\ &= 6i + 15 \\ &= 15 + 6i && \text{Commutative property}\end{aligned}$$

The last step gives the result in standard form.

(b) $(4 - 3i)(2 + 5i)$
Use FOIL.

$$\begin{aligned}(4 - 3i)(2 + 5i) &= 4(2) + 4(5i) + (-3i)(2) + (-3i)(5i) \\ &= 8 + 20i - 6i - 15i^2 \\ &= 8 + 14i - 15(-1) \\ &= 8 + 14i + 15 \\ &= 23 + 14i\end{aligned}$$

(c)
$$\begin{aligned}(1 + 2i)(1 - 2i) &= 1 - 2i + 2i - 4i^2 \\ &= 1 - 4(-1) \\ &= 1 + 4 \\ &= 5\end{aligned}$$

OBJECTIVE 4 **Write complex number quotients in standard form.** The quotient of two complex numbers is expressed in standard form by changing the denominator into a real number. For example, to write

$$\frac{8 + i}{1 + 2i}$$

in standard form, the denominator must be a real number. As seen in Example 3(c), the product $(1 + 2i)(1 - 2i)$ is 5, a real number. This suggests multiplying the numerator and denominator of the given quotient by $1 - 2i$ as follows.

$$\frac{8+i}{1+2i} = \frac{8+i}{1+2i} \cdot \frac{1-2i}{1-2i}$$

$$= \frac{8-16i+i-2i^2}{1-4i^2} \qquad \text{Multiply.}$$

$$= \frac{8-16i+i-2(-1)}{1-4(-1)} \qquad i^2 = -1$$

$$= \frac{10-15i}{5} \qquad \text{Combine terms.}$$

$$= \frac{5(2-3i)}{5} = 2-3i \qquad \text{Factor and write in standard form.}$$

Recall that this is the method used to rationalize some radical expressions in Chapter 9. The complex numbers $1 + 2i$ and $1 - 2i$ are *conjugates.* That is, the **conjugate** of the complex number $a + bi$ is $a - bi$. Multiplying the complex number $a + bi$ by its conjugate $a - bi$ gives the real number $a^2 + b^2$.

Product of Conjugates

$$(a+bi)(a-bi) = a^2 + b^2$$

The product of a complex number and its conjugate is the sum of the squares of the real and imaginary parts.

EXAMPLE 4 **Dividing Complex Numbers**

Write the following quotients in standard form.

(a) $\dfrac{-4+i}{2-i}$

Multiply numerator and denominator by $2 + i$, the conjugate of the denominator.

$$\frac{-4+i}{2-i} \cdot \frac{2+i}{2+i} = \frac{-8-4i+2i+i^2}{4-i^2}$$

$$= \frac{-8-2i-1}{4-(-1)} \qquad i^2 = -1$$

$$= \frac{-9-2i}{5}$$

$$= -\frac{9}{5} - \frac{2}{5}i \qquad \text{Standard form}$$

(b) $\dfrac{3+i}{-i}$

Here, the conjugate of $0 - i$ is $0 + i$, or i.

$$\frac{3+i}{-i} \cdot \frac{i}{i} = \frac{3i+i^2}{-i^2}$$

$$= \frac{-1+3i}{-(-1)} \qquad i^2 = -1;\ \text{commutative property}$$

$$= -1 + 3i$$

CONNECTIONS

The complex number $a + bi$ is also written with the notation $\langle a, b \rangle$. (Note the similarity to an ordered pair.) This notation suggests a way to graph complex numbers on a plane in a manner similar to the way we graph ordered pairs. For graphing complex numbers, the x-axis is called the *real axis* and the y-axis is called the *imaginary axis.* For example, we graph the complex number $2 + 3i$ or $\langle 2, 3 \rangle$ just as we would the ordered pair (2, 3), as shown in the figure. The figure also shows the graphs of the complex numbers $-1 - 4i$, $2i$, and -5.

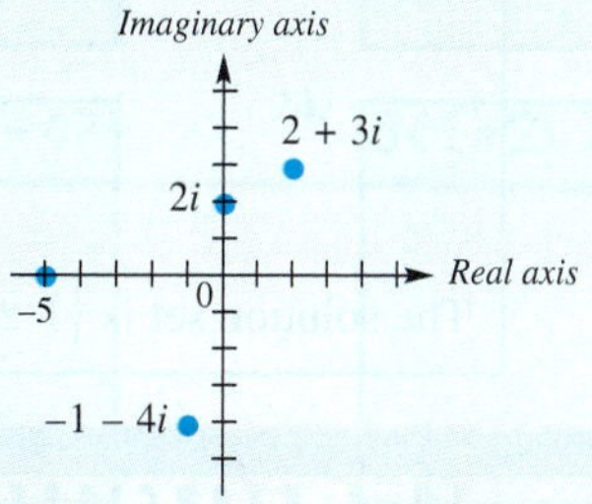

FOR DISCUSSION OR WRITING

1. Give the alternative notation for the last three complex numbers graphed above.
2. What is the real part of the complex number $-1 + 2i$? What is its imaginary part? Explain why we call the axes the real axis and the imaginary axis.

OBJECTIVE 5 **Solve quadratic equations with complex number solutions.** Quadratic equations that have no real solutions do have complex solutions, as shown in the next examples.

EXAMPLE 5 **Solving a Quadratic Equation with Complex Solutions (Square Root Method)**

Solve $(x + 3)^2 = -25$ for complex solutions.

Use the square root property.

$$(x + 3)^2 = -25$$

$$x + 3 = \sqrt{-25} \quad \text{or} \quad x + 3 = -\sqrt{-25}$$

Since $\sqrt{-25} = 5i$,

$$x + 3 = 5i \quad \text{or} \quad x + 3 = -5i$$

$$x = -3 + 5i \quad \text{or} \quad x = -3 - 5i.$$

The solution set is $\{-3 \pm 5i\}$.

EXAMPLE 6 **Solving a Quadratic Equation with Complex Solutions (Quadratic Formula)**

Solve $2p^2 = 4p - 5$ for complex solutions.

Write the equation as $2p^2 - 4p + 5 = 0$. Then $a = 2$, $b = -4$, and $c = 5$. The solutions are

$$p = \frac{-(-4) \pm \sqrt{(-4)^2 - 4(2)(5)}}{2(2)}$$

$$= \frac{4 \pm \sqrt{16 - 40}}{4}$$

$$= \frac{4 \pm \sqrt{-24}}{4}.$$

(c) Figure 10 shows the graph of $f(x) = x^2 + 2$. The equation $0 = x^2 + 2$ has no real solutions, since there are no x-intercepts. The solution set over the domain of real numbers is $\emptyset$. (The equation *does* have two imaginary solutions, $i\sqrt{2}$ and $-i\sqrt{2}$.)

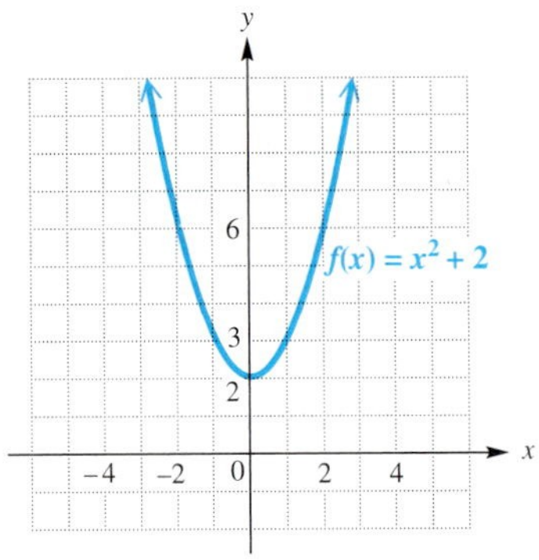

Figure 10

OBJECTIVE 4 **Solve applications using quadratic functions.** Because we can determine the coordinates of the vertex of the graph of a quadratic function, we are able to find the x-value that leads to the maximum or minimum y-value. This fact allows us to solve applications that lead to quadratic functions.

EXAMPLE 4 Solving a Problem Involving a Rectangular Region

A farmer wishes to enclose a rectangular region. He has 240 feet of fencing, and plans to use one side of his barn as part of the enclosure. See Figure 11. What dimensions should he use so that the enclosed region has maximum area? What will this maximum area be?

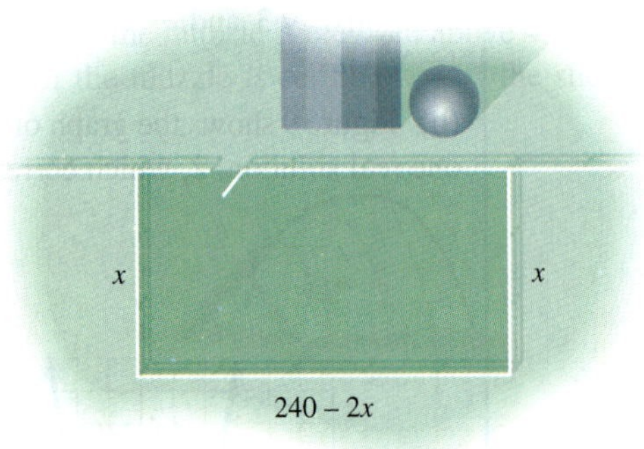

Figure 11

Let x represent the length of each of the two parallel sides of the enclosure. Then $240 - 2x$ represents the length of the third side formed from the fencing. Because area = length × width, the area of the region can be represented by the quadratic function

$$A(x) = x(240 - 2x)$$
$$A(x) = -2x^2 + 240x.$$

The graph of this function is a parabola that opens downward (because of the negative coefficient on x^2). The vertex is the highest point on the graph. Here, $a = -2$ and $b = 240$, so the x-coordinate of the vertex is

$$x = -\frac{b}{2a} = -\frac{240}{2(-2)} = 60.$$

So each of the two parallel sides of fencing should measure 60 feet. The remaining side will be $240 - 2(60) = 240 - 120 = 120$ feet long. The area of the region will be $60 \times 120 = 7200$ square feet. (This area can also be found by evaluating $A(60)$.)

Parabolic shapes are found all around us. Satellite dishes that deliver television signals are becoming more popular each year. Radio telescopes use parabolic reflectors to track incoming signals. The final example discusses how to describe a cross section of a parabolic dish using an equation.

EXAMPLE 5 Finding the Equation of a Parabolic Satellite Dish

The Parkes radio telescope has a parabolic dish shape with a diameter of 210 feet and a depth of 32 feet. (*Source:* J. Mar and H. Liebowitz, *Structure Technology for Large Radio and Radar Telescope Systems,* The MIT Press, Cambridge, MA, 1969.) Figure 12(a) shows a diagram of such a dish, and Figure 12(b) shows how a cross section of the dish can be modeled by a graph, with the vertex of the parabola at the origin of a coordinate system. Find the equation of this graph.

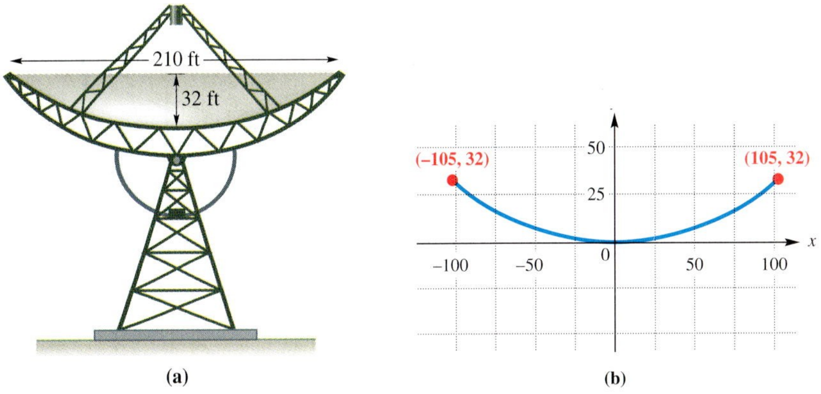

Figure 12

Because the vertex is at the origin, the equation will be of the form $y = ax^2$. As shown in Figure 12(b), one point on the graph has coordinates (105, 32). Letting $x = 105$ and $y = 32$, we can solve for a.

$y = ax^2$	General equation
$32 = a(105)^2$	Substitute for x and y.
$32 = 11{,}025a$	$105^2 = 11{,}025$
$a = \frac{32}{11{,}025}$	Divide by 11,025.

Thus the equation is $y = \frac{32}{11{,}025}x^2$.

Answers to Selected Exercises

TO THE STUDENT

If you need further help with algebra, you may want to obtain a copy of the *Student's Solution Manual* that goes with this book. You also may want the *Student's Study Guide.* It has extra examples and exercises to complete, corresponding to each learning objective of the book. In addition, there is a practice test for each chapter. Your college bookstore either has these books or can order them for you.

In this section we provide the answers that we think most students will obtain when they work the exercises using the methods explained in the text. If your answer does not look exactly like the one given here, it is not necessarily wrong. In many cases there are equivalent forms of the answer that are correct. For example, if the answer section shows $\frac{3}{4}$ and your answer is .75, you have obtained the right answer but written it in a different (yet equivalent) form. Unless the directions specify otherwise, .75 is just as valid an answer as $\frac{3}{4}$.

In general, if your answer does not agree with the one given in the text, see whether it can be transformed into the other form. If it can, then it is the correct answer. If you still have doubts, talk with your instructor.

CHAPTER 1 THE REAL NUMBER SYSTEM

SECTION 1.1 (PAGE 9)

CONNECTIONS **Page 2:** Answers will vary.

EXERCISES **1.** true **3.** false; The fraction $\frac{17}{51}$ can be reduced to $\frac{1}{3}$. **5.** false; *Product* refers to multiplication, so the product of 8 and 2 is 16. **7.** prime **9.** composite; $2 \cdot 2 \cdot 2 \cdot 2 \cdot 2 \cdot 2$ **11.** composite; $2 \cdot 7 \cdot 13 \cdot 19$ **13.** neither **15.** composite; $2 \cdot 3 \cdot 5$ **17.** composite; $2 \cdot 2 \cdot 5 \cdot 5 \cdot 5$ **19.** composite; $2 \cdot 2 \cdot 31$ **21.** prime **23.** $\frac{1}{2}$ **25.** $\frac{5}{6}$ **27.** $\frac{1}{3}$ **29.** $\frac{6}{5}$ **31.** (c) **33.** $\frac{24}{35}$ **35.** $\frac{6}{25}$ **37.** $\frac{6}{5}$ or $1\frac{1}{5}$ **39.** $\frac{232}{15}$ or $15\frac{7}{15}$ **41.** $\frac{10}{3}$ or $3\frac{1}{3}$ **43.** 12 **45.** $\frac{1}{16}$ **47.** $\frac{84}{47}$ or $1\frac{37}{47}$ **51.** $\frac{2}{3}$ **53.** $\frac{8}{9}$ **55.** $\frac{27}{8}$ or $3\frac{3}{8}$ **57.** $\frac{17}{36}$ **59.** $\frac{11}{12}$ **61.** $\frac{4}{3}$ or $1\frac{1}{3}$ **63.** 6 cups **65.** 34 dollars **67.** $\frac{9}{16}$ inch **69.** $618\frac{3}{4}$ feet **71.** $5\frac{5}{24}$ inches **73.** $\frac{1}{3}$ cup **75.** 650 **77.** **(a)** Crum **(b)** Jordan **(c)** Jordan **(d)** Baldock **(e)** Tobin and Perry; $\frac{1}{2}$ **79.** (b)

SECTION 1.2 (PAGE 19)

EXERCISES **1.** false; $4 + 3(8 - 2) = 4 + 3 \cdot 6 = 4 + 18 = 22$. The common error leading to 42 is adding 4 to 3 and then multiplying by 6. One must follow the rules for order of operations. **3.** false; The correct interpretation is $4 = 16 - 12$.

5. 49 **7.** 144 **9.** 64 **11.** 1000 **13.** 81 **15.** 1024 **17.** $\frac{16}{81}$ **19.** .000064 **23.** 32 **25.** $\frac{49}{30}$ or $1\frac{19}{30}$ **27.** 12 **29.** 23.01 **31.** 95 **33.** 90 **35.** 14 **37.** 9 **41.** true **43.** false **45.** true **47.** true **49.** false **51.** false **53.** true **55.** $15 = 5 + 10$ **57.** $9 > 5 - 4$ **59.** $16 \neq 19$ **61.** $2 \leq 3$ **63.** Seven is less than nineteen. True **65.** Three is not equal to six. True **67.** Eight is greater than or equal to eleven. False **69.** Answers will vary. One example is $5 + 3 \geq 2 \cdot 2$. **71.** $30 > 5$ **73.** $3 \leq 12$ **75.** is younger than **77.** The inequality symbol $\geq$ implies a true statement if 12 equals 12 *or* if 12 is greater than 12. **79.** December 1996, January 1997, May 1997, June 1997, November 1997 **81.** **(a)** .7 **(b)** 1.3%

SECTION 1.3 (PAGE 25)

EXERCISES **1.** 10 **3.** $12 + x$; 21 **5.** no **7.** $2x^3 = 2 \cdot x \cdot x \cdot x$, while $2x \cdot 2x \cdot 2x = (2x)^3$. **9.** The exponent 2 applies only to its base, which is x. (The expression $(4x)^2$ would require multiplying 4 by $x = 3$ first.) **11.** (Answers will vary.) Two such pairs are $x = 0$, $y = 6$ and $x = 1$, $y = 4$. To determine them, choose a value for x, substitute it into the expression $2x + y$, and then subtract the value of $2x$ from 6. **13.** **(a)** 13 **(b)** 15 **15.** **(a)** 20 **(b)** 30 **17.** **(a)** 64 **(b)** 144 **19.** **(a)** $\frac{5}{3}$ **(b)** $\frac{7}{3}$ **21.** **(a)** $\frac{7}{8}$ **(b)** $\frac{13}{12}$ **23.** **(a)** 52 **(b)** 114 **25.** **(a)** 25.836 **(b)** 38.754 **27.** **(a)** 24 **(b)** 28 **29.** **(a)** 12 **(b)** 33 **31.** **(a)** 6 **(b)** $\frac{9}{5}$ **33.** **(a)** $\frac{4}{3}$ **(b)** $\frac{13}{6}$ **35.** **(a)** $\frac{2}{7}$ **(b)** $\frac{16}{27}$ **37.** **(a)** 12 **(b)** 55 **39.** **(a)** 1 **(b)** $\frac{28}{17}$ **41.** **(a)** 3.684 **(b)** 8.841 **43.** $12x$ **45.** $x + 7$ **47.** $x - 2$ **49.** $7 - x$ **51.** $x - 6$ **53.** $\frac{12}{x}$ **55.** $6(x - 4)$ **57.** No, it is a connective word that joins the two factors: the number and 6. **59.** yes **61.** no **63.** yes **65.** yes **67.** yes **69.** $x + 8 = 18$; 10 **71.** $16 - \frac{3}{4}x = 13$; 4 **73.** $2x + 1 = 5$; 2 **75.** $3x = 2x + 8$; 8 **77.** expression **79.** equation **81.** equation **83.** \$10.50; less by \$.33 **85.** \$12.10; less by \$.27

SECTION 1.4 (PAGE 34)

EXERCISES **1.** 4 **3.** 0 **5.** One example is $\sqrt{12}$. There are others. **7.** true **9.** true **11.** **(a)** 3, 7 **(b)** 0, 3, 7 **(c)** $-9, 0, 3, 7$ **(d)** $-9, -1\frac{1}{4}, -\frac{3}{5}, 0, 3, 5.9, 7$ **(e)** $-\sqrt{7}, \sqrt{5}$ **(f)** All are real numbers. **15.** 93,000 **17.** $-30°$ **19.** $-31,532$ **21.** -8 **23.**

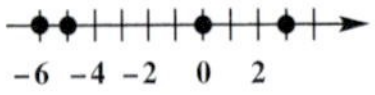

25.

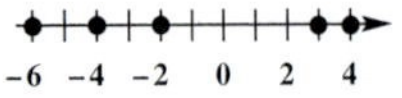

27.

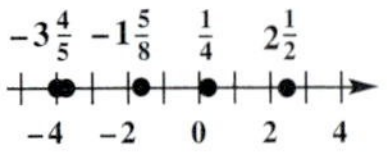

29. **(a)** A **(b)** A **(c)** B **(d)** B **31.** **(a)** 2 **(b)** 2 **33.** **(a)** -6 **(b)** 6 **35.** **(a)** -3 **(b)** 3 **37.** **(a)** 0 **(b)** 0 **39.** $a - b$ **41.** -12 **43.** -8 **45.** 3 **47.** $|-3|$ or 3 **49.** $-|-6|$ or -6 **51.** $|5 - 3|$ or 2 **53.** true **55.** true **57.** true **59.** false **61.** true **63.** false **65.** Softwood plywood from 1995 to 1996 represents the greatest drop. **67.** true

Answers will vary in Exercises 69–77. **69.** true: $a = 0$ or $b = 0$ or both $a = 0$ and $b = 0$; false: Choose any values for a and b so that neither a nor b is zero **71.** true: Choose a to be the opposite of b ($a = -b$); false: $a \neq -b$ **73.** $\frac{1}{2}, \frac{5}{8}, 1\frac{3}{4}$ **75.** $-3\frac{1}{2}, -\frac{2}{3}, \frac{3}{7}$ **77.** $\sqrt{5}, \pi, -\sqrt{3}$

SECTION 1.5 (PAGE 44)

EXERCISES **1.** negative **3.** -3; 5 **7.** 2 **9.** -3 **11.** -10 **13.** -13 **15.** -15.9 **17.** -1 **19.** 13 **21.** -3 **23.** -4 **25.** -10 **27.** -16 **29.** 11 **31.** 19 **33.** -4 **35.** 5 **37.** 0 **39.** $\frac{3}{4}$ **41.** -8 **43.** $\frac{15}{8}$ **45.** -6.3 **47.** -24 **49.** -16 **51.** no **53.** Answers will vary. One example is $-8 - (-2) = -6$. **55.** -1 **57.** $\frac{17}{9}$ or $1\frac{8}{9}$ **59.** $-5 + 12 + 6$; 13 **61.** $[-19 + (-4)] + 14$; -9 **63.** $[-4 + (-10)] + 12$; -2

65. $[8 + (-18)] + 4; -6$ **67.** $4 - (-8); 12$ **69.** $-2 - 8; -10$ **71.** $[9 + (-4)] - 7; -2$ **73.** $[8 - (-5)] - 12; 1$ **75.** \$13.2 billion **77.** $-$\$26.0 billion **79.** 50,395 feet **81.** 1345 feet **83.** **(a)** -10 **(b)** 5 **(c)** -12 **(d)** -31 **85.** 45°F **87.** -41°F **89.** 14,776 feet **91.** 365 pounds **93.** \$323.83

SECTION 1.6 (PAGE 57)

EXERCISES **1.** greater than 0 **3.** greater than 0 **5.** less than 0 **7.** greater than 0 **9.** equal to 0 **11.** 20 **13.** -28 **15.** 80 **17.** 0 **19.** $\frac{5}{6}$ **21.** $\frac{3}{2}$ **23.** $-32, -16, -8, -4, -2, -1, 1, 2, 4, 8, 16, 32$ **25.** $-40, -20, -10, -8, -5, -4, -2, -1, 1, 2, 4, 5, 8, 10, 20, 40$ **27.** $-31, -1, 1, 31$ **29.** -3 **31.** -2 **33.** 16 **35.** 0 **37.** 25.63 **39.** $\frac{3}{2}$ **43.** -11 **45.** -2 **47.** 35 **49.** 6 **51.** -18 **53.** 67 **55.** -8 **57.** 3 **59.** 7 **61.** 4 **63.** -3 **65.** First, substitute -3 for x and 4 for y to get $3(-3) + 2(4)$. Then perform the multiplications to get $-9 + 8$. Finally, add to get -1. **67.** 47 **69.** 72 **71.** $-\frac{78}{25}$ **73.** 0 **75.** -23 **77.** 2 **79.** $9 + (-9)(2); -9$ **81.** $-4 - 2(-1)(6); 8$ **83.** $(1.5)(-3.2) - 9; -13.8$ **85.** $12[9 - (-8)]; 204$ **87.** $\frac{-12}{-5 + (-1)}; 2$ **89.** $\frac{15 + (-3)}{4(-3)}; -1$ **91.** $\frac{\left(-\frac{1}{2}\right)\left(\frac{3}{4}\right)}{-\frac{2}{3}}; \frac{9}{16}$ **93.** $6x = -42; -7$ **95.** $\frac{x}{3} = -3; -9$ **97.** $x - 6 = 4; 10$ **99.** $x + 5 = -5; -10$ **101.** $8\frac{2}{5}$ **103.** 2 **105.** \$12.60 (to the nearest cent) **107.** 0 **109.** **(a)** 6 is divisible by 2. **(b)** 9 is not divisible by 2. **111.** **(a)** 64 is divisible by 4. **(b)** 35 is not divisible by 4. **113.** **(a)** 2 is divisible by 2 and $1 + 5 + 2 + 4 + 8 + 2 + 2 = 24$ is divisible by 3. **(b)** While 0 is divisible by 2, $2 + 8 + 7 + 3 + 5 + 9 + 0 = 34$ is not divisible by 3. **115.** **(a)** $4 + 1 + 1 + 4 + 1 + 0 + 7 = 18$ is divisible by 9. **(b)** $2 + 2 + 8 + 7 + 3 + 2 + 1 = 25$ is not divisible by 9.

SECTION 1.7 (PAGE 67)

EXERCISES **1.** B **3.** C **5.** B **7.** B **9.** G **11.** commutative property **13.** associative property **15.** associative property **17.** inverse property **19.** inverse property **21.** identity property **23.** commutative property **25.** distributive property **27.** identity property **29.** distributive property **31.** identity property **35.** $7 + r$ **37.** s **39.** $-6x + (-6) \cdot 7; -6x - 42$ **41.** $w + [5 + (-3)]; w + 2$ **43.** 11 **45.** 0 **47.** $-.38$ **49.** 1 **51.** Subtraction is not associative. **53.** The expression following the first equals sign should be $-3(4) -3(-6)$. The student forgot that 6 should be preceded by a $-$ sign. The correct work is $-3(4 - 6) = -3(4) - 3(-6) = -12 + 18 = 6$. **55.** $(5 + 1)x; 6x$ **57.** $4t + 12$ **59.** $-8r - 24$ **61.** $-5y + 20$ **63.** $-16y - 20z$ **65.** $8(z + w)$ **67.** $7(2v + 5r)$ **69.** $24r + 32s - 40y$ **71.** $(1 + 1 + 1)q; 3q$ **73.** $(-5 + 1)x; -4x$ **75.** $-4t - 3m$ **77.** $5c + 4d$ **79.** $3q - 5r + 8s$ **81.** for example, "putting on your socks" and "putting on your shoes" **83.** 0 **84.** $-3(5) + (-3)(-5)$ **85.** -15 **86.** We must interpret $(-3)(-5)$ as 15, since it is the additive inverse of -15.

SECTION 1.8 (PAGE 73)

EXERCISES **1.** false **3.** true **5.** (c) **7.** (a) **9.** $4r + 11$ **11.** $5 + 2x - 6y$ **13.** $-7 + 3p$ **15.** -12 **17.** 5 **19.** 1 **21.** -1 **23.** 74 **25.** Answers will vary. For example, $-3x$ and $4x$. **27.** like **29.** unlike **31.** like **33.** unlike **35.** Apples and oranges are examples of unlike fruits, just like x and y are unlike terms. We cannot add x and y to get an expression any simpler than $x + y$; we cannot add, for example, 2 apples and 3 oranges to obtain 5 fruits that are all alike. **37.** $9k - 5$ **39.** $-\frac{1}{3}t - \frac{28}{3}$ **41.** $-4.1r + 5.6$ **43.** $-2y^2 + 3y^3$ **45.** $-19p + 16$ **47.** $-4y + 22$ **49.** $-16y + 63$ **51.** $4k - 7$ **53.** $-23.7y - 12.6$ **55.** $(x + 3) + 5x; 6x + 3$ **57.** $(13 + 6x) - (-7x); 13 + 13x$ **59.** $2(3x + 4) - (-4 + 6x); 12$ **61.** Wording will vary. One example is "the difference between 9 times a number and the sum of the number and 2." **63.** 2, 3, 4, 5 **64.** 1 **65.** **(a)** 1, 2, 3, 4 **(b)** 3, 4, 5, 6 **(c)** 4, 5, 6, 7 **66.** The value of $x + b$ also increases by 1 unit. **67.** **(a)** 2, 4, 6, 8 **(b)** 2, 5, 8, 11 **(c)** 2, 6, 10, 14 **68.** m **69.** **(a)** 7, 9, 11, 13 **(b)** 5, 8, 11, 14 **(c)** 1, 5, 9, 13 In comparison, we see that while the values themselves are different, the number of units of increase is the same as the corresponding parts of Exercise 67. **70.** m

43. direct **45.** direct **47.** inverse **49.** 9 **51.** $\frac{16}{5}$ **53.** $\frac{4}{9}$ **55.** increases **57.** $106\frac{2}{3}$ miles per hour **59.** 25 kilograms per hour **61.** 20 pounds per square foot **63.** 15 feet **65.** 144 feet **67.** direct **69.** inverse **71.** approximately 2364 in 1995 and 716 in 1985 (The actual numbers were 2361 and 719.)

CHAPTER 6 REVIEW EXERCISES (PAGE 421)

1. 3 **3.** $-5, -\frac{2}{3}$ **5. (a)** $\frac{11}{8}$ **(b)** $\frac{13}{22}$ **7.** $\frac{b}{3a}$ **9.** $\frac{-(2x+3)}{2}$

Answers may vary in Exercise 11.

11. $\frac{-(4x-9)}{2x+3}, \frac{-4x+9}{2x+3}, \frac{4x-9}{-(2x+3)}, \frac{4x-9}{-2x-3}$ **13.** $\frac{72}{p}$ **15.** $\frac{5}{8}$ **17.** $\frac{3a-1}{a+5}$ **19.** $\frac{p+5}{p+1}$ **21.** $108y^4$ **23.** $\frac{15a}{10a^4}$ **25.** $\frac{15y}{50-10y}$ **27.** $\frac{15}{x}$ **29.** $\frac{4k-45}{k(k-5)}$ **31.** $\frac{-2-3m}{6}$ **33.** $\frac{7a+6b}{(a-2b)(a+2b)}$ **35.** $\frac{5z-16}{z(z+6)(z-2)}$ **37. (a)** $\frac{a}{b}$ **(b)** $\frac{a}{b}$ **(c)** Answers will vary. **39.** $\frac{4(y-3)}{y+3}$ **41.** $\frac{xw+1}{xw-1}$ **43.** It would cause the first and third denominators to equal 0. **45.** $\emptyset$ **47.** $t = \frac{Ry}{m}$ **49.** $m = \frac{4+p^2q}{3p^2}$ **51.** $\frac{2}{6}$ **53.** $3\frac{1}{13}$ hours **55.** inverse **57.** 4 centimeters **59. (a)** -3 **(b)** -1 **(c)** $-3, -1$ **60.** $\frac{15}{2x}$ **61.** If $x = 0$, the divisor R is equal to 0, and division by 0 is undefined. **62.** $(x+3)(x+1)$ **63.** $\frac{7}{x+1}$ **64.** $\frac{11x+21}{4x}$ **65.** $\emptyset$ **66.** We know that -3 is not allowed because P and R are undefined for $x = -3$. **67.** Rate is equal to distance divided by time. Here, distance is 6 miles and time is $x + 3$ minutes, so rate $= \frac{6}{x+3}$, which is the expression for P. **68.** $\frac{6}{5}, \frac{5}{2}$ **69.** $\frac{(5+2x-2y)(x+y)}{(3x+3y-2)(x-y)}$ **71.** $8p^2$ **73.** 3 **75.** $r = \frac{3kz}{5k-z}$ or $r = \frac{-3kz}{z-5k}$ **77.** approximately 7443 for hearts and 3722 for livers (The actual numbers were 7467 and 3698.) **79.** $\frac{36}{5}$

CHAPTER 6 TEST (PAGE 425)

[6.1] **1.** $-2, 4$ **2. (a)** $\frac{11}{6}$ **(b)** undefined **3.** (Answers may vary.) $\frac{-(6x-5)}{2x+3}, \frac{-6x+5}{2x+3}, \frac{6x-5}{-(2x+3)}, \frac{6x-5}{-2x-3}$ **4.** $-3x^2y^3$ **5.** $\frac{3a+2}{a-1}$ [6.2] **6.** $\frac{25}{27}$ **7.** $\frac{3k-2}{3k+2}$ **8.** $\frac{a-1}{a+4}$ [6.3] **9.** $150p^5$ **10.** $(2r+3)(r+2)(r-5)$ **11.** $\frac{240p^2}{64p^3}$ **12.** $\frac{21}{42m-84}$ [6.4] **13.** 2 **14.** $\frac{-14}{5(y+2)}$ **15.** $\frac{-x^2+x+1}{3-x}$ or $\frac{x^2-x-1}{x-3}$ **16.** $\frac{-m^2+7m+2}{(2m+1)(m-5)(m-1)}$ [6.5] **17.** $\frac{2k}{3p}$ **18.** $\frac{-2-x}{4+x}$ [6.6] **19.** $-1, 4$ **20.** $\left\{-\frac{1}{2}\right\}$ **21.** $D = \frac{dF-k}{F}$ or $D = \frac{k-dF}{-F}$ [6.7] **22.** 3 miles per hour **23.** $2\frac{2}{9}$ hours **24.** 27 days

CUMULATIVE REVIEW EXERCISES CHAPTERS 1–6 (PAGE 426)

[1.2, 1.5, 1.6] **1.** 2 [2.2] **2.** {17} [2.4] **3.** $b = \frac{2A}{h}$ [2.5] **4.** $\left\{-\frac{2}{7}\right\}$ [2.7] **5.** $[-8, \infty)$ **6.** $(4, \infty)$ [3.1] **7. (a)** $(-3, 0)$ **(b)** $(0, -4)$ [3.2] **8.**

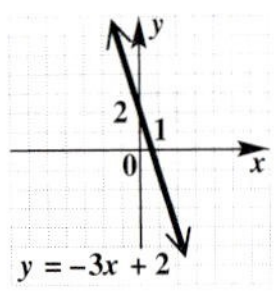

[4.1] **9.**

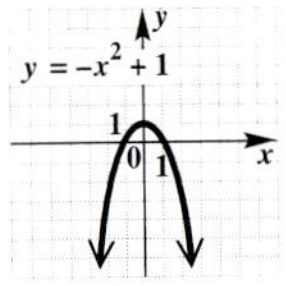

[4.2, 4.5] **10.** $\frac{1}{2^4x^7}$ **11.** $\frac{1}{m^6}$ **12.** $\frac{q}{4p^2}$ [4.1] **13.** $k^2 + 2k + 1$ [4.2] **14.** $72x^6y^7$ [4.4] **15.** $4a^2 - 4ab + b^2$

[4.3] **16.** $3y^3 + 8y^2 + 12y - 5$ [4.6] **17.** $6p^2 + 7p + 1 + \frac{3}{p-1}$ [4.7] **18.** 1.4×10^5 seconds

[5.3] **19.** $(4t + 3v)(2t + v)$ **20.** prime [5.4] **21.** $(4x^2 + 1)(2x + 1)(2x - 1)$ [5.5] **22.** $\{-3, 5\}$

23. $\left\{5, -\frac{1}{2}, \frac{2}{3}\right\}$ [5.6] **24.** -2 or -1 **25.** 6 meters **26.** 30 (The maximum percent is 30%.) [6.1] **27.** (a)

28. (d) [6.4] **29.** $\frac{4}{q}$ **30.** $\frac{3r + 28}{7r}$ **31.** $\frac{7}{15(q - 4)}$ **32.** $\frac{-k - 5}{k(k + 1)(k - 1)}$ [6.2] **33.** $\frac{7(2z + 1)}{24}$

[6.5] **34.** $\frac{195}{29}$ [6.6] **35.** 4, 0 **36.** $\left\{\frac{21}{2}\right\}$ **37.** $\{-2, 1\}$ [6.7] **38.** 150 miles **39.** $1\frac{1}{5}$ hours **40.** 20 pounds

CHAPTER 7 EQUATIONS OF LINES, INEQUALITIES, AND FUNCTIONS

SECTION 7.1 (PAGE 435)

CONNECTIONS **Page 429:** **1.** \$3500, \$5000; \$35,000, \$50,000 **2.** The loss in value each year; the depreciation when the item is brand new ($D = 0$)

EXERCISES **1.** D **3.** B **5.** The slope m of a vertical line is undefined, so it is not possible to write the equation of the line in the form $y = mx + b$. **7.** $y = 3x - 3$ **9.** $y = -x + 3$ **11.** $y = 4x - 3$ **13.** $y = 3$

15.
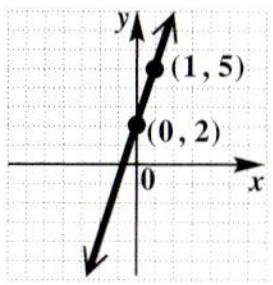

17.
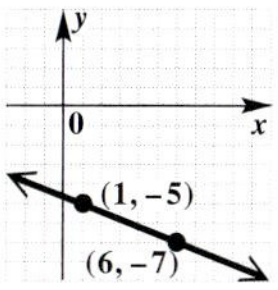

19.
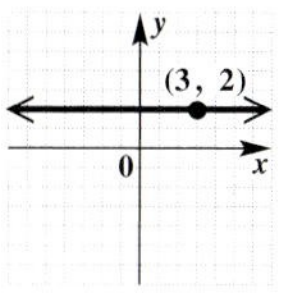

21.
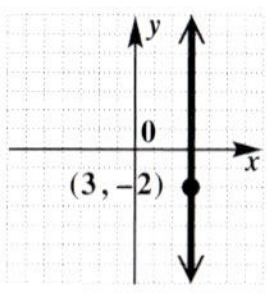

23. the y-axis

25. $y = 2x - 7$ **27.** $y = \frac{2}{3}x + \frac{19}{3}$ **29.** $y = -\frac{4}{5}x + \frac{9}{5}$ **31.** $4 = -3(-2) + b$ **32.** $b = -2$

33. $y = -3x - 2$ **34.** The equations are the same. **35.** $y = x$ (There are other forms as well.) **37.** $y = x - 3$

39. $y = -\frac{5}{7}x - \frac{54}{7}$ **41.** $y = -\frac{2}{3}x - 2$ **43.** $y = \frac{1}{3}x + \frac{4}{3}$ **45.** **(a)** (5, 42), (15, 61), (25, 76)

(b) yes
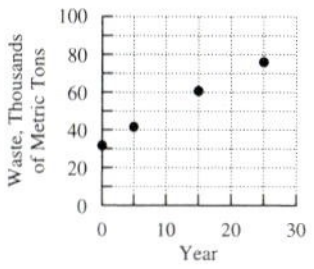

(c) $y = 1.76x + 32$ **(d)** $y = 49.6$, so there will be about 49,600 metric tons in 2005. **47.** **(a)** \$400 **(b)** \$.25 **(c)** $y = .25x + 400$ **(d)** \$425 **(e)** 1500 **49.** **(a)** (1, 822.3), (2, 774.9) **(b)** $y = -47.4x + 869.7$ **(c)** The slope represents the change in sales from 1996 to 1997. The negative slope indicates that sales *decreased*. **51.** $y = -3x + 6$

53. $Y_1 = \frac{3}{4}x + 1$ **55.** (0, 32); (100, 212) **56.** $\frac{9}{5}$ **57.** $F - 32 = \frac{9}{5}(C - 0)$ or $F - 212 = \frac{9}{5}(C - 100)$

58. $F = \frac{9}{5}C + 32$ **59.** $C = \frac{5}{9}(F - 32)$ or $C = \frac{5}{9}F - \frac{160}{9}$ **60.** 86° **61.** 10° **62.** $-40°$

24. $\frac{x + 1}{x}$ **25.** $\frac{(x + 3)^2}{3x}$ **26.** $\frac{4xy^4}{z^2}$ [6.5] **27.** $\frac{5}{8}$ **28.** 6 [3.3] **29.** $-\frac{4}{3}$ **30.** 0

[7.1] **31.** $y = -4x + 15$ **32.** $y = 4x$ [3.2] **33.**

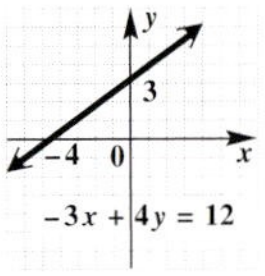

[7.2] **34.**

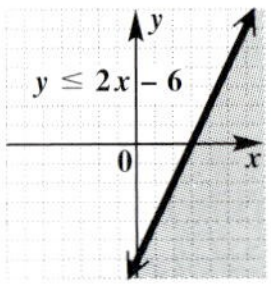

35.

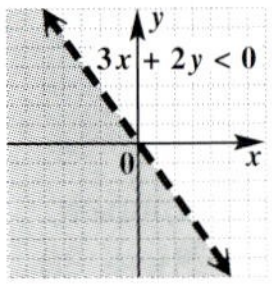

[2.3] **36.** corporate income taxes: \$171.8 billion; individual income taxes: \$656.4 billion [2.6] **37.** 15°, 35°, 130°
[5.6] **38.** 7 inches [2.5] **39.** 14 [6.7] **40.** 1 hour

CHAPTER 8 LINEAR SYSTEMS

SECTION 8.1 (PAGE 473)

Exercises **1. (a)** B **(b)** C **(c)** D **(d)** A **3.** yes **5.** no **7.** yes **9.** yes **11.** no **13.** (a); The ordered pair solution must be in quadrant II, and $(-4, -4)$ is in quadrant III. **15.** $\{(4, 2)\}$

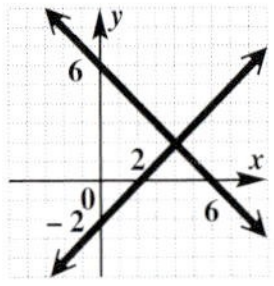

17. $\{(0, 4)\}$

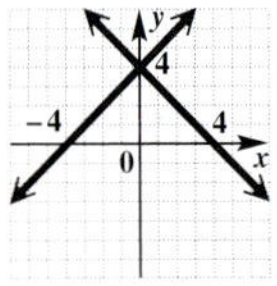

19. $\{(4, -1)\}$

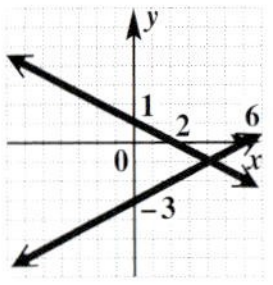

In Exercises 21–29, we do not show the graphs.

21. $\{(1, 3)\}$ **23.** $\{(0, 2)\}$ **25.** $\emptyset$ (inconsistent system) **27.** infinite number of solutions (dependent equations) **29.** $\{(4, -3)\}$ **33.** Yes, it is possible. For example, the system

$$\begin{aligned} x + y &= 5 \\ x - y &= -1 \\ 2x - y &= 1 \end{aligned}$$

has the single solution (2, 3).

35. about 350 million for each format **37.** 200 million **39.** between 1984 and 1986, and between 1988 and 1990 **41. (a)** neither **(b)** intersecting lines **(c)** one solution **43. (a)** dependent **(b)** one line **(c)** infinite number of solutions **45. (a)** inconsistent **(b)** parallel lines **(c)** no solution **47.** 40 **49.** (40, 30) **51.** B **53.** A **55.** $\{(-1, 5)\}$ **57.** $\{2\}$ **58.** 5 **59.** $\{(2, 5)\}$

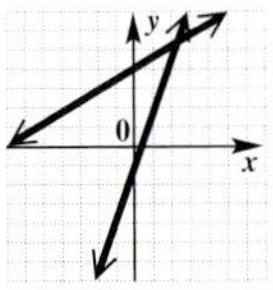

60. The x-coordinate, 2, is equal to the solution of the equation. **61.** The y-coordinate, 5, is equal to the value we obtained on both sides when checking. **62.** 5; 3; 5

SECTION 8.2 (PAGE 481)

Exercises **1.** No, it is not correct, because the solution set is $\{(3, 0)\}$. The y-value in the ordered pair must also be determined. **3.** $\{(3, 9)\}$ **5.** $\{(7, 3)\}$ **7.** $\{(0, 5)\}$ **9.** $\{(-4, 8)\}$ **11.** $\{(3, -2)\}$ **13.** infinite number of solutions **15.** $\left\{\left(\frac{1}{3}, -\frac{1}{2}\right)\right\}$ **17.** $\emptyset$ **19.** infinite number of solutions **21.** The first student had less work to do, because the coefficient of y in the first equation is -1. The second student had to divide by 2, introducing fractions into the expression for x. **23.** $\{(2, -3)\}$ **25.** $\{(3, 2)\}$ **27.** $\{(-2, 1)\}$ **29.** 1993 **31.** To find the total cost, multiply the number of bicycles (x) by the cost per bicycle (\$400), and add the fixed cost (\$5000). Thus, $y_1 = 400x + 5000$ gives this total cost (in dollars).

32. $y_2 = 600x$ **33.** $y_1 = 400x + 5000$; solution set: $\{(25, 15{,}000)\}$
$y_2 = 600x$ **34.** 25; 15,000; 15,000

35. $\{(2, 4)\}$

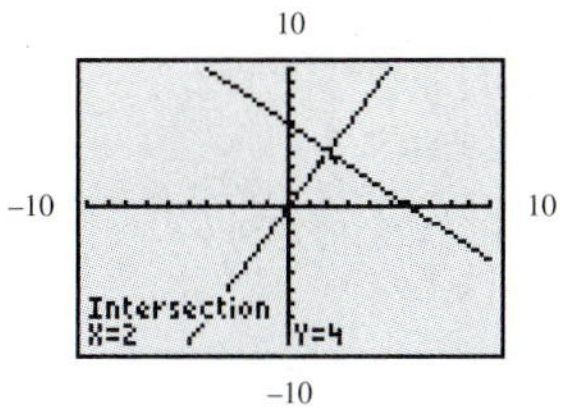

37. $\{(1, 5)\}$

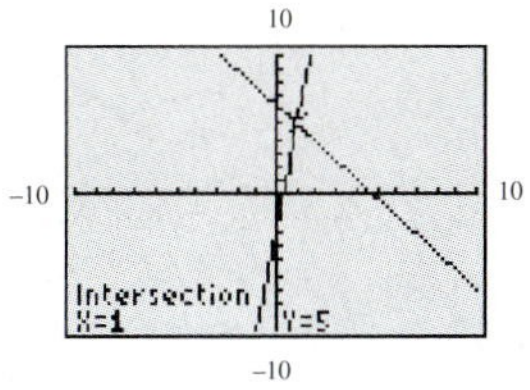

39. $\{(5, -3)\}$; The equations to input are $y_1 = \dfrac{5 - 4x}{5}$ and $y_2 = \dfrac{1 - 2x}{3}$.

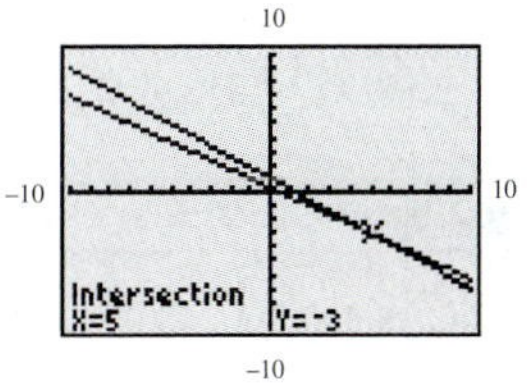

41. Adjust the viewing window so that it does appear.

SECTION 8.3 (PAGE 489)

EXERCISES **1.** true **3.** true **5.** $\{(4, 6)\}$ **7.** $\{(-1, -3)\}$ **9.** $\{(-2, 3)\}$ **11.** $\left\{\left(-\frac{2}{3}, \frac{17}{2}\right)\right\}$ **13.** $\{(3, -6)\}$ **15.** $\{(7, 4)\}$ **17.** $\{(0, 3)\}$ **19.** $\{(3, 0)\}$ **21.** $\left\{\left(-\frac{32}{23}, -\frac{17}{23}\right)\right\}$ **25.** $\{(-3, 4)\}$ **27.** infinite number of solutions **29.** $\{(0, 6)\}$ **31.** $\emptyset$ **33.** **(a)** $\{(1, 4)\}$ **(b)** $\{(1, 4)\}$ **(c)** Answers will vary. **35.** Yes, they should both get the same answer, since both procedures are mathematically valid. **37.** $\{(0, 3)\}$ **39.** $\{(24, -12)\}$ **41.** $\{(3, 2)\}$ **43.** $1141 = 1991a + b$ **44.** $1339 = 1996a + b$ **45.** $1991a + b = 1141$
$1996a + b = 1339$; solution set: $\{(39.6, -77{,}702.6)\}$
46. $y = 39.6x - 77{,}702.6$ **47.** 1220.2 (million); This is slightly less than the actual figure.
48. It is not realistic to expect the data to lie in a perfectly straight line; as a result, the quantity obtained from an equation determined in this way will probably be "off" a bit. One cannot put too much faith in models such as this one, because not all data points are linear in nature.

SECTION 8.4 (PAGE 496)

EXERCISES **1.** (d) **3.** (b) **5.** (c) **7.** the second number; $x - y = 48$; The two numbers are 73 and 25. **9.** Boyz II Men: 134; Bruce Springsteen & the E St. Band: 40 **11.** Terminal Tower: 708 feet; Society Center: 948 feet **13.** 46 ones; 28 tens **15.** 2 copies of *Godzilla*; 5 Aerosmith compact discs **17.** \$2500 at 4%; \$5000 at 5% **19.** Japan: \$17.19; Switzerland: \$13.15 **21.** 80 liters of 40% solution; 40 liters of 70% solution **23.** 30 pounds at \$6 per pound; 60 pounds at \$3 per pound **25.** 60 barrels at \$40 per barrel; 40 barrels at \$60 per barrel **27.** boat: 10 miles per hour; current: 2 miles per hour **29.** plane: 470 miles per hour; wind: 30 miles per hour **31.** car leaving Cincinnati: 55 miles per hour; car leaving Toledo: 70 miles per hour **33.** Roberto: 3 miles per hour; Juana: 2.5 miles per hour

SECTION 8.5 (PAGE 504)

EXERCISES **1.** C **3.** B **5.**

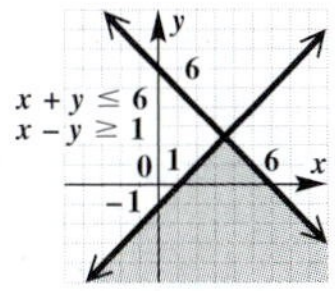

7.

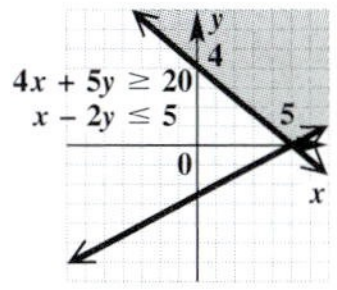

9.

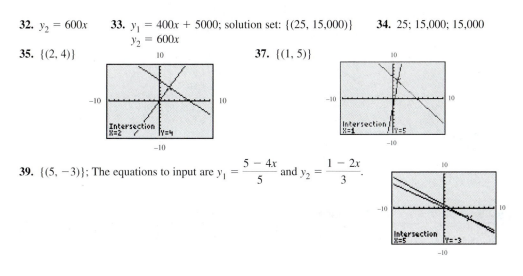

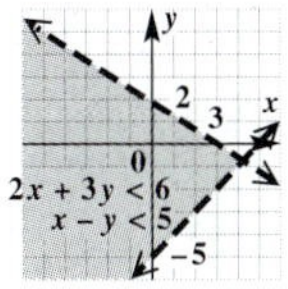

11.
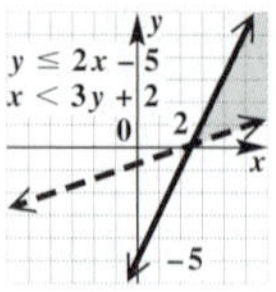

13.
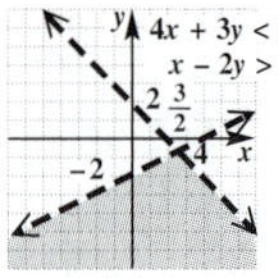

15.
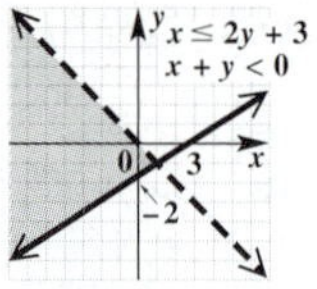

17.
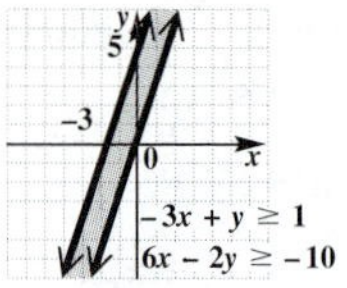

19.
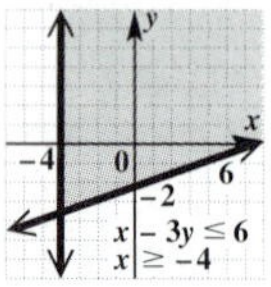

21.
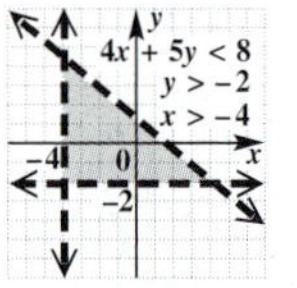

23.
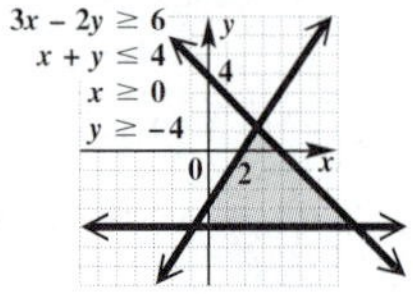

25. (4, 0), (4, 5), (9, 5)

27. (−3, 3), (5, 3), (5, −5) **29.** D **31.** A

CHAPTER 8 REVIEW EXERCISES (PAGE 510)

1. yes **3.** {(3, 1)} **5.** No, this is not correct. A false statement indicates that the solution set is ∅. **7.** {(2, 1)} **9.** infinite number of solutions **11.** His answer was incorrect since the system has infinitely many solutions (as indicated by the true statement $0 = 0$). **13.** (c) **15.** {(7, 1)} **17.** infinite number of solutions **19.** {(−4, 1)} **21.** {(9, 2)} **25.** *How Stella Got Her Groove Back*: 782,699; *The Deep End of the Ocean*: 840,263 **27.** Texas Commerce Tower: 75 stories; First Interstate Plaza: 71 stories **29.** length: 27 meters; width: 18 meters **31.** 25 pounds of \$1.30 candy; 75 pounds of \$.90 candy **33.** \$7000 at 3%; \$11,000 at 4% **35.** plane: 250 miles per hour; wind: 20 miles per hour

37.
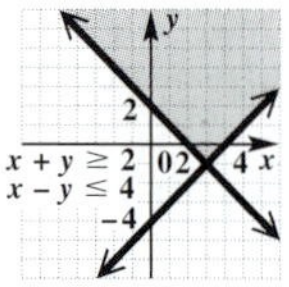

39.
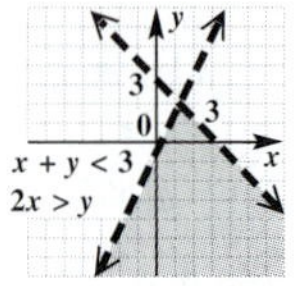

41. $\left\{\left(\frac{28}{5}, \frac{16}{5}\right)\right\}$ **42.** $\left\{\left(\frac{28}{5}, \frac{16}{5}\right)\right\}$ **43.** They are the same. It makes no difference which method we use. **44.** $y_1 = 2x - 8$ **45.** $y_2 = \frac{12 - x}{2}$ or $y_2 = 6 - \frac{1}{2}x$ **46.** $\left\{\frac{28}{5}\right\}$; The solution is the x-value found in Exercises 41 and 42. **47.** We get $\frac{16}{5} = \frac{16}{5}$. This value is the y-value found in Exercises 41 and 42.

48. 2 **49.** $-\frac{1}{2}$ **50.** They are perpendicular. **51.**
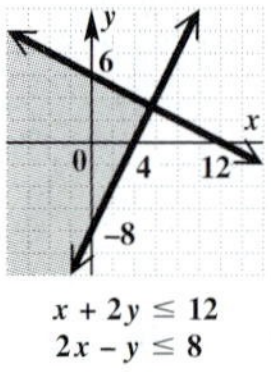

52. $\left\{\left(\frac{28}{5}, \frac{16}{5}\right)\right\}$ **53.** {(4, 8)}

55. {(2, 0)} **57.**
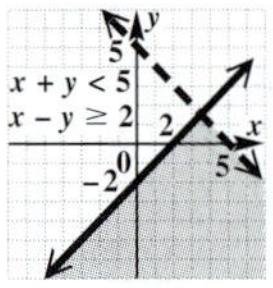

59. Great Smoky Mountains: 9.3 million; Rocky Mountain National Park: 2.9 million

61. slower car: 38 miles per hour; faster car: 68 miles per hour **63.** Yes. Let x represent each of the two equal side lengths. Then $x + 5$ is the length of the third side. The equation to solve is $x + x + (x + 5) = 29$, giving $x = 8$. The side lengths are 8, 8, and 13 inches.

CHAPTER 8 TEST (PAGE 514)

[8.1] **1. (a)** no **(b)** no **(c)** yes **2.** {(4, 1)} [8.2] **3.** {(1, −6)} **4.** {(−35, 35)} [8.3] **5.** {(5, 6)} **6.** {(−1, 3)} **7.** {(−1, 3)} **8.** ∅ **9.** {(0, 0)} **10.** {(−15, 6)} [8.1–8.3] **11.** infinite number of solutions **12.** It has no solution. [8.4] **13.** Memphis and Atlanta: 371 miles; Minneapolis and Houston: 671 miles

14. Disneyland: 15.0 million; Magic Kingdom: 13.8 million **15.** $33\frac{1}{3}$ liters of 25% solution; $16\frac{2}{3}$ liters of 40% solution **16.** slower car: 45 miles per hour; faster car: 60 miles per hour **17.** *The Lion King*: 30 million; *Snow White*: 28 million [8.5] **18.**

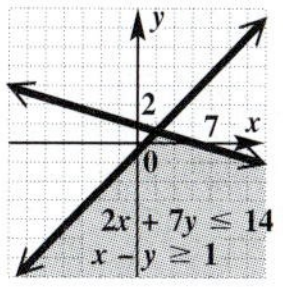

19.

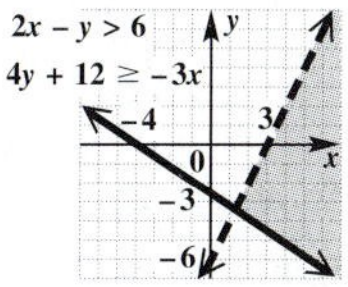

20. A

CUMULATIVE REVIEW EXERCISES CHAPTERS 1–8 (PAGE 515)

[1.7] **1.** −1, 1, −2, 2, −4, 4, −5, 5, −8, 8, −10, 10, −20, 20, −40, 40 [1.3] **2.** 1 [1.7] **3.** commutative property **4.** distributive property **5.** inverse property [1.6] **6.** 46 [2.2] **7.** $\left\{-\frac{13}{11}\right\}$ **8.** $\left\{\frac{9}{11}\right\}$ [2.4] **9.** width: $8\frac{1}{4}$ inches; length: $10\frac{3}{4}$ inches [2.7] **10.** $\left(-\frac{11}{2}, \infty\right)$ [3.1] **11.**

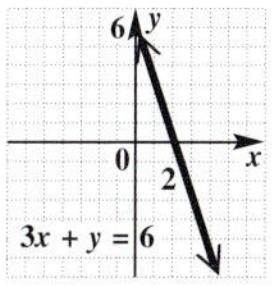

[3.3] **12.** $-\frac{4}{3}$ **13.** $-\frac{1}{4}$ [4.1] **14.** $14x^2 - 5x + 23$ [4.3] **15.** $6xy + 12x - 14y - 28$ [4.6] **16.** $3k^2 - 4k + 1$ [4.7] **17.** 3.65×10^{10} [4.5] **18.** x^6y [5.3] **19.** $(5m - 4p)(2m + 3p)$ [5.4] **20.** $(8t - 3)^2$ [5.5] **21.** $\left\{-\frac{1}{3}, \frac{3}{2}\right\}$ **22.** $\{-11, 11\}$ [6.1] **23.** $-1, \frac{5}{2}$ [6.4] **24.** $\frac{7}{x + 2}$ [6.2] **25.** $\frac{3}{4k - 3}$ [6.6] **26.** $\left\{-\frac{1}{4}, 3\right\}$ [5.7] **27.** $[-1, 6]$ [7.1] **28.** $y = 3x - 11$ **29.** $y = 4$ **30.** **(a)** $x = 9$ **(b)** $y = -1$ **31.** $y = 103.25x + 3502$; The slope represents the average yearly increase in health benefit cost during the period. [7.2] **32.**

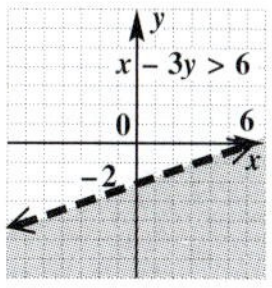

[7.3] **33.** 9 [8.1–8.3] **34.** $\{(-1, 6)\}$ **35.** $\{(3, -4)\}$ **36.** $\{(2, -1)\}$ [8.4] **37.** 405 adults and 49 children **38.** 19 inches, 19 inches, 15 inches **39.** 4 girls and 3 boys [8.5] **40.** (b)

CHAPTER 9 ROOTS AND RADICALS

SECTION 9.1 (PAGE 526)

CONNECTIONS **Page 523:** The area of the large square is $(a + b)^2$ or $a^2 + 2ab + b^2$. The sum of the areas of the smaller square and the four right triangles is $c^2 + 2ab$. Set these equal to each other and subtract $2ab$ from both sides to get $a^2 + b^2 = c^2$.

EXERCISES **1.** false; Zero has only one square root. **3.** true **5.** true **7.** −4, 4 **9.** −12, 12 **11.** $-\frac{5}{14}, \frac{5}{14}$ **13.** −30, 30 **15.** 7 **17.** −11 **19.** $-\frac{12}{11}$ **21.** not a real number **23.** 100 **25.** 19 **27.** $3x^2 + 4$ **29.** a must be positive. **31.** a must be negative. **33.** rational; 5 **35.** irrational; 5.385 **37.** rational; −8 **39.** not a real number **41.** The answer to Exercise 17 is the negative square root of a positive number. However, in Exercise 21, the square root of a negative number is not a real number. **43.** 23.896 **45.** 28.249 **47.** 1.985 **49.** 5 **51.** 17 **53.** 3.606 **55.** 2.289 **57.** 5.074 **59.** −4.431 **61.** $c = 17$ **63.** $b = 8$ **65.** $c = 11.705$ **67.** 24 centimeters **69.** 80 feet **71.** 195 miles **73.** 9.434 **75.** Answers will vary. For example, if $a = 2$ and

$b = 7$, $\sqrt{a^2 + b^2} = \sqrt{53}$, while $a + b = 9$. $\sqrt{53} \neq 9$. If $a = 0$ and $b = 1$, then $\sqrt{a^2 + b^2}$ is equal to $a + b$ (which is not true in general). **77.** $\sqrt{29}$ **79.** $\sqrt{2}$ **81.** 10 **83.** -3 **85.** 5 **87.** not a real number **89.** -3 **91.** 3 **93.** c^2 **94.** $(b - a)^2$ **95.** $2ab$ **96.** $(b - a)^2 = a^2 - 2ab + b^2$ **97.** $c^2 = 2ab + (a^2 - 2ab + b^2)$ **98.** $c^2 = a^2 + b^2$

SECTION 9.2 (PAGE 535)

CONNECTIONS **Page 530:** **1.** x; $\sqrt{x}$ **2.** The last part of it ($\sqrt{}$) is used as part of the radical symbol $\sqrt{\;}$.

EXERCISES **1.** false; $\sqrt{4}$ represents only the principal (positive) square root. **3.** false; $\sqrt{-6}$ is not a real number. **5.** true **7.** true **9.** 9 **11.** $3\sqrt{10}$ **13.** 13 **15.** $\sqrt{13r}$ **17.** (a) **19.** $3\sqrt{5}$ **21.** $5\sqrt{3}$ **23.** $5\sqrt{5}$ **25.** $-10\sqrt{7}$ **27.** $9\sqrt{3}$ **29.** $3\sqrt{6}$ **31.** 24 **33.** $6\sqrt{10}$ **37.** $\frac{4}{15}$ **39.** $\frac{\sqrt{7}}{4}$ **41.** 5 **43.** $\frac{25}{4}$ **45.** $6\sqrt{5}$ **47.** m **49.** y^2 **51.** $6z$ **53.** $20x^3$ **55.** $z^2\sqrt{z}$ **57.** x^3y^6 **59.** $2\sqrt[3]{5}$ **61.** $3\sqrt[3]{2}$ **63.** $2\sqrt[4]{5}$ **65.** $\frac{2}{3}$ **67.** $-\frac{6}{5}$ **69.** 5 **71.** $\sqrt[4]{12}$ **73.** $2x\sqrt[3]{4}$

In Exercise 75, the number of displayed digits will vary among calculator models. Also, less sophisticated models may exhibit round-off error in the final decimal place.

75. (a) 4.472135955 **(b)** 4.472135955 **(c)** The numerical results are not a proof because both answers are approximations and they might differ if calculated to more decimal places.

77. 6 centimeters **79.** 6 inches **81.** The product rule for radicals requires that both a and b must be nonnegative. Otherwise $\sqrt{a}$ and $\sqrt{b}$ would not be real numbers (except when $a = b = 0$). **83.** To verify the first length, we show, in the first triangle, that $1^2 + 1^2 = 2$, so the hypotenuse has length $\sqrt{2}$. Similarly, in the second triangle, $(\sqrt{2})^2 + 1^2 = 3$, so the hypotenuse has length $\sqrt{3}$, and so on. **84.** $\sqrt{4}$, $\sqrt{9}$, and so on; $\sqrt{16} = 4$ and $\sqrt{25} = 5$ **85.** Look at the radicands: $9 - 4 = 5$, $16 - 9 = 7$, $25 - 16 = 9$, and so on. **86.** The differences between consecutive whole number lengths increase by 2 each time, so we can predict that the next one will be $\sqrt{25 + 11} = \sqrt{36}$, and the one after that will be $\sqrt{36 + 13} = \sqrt{49}$.

SECTION 9.3 (PAGE 539)

EXERCISES **1.** distributive **3.** radicands **5.** $-5\sqrt{7}$ **7.** $5\sqrt{17}$ **9.** $5\sqrt{7}$ **11.** $11\sqrt{5}$ **13.** $15\sqrt{2}$ **15.** $-20\sqrt{2} + 6\sqrt{5}$ **17.** $17\sqrt{7}$ **19.** $-16\sqrt{2} - 8\sqrt{3}$ **21.** $20\sqrt{2} + 6\sqrt{3} - 15\sqrt{5}$ **23.** $4\sqrt{2}$ **25.** $2\sqrt{3} + 4\sqrt{3} = (2 + 4)\sqrt{3} = 6\sqrt{3}$ **27.** $11\sqrt{3}$ **29.** $5\sqrt{x}$ **31.** $3x\sqrt{6}$ **33.** 0 **35.** $-20\sqrt{2k}$ **37.** $42x\sqrt{5z}$ **39.** $-\sqrt[3]{2}$ **41.** $6\sqrt[3]{p^2}$ **43.** $21\sqrt[4]{m^3}$ **47.** $-6x^2y$ **48.** $-6(p - 2q)^2(a + b)$ **49.** $-6a^2\sqrt{xy}$ **50.** The answers are alike because the numerical coefficient of the three answers is the same: -6. Also, the first variable factor is raised to the second power, and the second variable factor is raised to the first power. The answers are different because the variables are different: x and y, then $p - 2q$ and $a + b$, and then a and $\sqrt{xy}$. **51.** 9.220 **53.** 5 **55.** $22\sqrt{2}$ **57. (a)** 1991 **(b)** 1997

SECTION 9.4 (PAGE 545)

EXERCISES **1.** radical **3.** fraction **5.** $4\sqrt{2}$ **7.** $\frac{-\sqrt{33}}{3}$ **9.** $\frac{7\sqrt{15}}{5}$ **11.** $\frac{\sqrt{30}}{2}$ **13.** $\frac{16\sqrt{3}}{9}$ **15.** $\frac{-3\sqrt{2}}{10}$ **17.** $\frac{21\sqrt{5}}{5}$ **19.** $\sqrt{3}$ **21.** $\frac{\sqrt{2}}{2}$ **23.** $\frac{\sqrt{65}}{5}$ **25.** 1; identity property for multiplication **27.** $\frac{\sqrt{21}}{3}$ **29.** $\frac{3\sqrt{14}}{4}$ **31.** $\frac{1}{6}$ **33.** 1 **35.** $\frac{\sqrt{7x}}{x}$ **37.** $\frac{2x\sqrt{xy}}{y}$ **39.** $\frac{x\sqrt{3xy}}{y}$ **41.** $\frac{3ar^2\sqrt{7rt}}{7t}$ **43.** (b) **45.** $\frac{\sqrt[3]{12}}{2}$ **47.** $\frac{\sqrt[3]{196}}{7}$ **49.** $\frac{\sqrt[3]{6y}}{2y}$ **51.** $\frac{\sqrt[3]{42mn^2}}{6n}$ **53. (a)** $\frac{9\sqrt{2}}{4}$ seconds **(b)** 3.182 seconds

SECTION 9.5 (PAGE 550)

EXERCISES **1.** 13 **3.** 4 **5.** 4 **7.** 5 **9.** $9\sqrt{5}$ **11.** $16\sqrt{2}$ **13.** $\sqrt{15}-\sqrt{35}$ **15.** $2\sqrt{10}+30$ **17.** $4\sqrt{7}$ **19.** $57+23\sqrt{6}$ **21.** $81+14\sqrt{21}$ **23.** $37+12\sqrt{7}$ **25.** 23 **27.** 1 **29.** $2\sqrt{3}-2+3\sqrt{2}-\sqrt{6}$ **31.** $15\sqrt{2}-15$ **33.** $87+9\sqrt{21}$ **35.** Because multiplication must be performed before addition, it is incorrect to add -37 and -2. Since $-2\sqrt{15}$ cannot be simplified, the expression cannot be written in a simpler form, and the final answer is $-37-2\sqrt{15}$. **37.** $\frac{3-\sqrt{2}}{7}$ **39.** $-4-2\sqrt{11}$ **41.** $1+\sqrt{2}$ **43.** $-\sqrt{10}+\sqrt{15}$ **45.** $3-\sqrt{3}$ **47.** $2\sqrt{5}+\sqrt{15}+4+2\sqrt{3}$ **49.** $\sqrt{11}-2$ **51.** $\frac{\sqrt{3}+5}{8}$ **53.** $\frac{6-\sqrt{10}}{2}$ **55.** $x\sqrt{30}+\sqrt{15x}+6\sqrt{5x}+3\sqrt{10}$ **57.** $6t-3\sqrt{14t}+2\sqrt{7t}-7\sqrt{2}$ **59.** $m\sqrt{15}+\sqrt{10mn}-\sqrt{15mn}-n\sqrt{10}$ **61.** $2-3\sqrt[3]{4}$ **63.** $12+10\sqrt[4]{8}$ **65.** $-1+3\sqrt[3]{2}-\sqrt[3]{4}$ **67.** 1 **69.** $30+18x$ **70.** They are not like terms. **71.** $30+18\sqrt{5}$ **72.** They are not like radicals. **73.** Make the first term $30x$, so that $30x+18x=48x$; make the first term $30\sqrt{5}$, so that $30\sqrt{5}+18\sqrt{5}=48\sqrt{5}$. **74.** When combining like terms, we add (or subtract) the coefficients of the common factors of the terms: $2xy+5xy=7xy$. When combining like radicals, we add (or subtract) the coefficients of the common radical: $2\sqrt{ab}+5\sqrt{ab}=7\sqrt{ab}$. **75.** 4 inches **76.** $\frac{\sqrt{AP}-P}{P}$; 8%

SECTION 9.6 (PAGE 557)

CONNECTIONS **Page 552: 1.** $6\sqrt{13}\approx 21.63$ (to the nearest hundredth) **2.** $h=\sqrt{13}$; $6\sqrt{13}\approx 21.63$

EXERCISES **1.** {49} **3.** {7} **5.** {85} **7.** {−45} **9.** $\left\{-\frac{3}{2}\right\}$ **11.** ∅ **13.** {121} **15.** {8} **17.** {1} **19.** {6} **21.** ∅ **23.** {5} **25.** Since $\sqrt{x}$ must be greater than or equal to zero for any replacement for x, it cannot equal −8, a negative number. **29.** {12} **31.** {5} **33.** {0, 3} **35.** {−1, 3} **37.** {8} **39.** {4} **41.** {8} **43.** {9} **45.** We cannot square term by term. The left side must be squared as a binomial in the first step. **47.** 158.6 feet **49. (a)** 70.5 miles per hour **(b)** 59.8 miles per hour **(c)** 53.9 miles per hour **51. (a)** 1991 **(b)** 1997; Here, $f(3)\approx 97.5$. Thus, 1997 is the year when about 3 million calls were made. **(c)** about 2.1 million **53.** 4 **55.** −2 **57.** −1

SECTION 9.7 (PAGE 565)

CONNECTIONS **Page 564: 1.** $(\sqrt{x}-3)(\sqrt{x}+1)$ **2.** $(x+\sqrt{10})(x-\sqrt{10})$ **3.** $(\sqrt{x}+2)^2$ **4.** $x\sqrt{x}+3\sqrt{x}+\frac{5\sqrt{x}}{x}$

EXERCISES **1.** (a) **3.** (c) **5.** 5 **7.** 4 **9.** 2 **11.** 2 **13.** 8 **15.** 9 **17.** 8 **19.** 4 **21.** −4 **23.** −4 **25.** $\frac{1}{343}$ **27.** $\frac{1}{36}$ **29.** $-\frac{1}{32}$ **31.** 2^3 **33.** $\frac{1}{6^{1/2}}$ **35.** $\frac{1}{15^{1/2}}$ **37.** $11^{1/7}$ **39.** 8^3 **41.** $6^{1/2}$ **43.** $\frac{5^3}{2^3}$ **45.** $\frac{1}{2^{8/5}}$ **47.** $6^{2/9}$ **49.** z **51.** $m^2n^{1/6}$ **53.** $\frac{a^{2/3}}{b^{4/9}}$ **55.** 2 **57.** 2 **59.** $\sqrt{a}$ **61.** $\sqrt[3]{k^2}$ **63.** $\sqrt{2}=2^{1/2}$ and $\sqrt[3]{2}=2^{1/3}$ **64.** $2^{1/2}\cdot 2^{1/3}$ **65.** 6 **66.** $2^{3/6}\cdot 2^{2/6}$ **67.** $2^{5/6}$ **68.** $\sqrt[6]{2^5}$ or $\sqrt[6]{32}$ **69.** 2 **71.** 1.883 **73.** 3.971 **75.** 9.100 **81. (a)** $d=1.22x^{1/2}$ **(b)** 211.31 miles **83.** Because $(7^{1/2})^2=7$ and $(\sqrt{7})^2=7$, they should be equal, so we define $7^{1/2}$ to be $\sqrt{7}$.

CHAPTER 9 REVIEW EXERCISES (PAGE 571)

1. −7, 7 **3.** −14, 14 **5.** −15, 15 **7.** 4 **9.** 10 **11.** not a real number **13.** $\frac{7}{6}$ **15.** a must be negative. **17.** irrational; 4.796 **19.** rational; −5 **21.** $5\sqrt{3}$ **23.** $4\sqrt{10}$ **25.** 12 **27.** $16\sqrt{6}$ **29.** $-\frac{11}{20}$ **31.** $\frac{\sqrt{7}}{13}$

G

H

I

K

L

S

T

U

Index of Applications

Astronomy/Aerospace

Biology

Business

Chemistry

Construction

Consumer Applications

Economics

Education

Environment

Finance

Geometry

Government

Health/Life Sciences

Labor

Miscellaneous

Physics

Sports/Entertainment

Statistics/Demographics

Transportation